Introduction to algebra

INTRODUCTION TO ALGEBRA

R. Kochendörffer
Professor of Mathematics, University of Dortmund

WOLTERS-NOORDHOFF PUBLISHING GRONINGEN
THE NETHERLANDS

Softcover reprint of the hardcover 1st edition 1972

Library of Congress Catalog Card No. 75-119885

ISBN-13: 978-94-009-8181-2 e-ISBN-13: 978-94-009-8179-9
DOI: 10.1007/978-94-009-8179-9

Contents

Preface IX

Chapter 1. Basic concepts 1

1.1. Sets 1
1.2. Relations 3
1.3. Mappings 6
1.4. Operations 10
1.5. Algebraic systems 11

Exercises 15

Chapter 2. The integers 20

2.1. The natural numbers and the integers 20
2.2. Divisibility. Prime numbers 22
2.3. The greatést common divisor 24
2.4. Prime factorization 27
2.5. Congruences. Residue classes 29
2.6. The residue class ring 32
2.7. Simultaneous congruences. Euler's function 35

Exercises 37

Chapter 3. Groups 40

3.1. Semigroups 40
3.2. Groups 42

3.3. Isomorphisms. Automorphisms 50
3.4. Embedding of abelian semigroups in groups 53
3.5. Subgroups 56
3.6. Cyclic groups 61
3.7. Homomorphisms 64
3.8. Subnormal series 72
3.9. Direct products 76
3.10. Permutation groups 81
3.11. Sylow subgroups and p-groups 89
3.12. Endomorphisms and operators 94
3.13. Vector spaces. Modules 98
Exercises 109

Chapter 4. Rings. Integral domains 113

4.1. Definitions and examples 113
4.2. Homomorphisms 120
4.3. Commutative rings. Integral domains 127
4.4. Principal ideal rings 131
4.5. Euclidean rings 137
4.6. Fields of quotients 142
4.7. Prime fields. Characteristic 144
Exercises 147

Chapter 5. Polynomials 151

5.1. Polynomials in one indeterminate 151
5.2. Polynomials over fields 152
5.3. Polynomials over integral domains 155
5.4. Roots. The derivative 160
5.5. Polynomials in several indeterminates 168
5.6. Symmetric polynomials 171
5.7. The resultant and the discriminant 176
Exercises 180

Chapter 6. Fields 183

6.1. Adjunction 183
6.2. Algebraic extension fields 187

6.3. Construction of extension fields 191
6.4. Normal extensions 200
6.5. Separable and inseparable extensions 202
6.6. Galois theory 211
6.7. Cyclotomic fields 225
6.8. Galois fields 235
Exercises 240

Chapter 7. Galois theory of equations 243

7.1. The Galois group of a polynomial 243
7.2. Solubility of equations by radicals 251
7.3. Quadratic, cubic, and quartic equations 255
7.4. Constructions by ruler and compass 261
Exercises 265

Chapter 8. Order and valuations 267

8.1. Ordered fields 267
8.2. Formally real fields 269
8.3. Valuations 275
Exercises 283

Chapter 9. Modules 285

9.1. Elementary divisors 285
9.2. Modules over principal ideal rings 291
9.3. Endomorphisms of vector spaces 303
9.4. Finiteness conditions 315
9.5. Algebraic integers 321
Exercises 331

Chapter 10. Algebras 336

10.1. Basic definitions 336
10.2. The radical 338
10.3. Semi-simple rings 341
10.4. Simple rings 347

10.5. Division algebras over the field of the real numbers . . 352
10.6. Representation modules 356
10.7. Representations of semi-simple algebras 361
Exercises 375

Chapter 11. Lattices 379

11.1. Lattices and partially ordered sets 379
11.2. Modular lattices 385
11.3. Distributive lattices 392
Exercises 404

Bibliography 407

Index 410

Preface

This book is intended as a textbook for an undergraduate course on algebra. In most universities a detailed study of abstract algebraic systems commences in the second year. By this time the student has gained some experience in mathematical reasoning so that a too elementary book would rob him of the joy and the stimulus of using his ability. I tried to make allowance for this when I chose the level of presentation. On the other hand, I hope that I also avoided discouraging the reader by demands which are beyond his strength. So, the first chapters will certainly not require more mathematical maturity than can reasonably be expected after the first year at the university.

Apart from one exception the formal prerequisites do not exceed the syllabus of an average high school. As to the exception, I assume that the reader is familiar with the rudiments of linear algebra, i.e. addition and multiplication of matrices and the main properties of determinants. In view of the readers for whom the book is designed I felt entitled to this assumption. In the first chapters, matrices will almost exclusively occur in examples and exercises providing non-trivial instances in the theory of groups and rings. In Chapters 9 and 10 only, vector spaces and their properties will form a relevant part of the text. A reader who is not familiar with these concepts will have no difficulties in acquiring these prerequisites by any elementary textbook, e.g. [10]. (Numbers in brackets refer to the Bibliography at the end of this volume.)

There is certainly some basic material which must be contained in any undergraduate textbook on algebra, but apart from this there is still much room for choice. The list of contents will indicate my choice but I would like to add a few comments. The fundamental concepts of algebra were originally introduced as tools to deal with special problems. In my opinion it would give the student a false impression of the development of mathematical ideas if he were to indulge in abstract concepts only and forget their origin.

A great, if not the greatest, part of the students of mathematics are future teachers. These students should not only become familiar with the fundamental concepts of algebra but they should also realize that these concepts are indispensable tools for a deeper study of problems which are closely related to the high school syllabus. So, when I had to decide whether a certain topic should be included in this book and to which extent it should be persued, an important criterion, though not the only one, was its relevance for a student who will later teach at a high school.

I wish to acknowledge the kind cooperation of both Wolters-Noordhoff Publishing and the VEB Deutscher Verlag der Wissenschaften. I also thank my wife for patiently and skillfully typing the manuscript.

January 1971 Rudolf Kochendörffer

1

Basic concepts

This introductory chapter contains a survey of some basic concepts which will be used throughout the book. The reader will probably be acquainted with most of these concepts, but we mention them to fix the notations and to indicate the prerequisites.

1.1. Sets

With regard to sets we adopt the naive standpoint. The following sets of numbers will frequently occur:

(a) The set $\boldsymbol{N}$ of the *natural numbers*, viz. 1, 2, 3, ...

(b) The set $\boldsymbol{Z}$ of the *integers*, viz. 0, ± 1, ± 2, ± 3, ...

(c) The set $\boldsymbol{Q}$ of the *rational numbers* a/b where a and b are integers and $b \neq 0$. Two rational numbers a/b and a'/b' are regarded as equal if and only if $ab' = a'b$.

(d) The set $\boldsymbol{R}$ of the *real numbers*.

(e) The set $\boldsymbol{C}$ of all *complex numbers*. Note that we include the real numbers among this set.

In order to indicate that α is an element of the set M (or belongs to M) we write $\alpha \in M$, whereas $\alpha \notin M$ means that α does not belong to M. For reasons of expedience we introduce the *empty set* $\emptyset$ that contains no element.

If the set M consists of the elements α, β, γ, ... we write

$$M = \{\alpha, \beta, \gamma, \ldots\}.$$

Of course, the order of the elements is irrelevant. The set of all elements of M that satisfy a condition C will be denoted by

$$\{\xi: \xi \in M,\ \xi \text{ satisfies C}\}.$$

For instance $\{\xi: \xi \in \mathbf{Z},\ 2 \text{ divides } \xi\}$ is the set of all even integers.

Two sets M and N are said to be equal, $M = N$, if they consist of the same elements. If M and N are not equal we write $M \neq N$. The symbols

$$M \subseteq N \text{ or } N \supseteq M$$

mean that every element of M belongs to N. The set M is then called a *subset* of N, or we say that M is contained in N or that N contains M. If $M \subseteq N$ and $M \supseteq N$ hold simultaneously, then obviously $M = N$. We say that M is a *proper subset* of N and write

$$M \subset N \text{ or } N \supset M$$

if $M \subseteq N$ and $M \neq N$. Among the subsets of any set we include the empty set.

The set D of all elements that are contained both in M and N is called the *intersection* of M and N. It is denoted by

$$D = M \cap N.$$

In case $M \cap N = \emptyset$ the sets M and N are said to be *disjoint*.

The *union* $M \cup N$ consists of all elements that are contained in at least one of the sets M and N.

The intersection and the union of more than two sets are defined analogously. Suppose that to every element λ of some index set Λ there is assigned a set M_λ. Then the intersection and the union of the sets M_λ, $\lambda \in \Lambda$ are denoted by

$$\bigcap_{\lambda\in\Lambda} M_\lambda \quad \text{and} \quad \bigcup_{\lambda\in\Lambda} M_\lambda,$$

respectively. Hence

$$x \in \bigcap_{\lambda\in\Lambda} M_\lambda$$

if and only if $x \in M_\lambda$ for all $\lambda \in \Lambda$, and

$$y \in \bigcup_{\lambda\in\Lambda} M_\lambda$$

if and only if $y \in M_\lambda$ for at least one $\lambda \in \Lambda$.

1.2. Relations

Let us consider some examples of binary relations.

(a) Equality, $\alpha = \beta$, for elements α, β of a given set.

(b) α similar to β for triangles α, β in a given plane.

(c) $|\alpha| = |\beta|$ for $\alpha, \beta \in \boldsymbol{C}$ where $|\alpha|$ denotes the absolute value of α.

(d) For $\alpha, \beta \in \boldsymbol{Z}$, α is said to be of the same parity as β if α and β are both even or both odd.

(e) $\alpha \leqq \beta$ for $\alpha, \beta \in \boldsymbol{R}$.

(f) α divides β for $\alpha, \beta \in \boldsymbol{N}$.

(g) $A \subseteq B$ for subsets A and B of a given set (inclusion relation).

A *binary relation* R on a set M is a condition on the ordered pairs (ξ, η) of elements of M. We write $\xi R \eta$ if the pair (ξ, η) satisfies the condition R. It is necessary to consider ordered pairs since, in general, $\xi R \eta$ does not imply that $\eta R \xi$, cf. examples (e), (f), (g). To define a binary relation in set theoretical terms we form the so-called *Cartesian product* $M \times M$ of M by itself, i.e. the set of all ordered pairs (ξ, η) of elements of M. The relation R then defines a subset $S(R)$ of $M \times M$, namely

$$S(R) = \{(\xi, \eta) \colon \xi, \eta \in M; \xi R \eta\}.$$

Conversely, every subset S of $M \times M$ gives rise to a binary relation R on M if we define $\xi R \eta$ if and only if $(\xi, \eta) \in S$. Thus we may define:

A binary relation on M is a subset of the Cartesian product $M \times M$.

In a similar way one can also define ternary, quaternary etc. relations. Since only binary relations will occur we shall from now on omit the word "binary". Moreover, we are not concerned with arbitrary relations but only with those that have some additional properties. For a relation R on M the following properties will play a role:

R is *reflexive* if $\xi R \xi$ for every $\xi \in M$.

R is *symmetric* if $\xi R \eta$ implies that $\eta R \xi$.

R is *transitive* if $\xi R \eta$ and $\eta R \zeta$ imply that $\xi R \zeta$.

R is *antisymmetric* if $\xi R \eta$ and $\eta R \xi$ imply that $\xi = \eta$.

All the above relations are reflexive and transitive, but $\alpha < \beta$ for $\alpha, \beta \in \mathbf{R}$ is an example of a non-reflexive relation. (a), (b), (c), (d) are symmetric, whereas (e), (f), (g) are antisymmetric.

An *equivalence relation* is one that is reflexive, symmetric, and transitive. Equivalence relations play an important role in mathematics as well as in many branches of science. Due to the symmetry of an equivalence relation we may say that two elements are equivalent to each other. (a), (b), (c), (d) are equivalence relations.

We shall now derive an important theorem on equivalence relations. Let R be an equivalence relation on M. We write $\xi \sim \eta$ and say that ξ is equivalent to η if $\xi R \eta$. For $\alpha \in M$ we define the *equivalence class* $C(\alpha)$ of α as follows:

$$C(\alpha) = \{\xi : \xi \in M, \xi \sim \alpha\}.$$

Since R is reflexive it follows that $\alpha \in C(\alpha)$ so that $C(\alpha)$ is not empty.

Let ξ and η be two elements of $C(\alpha)$. Then $\xi \sim \alpha$ and $\eta \sim \alpha$ and hence by the symmetry of R also $\alpha \sim \eta$. Since R is transitive it follows that $\xi \sim \eta$. Hence any two elements of an equivalence class are equivalent. Moreover, every element of M that is equivalent to an element of $C(\alpha)$ is contained in $C(\alpha)$. Indeed, if $\eta \sim \xi$ and $\xi \in C(\alpha)$, then $\xi \sim \alpha$ and, by the transitivity of R, $\eta \sim \alpha$, which means $\eta \in C(\alpha)$. This shows that two elements of M are equivalent if and only if they belong to the same equivalence class. In particular, two distinct equivalence classes are disjoint.

Since every element of M is contained in its equivalence class it follows that the equivalence classes form a partition of M into mutually disjoint subsets.

Conversely, any partition of M into mutually disjoint subsets gives rise to a relation R on M if we define $\xi R \eta$ if and only if ξ and η belong to the same subset of the partition. It is obvious that R is reflexive, symmetric, and transitive, i.e. an equivalence relation.

Thus we proved the following theorem:

1.2.1. THEOREM. *To any equivalence relation R on a set M there belongs a partition of M into mutually disjoint subsets, the equivalence classes of R. Two elements ξ, η of M belong to the same equivalence* *class if and only if* $\xi R \eta$. *Conversely, to any partition of* M *into mutually disjoint subsets there belongs an equivalence relation* R *on* M *such that the subsets of the partition are the equivalence classes of* R.

A subset T of M that contains one and only one element of each equivalence class is called a *transversal* or a set of *representatives.* The partition of M into equivalence classes can be written as follows:

$$M = \bigcup_{\tau \in T} C(\tau).$$

An *order relation* is one that is reflexive, transitive, and anti-symmetric. In the above list, (e), (f), (g) are order relations.

An arbitrary order relation on a set M is frequently denoted by $\leqq$ but we emphasize that this symbol does not imply any allusion to real numbers. In particular, there may exist so-called incomparable pairs $x, y \in M$, i.e. pairs for which neither $x \leqq y$ nor $y \leqq x$ holds. Therefore M is said to be *partially ordered.*

An order relation $\leqq$ on M is called a *complete* or *linear* order if, for any two elements x, y of M, we have $x \leqq y$ or $y \leqq x$ (or both if $x = y$). In this case, M is said to be *completely ordered* and is called a *chain.*

Clearly, an order relation on a partially ordered set M induces an order relation on any non-empty subset of M. If a subset C is completely ordered with respect to that induced order relation, then C is called a chain in M.

An element m of an ordered set M is called a *maximal* element of M if $x \in M$ and $m \leqq x$ imply that $m = x$.

A partially ordered set may contain several, possibly infinitely many maximal elements. For example, let S be any set with at least two elements and let M denote the set of all subsets of S that are distinct from S. By the set theoretical inclusion, M is partially ordered. The maximal elements of M are precisely those subsets of S that are obtained from S by removing a single element. On the other hand, even a completely ordered set need not have a maximal element, e.g. the real numbers x such that $0 \leqq x < 1$ with $<$ in the usual meaning.

Let T be a subset of an ordered set M. If there exists an element b in M such that $x \leqq b$ for all $x \in T$, then T is said to be *bounded above*, and any element $b \in M$ with this property is called an *upper bound* of T. Note that b need not belong to T.

We now state a proposition about the existence of maximal elements.

1.2.2. Zorn's Lemma. *If every chain in a partially ordered set M is bounded above, then M contains a maximal element.*

We regard Zorn's Lemma as an axiom. It turns out that it is equivalent to the so-called *Axiom of Choice*. This is, roughly speaking, the following proposition: If any collection of non-empty sets is given, then we may assume that there exists a function that assigns to each set one of its elements. It is a matter of taste and expedience which of these two (or of several other) equivalent propositions is taken as an axiom. We choose Zorn's Lemma because it is best suitable for our purpose.

1.3. Mappings

Let M and N be two sets. A *mapping f of M into N* is a rule that assigns to each element ξ of M an element ξf of N. We write

$$f\colon M \to N \quad \text{or} \quad M \xrightarrow{f} N.$$

We also say that f is a *function* defined for $\xi \in M$ whose values ξf belong to N. The reader will probably be more accustomed to the notation $f(\xi)$ instead of ξf, and occasionally we shall use this notation. In general, however, the notation ξf is more convenient for our purposes. Sometimes we shall write ξ^f instead of ξf. To indicate the effect of f on the elements of M we use the notation

$$f\colon \xi \to \xi f \quad (\xi \in M).$$

For instance the mapping

$$f\colon \xi \to \xi^2 \quad (\xi \in \boldsymbol{Z})$$

is a mapping of $\boldsymbol{Z}$ into the set of all non-negative integers.

The element ξf is called the *image* of ξ under f, and ξ is said to be a *preimage* of ξf. (Note that we have to use the indefinite article since an element of N may have several or even infinitely many preimages in M.)

The subset Mf of N that consists of all elements ξf where $\xi \in M$ is called the image of M under f, and similarly the image of a subset A of M under f is defined by $Af = \{\xi f: \xi \in A\}$. For a subset N_0 of Mf we put $N_0 f^{-1} = \{\xi: \xi \in M, \xi f \in N_0\}$ and call $N_0 f^{-1}$ the *inverse image* of N_0. In particular, for an element $\eta \in Mf$ the inverse image ηf^{-1} consists of all preimages of η.

The concept of a mapping can also be described in set theoretical terms. Let $M \times N$ denote the Cartesian product of M and N, i.e. the set of all ordered pairs (ξ, η) with $\xi \in M$, $\eta \in N$. Then a mapping f of M into N is uniquely determined by the subset $S(f)$ of $M \times N$ that consists of all pairs $(\xi, \xi f)$, $\xi \in M$. Since every element of M has a unique image the set $S(f)$ has the following property:

If $(\xi, \eta_1) \in S(f)$ and $(\xi, \eta_2) \in S(f)$, then $\eta_1 = \eta_2$.

Conversely, let S be a subset of $M \times N$ with this property and such that for each element $\xi \in M$ at least one (and hence exactly one) pair (ξ, η) belongs to S. Then we obviously obtain a mapping f of M into N if we put $\xi f = \eta$ if $(\xi, \eta) \in S$.

A mapping f is said to be *injective* or *one-to-one* if $\xi_1 \neq \xi_2$ implies that $\xi_1 f \neq \xi_2 f$.

We say that f is a *surjective* mapping or a mapping of M *onto* N if $Mf = N$, i.e. if every element of N is the image of at least one element of M. When we say that f maps M into N, then this does not exclude that f is actually a mapping onto N.

A mapping that is both injective and surjective is called *bijective*. An injective mapping f of M into N yields a bijective mapping of M onto Mf.

If f is a bijective mapping of M onto N, then each element η of N has a unique inverse image ηf^{-1} in M. In this case, therefore,

$$f^{-1}: \eta \to \eta f^{-1} \quad (\eta \in N)$$

is a bijective mapping of N onto M. We call f^{-1} the *inverse* of f. Obviously, f is the inverse of f^{-1}.

Suppose that we are given two mappings

$$f: M \to N, \quad g: N \to P.$$

Then we may combine these two mappings to a mapping of M into P.

To any element $\xi \in M$ we first apply f to obtain the element ξf of N, and then we apply g to ξf which yields the element $(\xi f)\, g$ of P. The combined mapping of M into P is denoted by fg and is called the *product* of f and g. So we have

$$\xi(fg) = (\xi f)\, g \quad (\xi \in M). \tag{1}$$

It is clear that the product fg is only defined if Mf is contained in the set on which g is defined. Hence the existence of fg does not imply that the converse product gf exists. Moreover, we shall see that fg and gf need not coincide if both are defined.

Though, in contrast to the product of numbers, the order of the factors is in general relevant for products of mappings there is another property of the products of numbers which is shared by products of mappings.

Let us consider a third mapping

$$h: P \to Q.$$

Then we may combine the mappings f, g, and h in this order. The combined mapping maps M into Q, and the image of any element ξ of M is

$$\big((\xi f)\, g\big)\, h.$$

Due to (1) we obtain

$$\big((\xi f)\, g\big)\, h = (\xi fg)\, h = \xi(fg)\, h$$

and on the other hand

$$\big((\xi f)\, g\big)\, h = (\xi f)\, (gh) = \xi f(gh).$$

Since ξ is an arbitrary element of M it follows that

$$(fg)\, h = f(gh).$$

This rule is called the *associative law*. Due to this law we may omit the parentheses and denote the combined mapping without ambiguity by fgh.

A bijective mapping of a set M onto itself is called a *permutation* of M. A permutation

$$a: \xi \to \xi a \quad (\xi \in M)$$

is frequently denoted as follows:

$$a = \begin{pmatrix} \xi \\ \xi a \end{pmatrix} \quad (\xi \in M).$$

It is obvious that the inverse mapping

$$a^{-1} = \begin{pmatrix} \xi a \\ \xi \end{pmatrix} \quad (\xi \in M)$$

is also a permutation of M. If b is another permutation of M, then the product ab is also a permutation of M. Indeed, since a and b are surjective we have $Ma = M$, $Mb = M$ and hence $M(ab) = (Ma)\,b = Mb = M$ so that ab is surjective. For two distinct elements ξ, η of M we have $\xi a \neq \eta a$ since a is injective. Because b is injective it follows that $\xi(ab) = (\xi a)\,b \neq (\eta a)\,b = \eta(ab)$, hence ab is surjective. Therefore ab is a permutation of M.

Clearly, both products aa^{-1} and $a^{-1}a$ are equal to the *identity permutation* that maps each element of M onto itself. The identity permutation will frequently be denoted by e.

In case M is a finite set, $M = \{\xi_1, \xi_2, \ldots, \xi_n\}$ say, we frequently use the more explicit notation

$$a = \begin{pmatrix} \xi_1 & \xi_2 & \ldots & \xi_n \\ \xi_1 a & \xi_2 a & \ldots & \xi_n a \end{pmatrix}. \tag{2}$$

Here the elements $\xi_1 a, \xi_2 a, \ldots, \xi_n a$ coincide, except for the order, with $\xi_1, \xi_2, \ldots, \xi_n$ so that the second row of (2) is a permutation of the first row in the sense of combinatorics. This accounts for the term permutation. As is well known there are $n!$ distinct permutations of a set with n elements.

As an example we consider the permutations of the set $M = \{1, 2, 3\}$. Let

$$a = \begin{pmatrix} 1 & 2 & 3 \\ 2 & 3 & 1 \end{pmatrix}, \quad b = \begin{pmatrix} 1 & 2 & 3 \\ 1 & 3 & 2 \end{pmatrix}.$$

The permutation a carries 1 into 2, and b carries 2 into 3. Hence the product ab maps 1 onto 3. Similarly we obtain

$$2(ab) = (2a)b = 3b = 2,$$

$$3(ab) = (3a)b = 1b = 1.$$

This gives

$$ab = \begin{pmatrix} 1 & 2 & 3 \\ 3 & 2 & 1 \end{pmatrix}.$$

We find

$$ba = \begin{pmatrix} 1 & 2 & 3 \\ 2 & 1 & 3 \end{pmatrix}.$$

This shows that $ab \neq ba$. Further

$$a^{-1} = \begin{pmatrix} 2 & 3 & 1 \\ 1 & 2 & 3 \end{pmatrix} = \begin{pmatrix} 1 & 2 & 3 \\ 3 & 1 & 2 \end{pmatrix},$$

$$aa = \begin{pmatrix} 1 & 2 & 3 \\ 3 & 1 & 2 \end{pmatrix} = a^{-1}, \quad b = b^{-1}.$$

1.4. Operations

A *binary operation* Φ in a set M is a rule that assigns to each ordered pair ξ, η of elements of M an element $\Phi(\xi, \eta)$ of M. In other words, a binary operation is a mapping of $M \times M$ into M.

Since we shall always deal only with a small number of distinct operations simultaneously, it is convenient to denote operations by special symbols. We might for instance use the symbol $\circ$ to denote the operation Φ and write $\xi \circ \eta$ instead of $\Phi(\xi, \eta)$. The symbols $+$, $-$, and $\times$ are symbols for familiar operations in $\boldsymbol{C}$, namely addition, subtraction and multiplication. If there is no danger of confusion we shall use these symbols also for other binary operations in other sets. Very frequently we shall denote a binary operation simply by juxtaposition, i.e. we write $\xi\eta$ to denote the element that a given binary operation assigns to the pair ξ, η. The product of two mappings is an example of this convenient notation.

It is necessary to generalize the concept of a binary operation to pairs of elements of different sets. In the most general case we have three sets M, N, and P and a mapping of $M \times N$ into P. For our purposes it is sufficient to deal with the following generalization: For two sets Ω and M we consider a mapping of $\Omega \times M$ into M. For example let $\Omega = \boldsymbol{R}$ and let M be the set of all vectors in a given plane; for a real number α and a vector $v \in M$ we then define αv as the product of the vector v by α.

Occasionally ternary operations will occur. It is sufficient to consider the following case. Let M, N, and P be three sets. Then a mapping of $(M \times N) \times P$ into N is a ternary operation which assigns to each triplet α, x, β ($\alpha \in M$, $x \in N$, $\beta \in P$) an element of N.

1.5. Algebraic systems

An *algebraic system* is a set in which one or several operations and possibly also one or several relations are defined and satisfy certain postulates. In this section we introduce some of the most important algebraic systems.

Rings

Let M_n be the set of all $n \times n$ matrices whose elements are real numbers. We assume that the reader is acquainted with two operations in M_n, namely addition and multiplication of matrices. Addition and multiplication are also defined in the set $\boldsymbol{Z}$ of the integers. In both cases the following laws are satisfied for arbitrary elements a, b, c of M_n or $\boldsymbol{Z}$, respectively:

(R_1) $a + b = b + a$ (*commutative law of addition*).

(R_2) $(a + b) + c = a + (b + c)$ (*associative law of addition*).

(R_3) There exists an element 0 such that

$$a + 0 = a \text{ for every } a.$$

(R_4) For any a there exists an element $-a$ such that

$$a + (-a) = 0.$$

(R_5) $(ab)\,c = a(bc)$ (*associative law of multiplication*).

(R_6) $(a + b)\,c = ac + bc, \quad c(a + b) = ca + cb$ (*distributive laws*).

Any algebraic system with two operations, addition and multiplication, in which these laws hold for arbitrary elements is called a *ring*.

The element whose existence is postulated by (R_3) is called the *zero element*. We shall prove later that a ring contains only a single zero element. It will also turn out that the element $-a$ in (R_4) is uniquely determined by a; it is called the additive inverse of a. A sum $b + (-a)$ is simply denoted by $b - a$.

In contrast to M_n, in the ring $\boldsymbol{Z}$ of the integers another law is satisfied, namely:

(R_7) $ab = ba$ (*commutative law of multiplication*).

Every ring in which (R_7) holds is called a *commutative ring.* In any commutative ring each of the two distributive laws is obviously a consequence of the other. Another example of a commutative ring is the set of all complex numbers $a + bi$ with $a, b \in \boldsymbol{Z}$.

So far all rings had the following property:

(R_8) There exists a *unit element*, i.e. an element e such that

$$ea = ae \text{ for every } a.$$

The ring of the even integers shows that a ring need not contain a unit element. We shall see later that a ring cannot contain more than one unit element. The unit element will frequently be denoted by 1 also when the ring does not consist of numbers.

The rational numbers too form a commutative ring. But they have the following additional property:

(R_9) For every $a \neq 0$ there exists an element a^{-1} such that

$$aa^{-1} = a^{-1}a = 1.$$

Any algebraic system that contains more than one element and satisfies (R_1) through (R_9) is called a *field.*

So the rational numbers form a field. Other examples of fields are the real numbers and the complex numbers. The fields of the rational numbers, the real numbers, and the complex numbers will be denoted by $\boldsymbol{Q}$, $\boldsymbol{R}$, and $\boldsymbol{C}$, respectively as in § 1.1.

Finally, the complex numbers $a + bi$ with $a, b \in \boldsymbol{Q}$ also form a field.

Groups

Let S be a given non-empty set. We consider the set $P(S)$ of all permutations of S. As we proved in § 1.3 the product of two permutations of S is again a permutation of S. Hence $P(S)$ can be regarded as an algebraic system with one operation. As we saw in § 1.3 the following laws are satisfied for arbitrary elements a, b, c of $P(S)$:

(G_1) $(ab)\,c = a(bc)$ (*associative law*).

(G_2) There exists an element e such that

$$ae = ea = a \text{ for every } a\,.$$

(G_3) For every a there exists an element a^{-1} such that

$$aa^{-1} = a^{-1}a = e\,.$$

Any algebraic system with one operation in which (G_1), (G_2), (G_3) are satisfied is called a *group*.

Later we shall prove that a group cannot contain more than one element e with the property (G_2). This element is called the *unit element* or the *identity* of the group. It will turn out that for a given a the element a^{-1} is uniquely determined. It is called the *inverse* of a.

Let us consider the set of all non-singular $n \times n$ matrices whose elements belong to a given field $\mathbf{K}$. The product of two non-singular matrices is again non-singular and the inverse of a non-singular matrix with elements in $\mathbf{K}$ has also elements in $\mathbf{K}$. Moreover, multiplication of matrices is associative. Finally the $n \times n$ identity matrix has the properties of the unit element. Therefore the non-singular $n \times n$ matrices with elements in $\mathbf{K}$ form a group under matrix multiplication. This group is denoted by $\mathrm{GL}(n, \mathbf{K})$ and is called the *general linear group* of degree n over $\mathbf{K}$.

A group is said to be *abelian* if the *commutative law* holds, i.e. if

$$ab = ba \text{ for any two elements } a, b\,.$$

The group operation has been denoted by juxtaposition. We refer to this notation as the multiplicative notation. We also say that ab is the *product* of a and b. However, we may just as well adopt the additive notation and write $a + b$ instead of ab. In this case $a + b$ is called the *sum* of a and b. The additive notation suggests to denote the identity by 0 and the inverse of a by $-a$ so that

$$a + 0 = a, \quad a + (-a) = 0\,.$$

Additive notation is mainly used for abelian groups.

The four laws (R_1) through (R_4) for addition in a ring can be summarized in the statement that the elements of a ring form an abelian group with respect to addition. Similarly, (R_5) and (R_7) through (R_9) mean that the non-zero elements of a field form an abelian group under multiplication.

Vector spaces

Let $\mathbf{K}^n$ denote the set of all $n \times 1$ matrices or columns

$$x = \begin{bmatrix} \xi_1 \\ \vdots \\ \xi_n \end{bmatrix}$$

whose elements $\xi_1, \ldots, \xi_n$ belong to a given field $\mathbf{K}$. Addition of columns and multiplication of a column by an element α of $\mathbf{K}$ are defined in the usual way, namely

$$x + y = \begin{bmatrix} \xi_1 \\ \vdots \\ \xi_n \end{bmatrix} + \begin{bmatrix} \eta_1 \\ \vdots \\ \eta_n \end{bmatrix} = \begin{bmatrix} \xi_1 + \eta_1 \\ \vdots \\ \xi_n + \eta_n \end{bmatrix}, \quad \alpha x = \begin{bmatrix} \alpha\xi_1 \\ \vdots \\ \alpha\xi_n \end{bmatrix}.$$

In case that $\mathbf{K} = \boldsymbol{R}$ and $n = 2$ or $n = 3$ we may think of $\xi_1, \ldots, \xi_n$ as coordinates of a vector in a plane or a three-dimensional space with respect to a given coordinate system. Then addition of columns corresponds to addition of vectors, and multiplication of a column by an element of $\mathbf{K}$ corresponds to multiplication of a vector by a scalar. In view of this special case any algebraic system with properties similar to those of $\mathbf{K}^n$ is called a vector space. This leads to the following general definition.

Let V be a non-empty set for whose elements addition is defined. Let further $\mathbf{K}$ be any field and assume that for any element α of $\mathbf{K}$ and any element x of V there is defined an element αx of V. Then V is said to be a *vector space over* $\mathbf{K}$ if the following postulates are satisfied:

(V_1) $x + y = y + x \quad (x, y \in V)$.

(V_2) $(x + y) + z = x + (y + z) \quad (x, y, z \in V)$.

(V_3) V contains an element 0 such that

$$x + 0 = x \text{ for every } x \in V.$$

(V_4) For every $x \in V$ there exists an element $-x$ in V such that

$$x + (-x) = 0.$$

(V_5) $\alpha(x + y) = \alpha x + \alpha y \quad (\alpha \in \mathbf{K};\ x, y, \in V).$

(V_6) $(\alpha + \beta)\, x = \alpha x + \beta x \quad (\alpha, \beta \in \mathbf{K};\ x \in V).$

(V_7) $(\alpha\beta)\, x = \alpha(\beta x) \quad (\alpha, \beta \in \mathbf{K};\ x \in V).$

(V_8) $1x = x$ for the unit element 1 of $\mathbf{K}$.

The postulates (V_1) through (V_4) mean that V is an abelian group with respect to addition.

It is obvious that these laws hold in $\mathbf{K}^n$.

Exercises

1. Prove that for any three subsets A, B, C of a given set

$$A \cap (B \cup C) = (A \cap B) \cup (A \cap C),$$

$$A \cup (B \cap C) = (A \cup B) \cap (A \cup C).$$

2. Prove that for two sets A, B the following propositions are equivalent (i.e. each is a consequence of any other):

   (i) $A \subseteq B$, (ii) $A \cap B = A$, (iii) $A \cup B = B$.

3. For any subset S of a set M let S' denote the set of all those elements of M that are not contained in S. (S' is called the *complement* of S in M.) Prove that for any two subsets A, B of M

$$(A \cap B)' = A' \cup B', \qquad (A \cup B)' = A' \cap B'$$

   (De Morgan's laws).

4. Give an example of a relation that is reflexive and symmetric but not transitive.
5. What is wrong with the following argument to show that every symmetric and transitive relation is reflexive? If $\alpha \sim \beta$ then by the symmetriy $\beta \sim \alpha$ and hence by the transitivity $\alpha \sim \alpha$. Give an example of a relation that is symmetric and transitive but not reflexive.
6. Find the unique relation that is reflexive, symmetric, transitive, and antisymmetric.

7. Let M be a non-empty set and $\gamma \in M$. Define a binary relation $*$ on M such that $\xi * \eta$ if and only if $\xi = \gamma$ or $\eta = \gamma$. When is this relation transitive?
8. Describe the equivalence classes of example (c) in § 1. 2 as sets of points in the complex plane. Find a system of representatives for the equivalence classes.
9. Let R_1 and R_2 be relations on a set M and let $S(R_i)$ denote the subset of $M \times M$ that defines R_i in the sense of § 1. 2. Define relations $R_1 \wedge R_2$ and $R_1 \vee R_2$ as follows:

$$\xi(R_1 \wedge R_2)\,\eta \text{ if and only if } \xi R_1 \eta \textit{ and } \xi R_2 \eta,$$

$$\xi(R_1 \vee R_2)\,\eta \text{ if and only if } \xi R_1 \eta \textit{ or } \xi R_2 \eta.$$

(Here $\xi R_1 \eta$ or $\xi R_2 \eta$ does not exclude that both relations hold simultaneously.)

(i) Express $S(R_1 \wedge R_2)$ and $S(R_1 \vee R_2)$ in terms of $S(R_1)$ and $S(R_2)$.

(ii) Let R_1 and R_2 be equivalence relations. Prove that $R_1 \wedge R_2$ is also an equivalence relation and describe the equivalence classes of $R_1 \wedge R_2$ in terms of those of R_1 and R_2. Is $R_1 \vee R_2$ also an equivalence relation?

10. Let R, R_1, R_2, R_3 be relations on a set M. Write $R \subseteq R_1$ to mean that $\xi R \eta$ implies $\xi R_1 \eta$. Define a new relation $R_1 \circ R_2$ as follows: $\xi(R_1 \circ R_2)\,\eta$ if and only if there exists at least one $\zeta \in M$ such that $\xi R_1 \zeta$ and $\zeta R_2 \eta$. Moreover, define R^{-1} as follows: $\xi R^{-1} \eta$ if and only if $\eta R \xi$. Finally, let E denote the identity relation, i.e. $\xi E \eta$ if and only if $\xi = \eta$.

(i) Explain the notation $R \subseteq R_1$ in terms of $S(R)$ and $S(R_1)$.

(ii) Prove that

$$(R_1 \circ R_2) \circ R_3 = R_1 \circ (R_2 \circ R_3).$$

Give an example to show that the relations $R_1 \circ R_2$ and $R_2 \circ R_1$ need not coincide.

(iii) Show that

R is reflexive if and only if $E \subseteq R$,
R is symmetric if and only if $R^{-1} = R$,
R is transitive if and only if $R \circ R \subseteq R$,
R is antisymmetric if and only if $R \wedge R^{-1} \subseteq E$.

(iv) Prove that

$$(R_1 \circ R_2)^{-1} = R_2^{-1} \circ R_1^{-1},$$
$$(R_1 \wedge R_2)^{-1} = R_1^{-1} \wedge R_2^{-1},$$
$$(R_1 \vee R_2)^{-1} = R_1^{-1} \vee R_2^{-1}.$$

(v) Suppose that R_1 and R_2 are equivalence relations. Prove that $R_1 \circ R_2$ is an equivalence relation if and only if $R_1 \circ R_2 = R_2 \circ R_1$.

(vi) Prove that

$$(R_1 \wedge R_2) \circ R \subseteq (R_1 \circ R) \wedge (R_2 \circ R),$$
$$R \circ (R_1 \wedge R_2) \subseteq (R \circ R_1) \wedge (R \circ R_2),$$
$$(R_1 \vee R_2) \circ R = (R_1 \circ R) \vee (R_2 \circ R),$$
$$R \circ (R_1 \vee R_2) = (R \circ R_1) \vee (R \circ R_2).$$

11. How many relations can be defined on a set of n elements (n a natural number)?
12. Let m be a natural number. For $x, y \in \boldsymbol{Z}$ let $x \sim y$ if and only if $x - y$ is divisible by m. Show that this is an equivalence relation.
13. For $\alpha, \beta \in \boldsymbol{N}$ verify that the relation "α divides β" is an order relation. Show that $\boldsymbol{N}$ is not completely ordered by this relation. Find a chain in $\boldsymbol{N}$.
14. Define a finite set to be one that admits no bijective mapping onto a proper subset. Show that every subset of a finite set is finite or empty.
15. Let c be a given integer and define for $x \in \boldsymbol{Z}$ the mapping

$$f_c : x \to x + c - xc$$

of $\boldsymbol{Z}$ into itself. For what values of c is f_c injective and for what values of c is f_c surjective? Answer the same questions for $\boldsymbol{Q}$ instead of $\boldsymbol{Z}$ where c is a given rational number.

16. Let

$$f : M \to N, \quad g : N \to P$$

be two bijective mappings.

(i) Show that fg is a bijective mapping of M onto P.
(ii) Prove that $(fg)^{-1} = g^{-1}f^{-1}$.

17. Let

$$f: M \to N, \quad g: N \to P$$

be two mappings.

(i) Suppose that fg is surjective. Show that g is surjective. Is f necessarily surjective?
(ii) Suppose that fg is injective. Show that f is injective. Is g necessarily injective?

18. A ternary relation on a set M is defined by a subset of the Cartesian product $(M \times M) \times M$. Describe the subsets which belong to those ternary relations that are binary operations in M.

19. In $\boldsymbol{Q}$ define the operation $a \circ b = a + b - ab$. Is $\boldsymbol{Q}$ a group with respect to this operation?

20. Define addition of complex numbers as usual and replace multiplication by the operation $a * b = \overline{ab}$. (Here $\bar{a}$ denotes the complex conjugate of a). Which of the axioms (R_1) through (R_9) are valid for the operations $+$ and $*$?

21. Let S be a non-empty set and G a set of permutations of S. In S define the relation $\xi \sim \eta$ to hold if and only if G contains at least one permutation that carries ξ into η. Show that $\sim$ is an equivalence relation if G is a group.

22. Let $\mathbf{K}^n$ denote the vector space of all $n \times 1$ matrices

$$x = \begin{bmatrix} \xi_1 \\ \xi_2 \\ \vdots \\ \xi_n \end{bmatrix}$$

with elements ξ_i in a field $\mathbf{K}$. Define addition as in § 1.5 and put

$$\alpha x = \begin{bmatrix} 2\alpha\xi_1 \\ \alpha\xi_2 \\ \vdots \\ \alpha\xi_n \end{bmatrix}$$

for $\alpha \in \mathbf{K}$. Which of the axioms (V_1) through (V_8) are satisfied?

23. Let a be a fixed vector of the vector space $\boldsymbol{R}^3$.
    (i) For any $x \in \boldsymbol{R}^3$ let $a \times x$ denote the vector product. Is the mapping $x \to a \times x$ injective and surjective?
    (ii) Answer the same question for the mapping $x \to (a, x)\, x$ where (a, x) denotes the scalar product.

2

The integers

In this chapter we shall derive some results of elementary number theory which will be used later. Moreover the ring of the integers will provide examples of some basic concepts which play a role in the study of more general rings.

Throughout this chapter lower case Latin letters will denote integers.

2.1. The natural numbers and the integers

From long experience at school the reader will certainly be familiar with the following facts about integers:

(a) *The integers form a commutative ring with the unit element* 1.

(b) *The integers are completely ordered, i.e. for any two distinct integers a, b we have*

$$\textit{either } a > b \textit{ or } b < a.$$

(c) *If* $a > 0$ *and* $b > 0$, *then* $a + b > 0$ *and* $ab > 0$.

(d) *Every non-empty set of natural numbers contains a least element.*

In our approach these facts play the role of axioms. These properties of the integers can actually be derived from fewer and simpler axioms. The best known system of axioms for the integers is due to G. Peano. The way from Peano's axiom to (a), (b), (c),

(d) is not difficult but rather long. Therefore we do not present it here but refer to [3].

It is easy to derive all the familiar properties of the order from (c). In particular, if $a > 0$, $b > 0$ or $0 > a$, $0 > b$, then $ab > 0$; and if $a > 0$, $0 > b$, then $0 > ab$. This implies that the product of two non-zero integers is distinct from zero, a property which is not at all shared by all rings.

In view of (d) the natural numbers are said to be *well-ordered.* We leave it as an exercise to the reader to prove that every non-empty set of integers is well ordered if all its elements are greater than some given number.

The well-ordering property of the natural numbers is the basis for proofs by induction.

Let $P(n)$ be a proposition that involves a natural number n, e.g.

$$P(n)\colon 1 + 2 + \ldots + n = 1/2\, n(n + 1).$$

In order to prove that $P(n)$ is true for all natural numbers n it is sufficient to prove the following two propositions:

(i) $P(1)$ is true.

(ii) If k is an arbitrary natural number and if $P(x)$ is true for all natural numbers with $1 \leqq x \leqq k$, then $P(k + 1)$ is true.

Indeed, let F be the set of all m for which $P(m)$ is false. We have to show that F is empty. Suppose that F is not empty. Then, by (d), F contains a least element l. From (i) we conclude that $l > 1$. Therefore we can write $l = k + 1$ for some natural number k. By the choice of l, $P(x)$ is true for all x with $1 \leqq x \leqq k$. Hence it follows from (ii) that $P(l) = P(k + 1)$ is true, a contradiction.

In the proof of (ii) one can frequently verify $P(k + 1)$ by the assumption that $P(k)$ is true without considering $P(1), \ldots, P(k - 1)$. Due to the above remark about well-ordered sets of integers one can start a proof by induction with $P(0)$ instead of $P(1)$ or with $P(g)$ for some integer g.

Another familiar property of the integers is also a consequence of the well-ordering.

2.1.1. Division Algorithm. *Let a and m be integers and $m > 0$. Then there exists one and only one pair q, r of integers such that*

$$a = qm + r, \quad 0 \leqq r < m. \tag{1}$$

The integer r is called the *remainder* in the division of a by m.

Proof. Let D be the set of all integers of the form $a - xm$ where x ranges over all integers. For the particular choice $x = 0$ if $a \geqq 0$ and $x = a$ if $a < 0$ the number $a - xm$ is non-negative. Hence the subset D_0 of the non-negative integers in D is not empty. Due to the well-ordering of the non-negative integers, D_0 contains a least element, $r = a - qm$ say. Since all integers in D_0 are non-negative we obtain $r = a - qm \geqq 0$. Moreover, since $a - qm$ is the least integer in D_0 we conclude that $a - (q + 1)\, m < 0$. This gives

$$r - m = a - qm - m = a - (q + 1)\, m < 0,$$

i.e. $r < m$. Hence the integers q and r have the desired property.

To prove the uniqueness, let q', r' be any pair of integers such that

$$a = q'm + r', \quad 0 \leqq r' < m.$$

Suppose that $q' < q$. Then we obtain

$$a - q'm = r' \geqq a - (q - 1)\, m = r + m \geqq m$$

which contradicts the condition $r' < m$. Hence $q' < q$ is impossible. Similarly, $q < q'$ leads to a contradiction. It follows that $q = q'$ and hence $r = r'$. This completes the proof.

2.2. Divisibility. Prime numbers

An integer $a \neq 0$ is called a *divisor* of the integer b if there exists an integer g such that $ag = b$. To indicate that a is a divisor of b we write $a \mid b$. We also say that a divides b, that b is divisible by a, or that b is a multiple of a.

The following propositions are immediate consequences of this definition so that the proof can be left to the reader:

If $a \mid b$ and $b \mid c$, then $a \mid c$.

If $a \mid b$ and $a \mid b'$, then $a \mid (bx + b'x')$ for any two integers x, x'.

If $a \mid b$ and $a > 0$, $b > 0$, then $a \leqq b$.

If $a \mid b$, then $(\pm a) \mid (\pm b)$ for every choice of the signs.

Due to the last proposition it is frequently sufficient to study divisibility only for natural numbers.

Any integer $b \neq 0$ has the so-called *trivial divisors* $\pm b$ and ± 1.

A *prime number* is a natural number that is greater than 1 and has only trivial divisors. The first prime numbers are 2, 3, 5, 7, 11, 13, 17, 19, 23, ...

Any prime number that divides an integer b is called a *prime divisor* of b.

2.2.1. Lemma. *Every natural number $b > 1$ has at least one prime divisor.*

Proof. Let D be the set of all natural numbers that divide b and are greater than 1. The set D is not empty since $b \mid b$ and $b > 1$. Hence D contains a least number p. We show that p is a prime number. Suppose that q is a non-trivial positive divisor of p so that $q \mid b$ and $1 < q < p$. From $q \mid p$ and $p \mid b$ it follows that $q \mid b$ and hence $q \in D$ which contradicts the choice of p. Hence the theorem is proved.

2.2.2. Theorem. *There are infinitely many prime numbers.*

Proof. We shall prove the following propositions. Let $\{p_1, p_2, \ldots, p_r\}$ be a finite set of prime numbers. Then there exists a prime number that is not contained in this set. Put

$$m = p_1 p_2 \cdots p_r + 1 .$$

By the previous theorem, m has a prime divisor p, possibly $p = m$. The prime number p is distinct from $p_1, p_2, \ldots, p_r$ since m is divisible by p whereas m has the remainder 1 under division by $p_1, p_2, \ldots, p_r$. This completes the proof. (This proof is contained in Euclid's Elements, about 300 B.C.)

2.3. The greatest common divisor

Let $a_1, a_2, \ldots, a_n$ be integers and $a_i \neq 0$ for $i = 1, 2, \ldots, n$. We may assume that $n \geqq 2$ since for $n = 1$ the following is trivial. An integer t such that

$$t \mid a_i \text{ for } i = 1, 2, \ldots, n$$

is called a *common divisor* of $a_1, a_2, \ldots, a_n$.

We now come to one of the most important theorems of number theory.

2.3.1. THEOREM ON THE GREATEST COMMON DIVISOR. *Let $a_1, a_2, \ldots, a_n$ be non-zero integers. There exists one and only one positive common divisor d of $a_1, a_2, \ldots, a_n$ that satisfies the condition:*

(i) *d is divisible by every common divisor of $a_1, a_2, \ldots, a_n$.*

d is called the greatest common divisor of $a_1, a_2, \ldots, a_n$ and is denoted by

$$d = (a_1, a_2, \ldots, a_n).$$

(ii) *The greatest common divisor can be represented in the form*

$$d = a_1x_1 + a_2x_2 + \ldots + a_nx_n \tag{1}$$

where $x_1, x_2, \ldots, x_n$ are suitable integers.

Conversely, every common divisor of $a_1, a_2, \ldots, a_n$ that can be represented in the form (1) *satisfies* (i).

Proof. Let d and d' be positive common divisors of $a_1, a_2, \ldots, a_n$ that satisfy (i). It follows that $d \mid d'$ and $d' \mid d$ so that $d = d'$. Hence there exists at most one positive common divisor of $a_1, a_2, \ldots, a_n$ that satisfies (i).

Every common divisor d that can be represented in the form (1) satisfies (i) because every common divisor of $a_1, a_2, \ldots, a_n$ divides the right-hand side of (1).

Therefore it remains to prove that there exists a positive common divisor d of $a_1, a_2, \ldots, a_n$ that admits a representation (1). The proof will be constructive, i.e. it provides an algorithm to compute d and the representation (1).

We first consider the case $n = 2$. We may obviously assume that a_1 and a_2 are natural numbers. For $a_1 = a_2$ the theorem is certainly true so that we may assume that $a_1 > a_2$. In case $a_2 \mid a_1$ it is clear that $a_2 = d$ can be represented in the form (1) and the theorem follows. Otherwise repeated application of the division algorithm gives

(2.1) $a_1 = a_2q_1 + r_1, \quad 0 < r_1 < a_2,$

(2.2) $a_2 = r_1q_2 + r_2, \quad 0 < r_2 < r_1,$

(2.3) $r_1 = r_2q_3 + r_3, \quad 0 < r_3 < r_2,$

(2.4) $r_2 = r_3q_4 + r_4, \quad 0 < r_4 < r_3,$

....

The remainders $r_1, r_2, r_3, \ldots$ form a strictly decreasing sequence of non-negative integers. Therefore we must arrive at the remainder 0 after a finite number of steps. Hence, for some natural number k we obtain

(2.k − 2) $r_{k-4} = r_{k-3}q_{k-2} + r_{k-2}, \quad 0 < r_{k-2} < r_{k-3},$

(2.k − 1) $r_{k-3} = r_{k-2}q_{k-1} + r_{k-1}, \quad 0 < r_{k-1} < r_{k-2},$

(2.k) $r_{k-2} = r_{k-1}q_k + r_k, \quad 0 < r_k < r_{k-1},$

(2.k + 1) $r_{k-1} = r_kq_{k+1}.$

We will show that r_k, the last non-zero remainder, is a common divisor of a_1 and a_2 and can be represented in the form $r_k = a_1x_1 + a_2x_2$ with suitable integers x_1, x_2.

From (2.k + 1) it follows that $r_k \mid r_{k-1}$. Hence the right-hand side of (2.k) is divisible by r_k so that $r_k \mid r_{k-2}$. From $r_k \mid r_{k-1}$ and $r_k \mid r_{k-2}$ it follows that the right-hand side of (2.k − 1) is divisible by r_k and hence $r_k \mid r_{k-3}$. When we proceed in this way and arrive at (2.2) we know that r_k divides both r_1 and r_2 so that $r_k \mid a_2$. Finally, (2.1) shows that $r_k \mid a_1$. Therefore r_k is a common divisor of a_1 and a_2.

If a number d can be represented in the form (1) we shall say for short that d can be represented as a linear combination of $a_1, a_2, \ldots, a_n$. We have to show that r_k can be represented as a linear combination of a_1 and a_2. From (2.k) we obtain

$$r_k = r_{k-2} - r_{k-1}q_k$$

so that r_k is a linear combination of r_{k-1} and r_{k-2}. By (2.k − 1), r_{k-1} can be replaced by a linear combination of r_{k-2} and r_{k-3}.

Hence we obtain a representation of r_k as a linear combination of r_{k-2} and r_{k-3}. Then we use $(2.k-2)$ to obtain a representation of r_k as a linear combination of r_{k-3} and r_{k-4}. Proceeding in this way (2.1) finally yields a representation of r_k as a linear combination of a_1 and a_2. This proves the theorem for $n=2$.

We now complete the proof by induction on n. Suppose that the theorem is true for all sets of l integers where $2 \leq l < n$. If $a_1, a_2, \ldots, a_n$ are given we choose an arbitrary integer l with $2 \leq l < n$. From the induction hypothesis it follows that the following greatest common divisors exist and can be represented as linear combinations:

$$d_1 = (a_1, \ldots, a_l), \quad d_2 = (a_{l+1}, \ldots, a_n), \quad d = (d_1, d_2).$$

Since d is a common divisor of d_1 and d_2 we conclude that d is a common divisor of $a_1, \ldots, a_l, a_{l+1}, \ldots, a_n$. Moreover, d can be expressed in the form $d = d_1y_1 + d_2y_2$ with suitable integers y_1 and y_2. In this expression we substitute d_1 and d_2 by their representations as linear combinations of $a_1, \ldots, a_l$ and $a_{l+1}, \ldots, a_n$, respectively. This gives a representation of d as a linear combination of $a_1, \ldots, a_n$.

The theorem is now proved.

The above division algorithm for the greatest common divisor of two integers is called the *Euclidean algorithm*.

The integers $a_1, a_2, \ldots, a_n$ are said to be *relatively prime* if $(a_1, a_2, \ldots, a_n) = 1$. By the last theorem, $a_1, a_2, \ldots, a_n$ are relatively prime if and only if

$$a_1x_1 + a_2x_2 + \ldots. + a_nx_n = 1$$

with suitable integers $x_1, x_2, \ldots, x_n$.

It is clear that the greatest common divisor as defined in the previous theorem is the greatest positive common divisor with regard to the order of the integers. This accounts for its name. In our definition, however, we did not refer to the order. This has been done on purpose because our definition also applies to rings in which no order relation can be defined. Such rings will be studied in Chapter 4.

2.4. Prime factorization

Let m be a natural number, $m > 1$. We shall prove that m can be represented as a product of prime numbers. This is certainly true if m itself is a prime number. Hence it is in particular true for $m = 2$. We apply induction and assume that every natural number that is greater than 1 and less than m can be represented as a product of prime numbers. By Lemma 2.2.1, m has at least one prime divisor, p_1 say. This gives $m = p_1 m_1$ where m_1 is a natural number and less than m. In case $m_1 = 1$ we obtain $m = p_1$ and our proposition is true. Otherwise we have $1 < m_1 < m$ so that by the induction hypothesis m_1 is a product of prime numbers, $m_1 = p_2 \dots p_r$ say. This gives $m = p_1 p_2 \dots p_r$ which proves our proposition.

Our aim is to prove that the factorization of a natural number $m > 1$ is unique but for the order of the prime factors. The proof is based on the following lemma:

2.4.1. Lemma. *If a product ab of two integers a and b is divisible by a prime number p, than at least one factor is divisible by p.*

Proof. Suppose that p does not divide a. Then we have to prove that p divides b. If p does not divide a, then $(a, p) = 1$. Therefore 1 can be represented as a linear combination of a and p, i.e.

$$1 = ax + py, \quad x, y \in \mathbf{Z}.$$

Multiplication by b gives

$$b = abx + pby.$$

Due to $p \mid ab$ the right-hand side is divisible by p and hence $p \mid b$. This completes the proof of the lemma.

By induction the lemma can obviously be generalized to the following proposition:

If a product $a_1 a_2 \dots a_k$ of integers a_i is divisible by a prime number p, than at least one factor a_i is divisible by p.

2.4.2. Unique Factorization Theorem. *Every natural number $m > 1$ can be represented as a product*

$$m = \prod_{p|m} p^{\mu_p}$$

of powers of distinct prime numbers. The prime powers are uniquely determined except for their order.

Proof. We saw that m can be expressed as a product of prime numbers. Let

$$m = p_1 p_2 \dots p_r = q_1 q_2 \dots q_s$$

be two factorizations of m into prime numbers p_i and q_j. We shall prove that $r = s$ and that $p_1, p_2, \dots, p_r$ coincide with $q_1, q_2, \dots, q_s$ except possibly for the order. We proceed by induction on r. For $r = 1$ our proposition is obviously true since a single prime number has no further decomposition so that $m = p_1 = q_1$. We assume that in a product of less than r prime numbers the prime factors are uniquely determined but for their order. In the equation

$$p_1 p_2 \dots p_r = q_1 q_2 \dots q_s \tag{1}$$

the right-hand side is divisible by p_1. By the lemma, at least one of the factors on the right-hand side is divisible by p_1. Under suitable notation we may assume that p_1 divides q_1. Since q_1 is a prime number this gives $p_1 = q_1$. By cancelling this factor on both sides of (1) we obtain

$$p_2 \dots p_r = q_2 \dots q_s.$$

Since the left-hand side is a product of less than r prime numbers the induction hypothesis implies that $r - 1 = s - 1$ and that $p_2, \dots, p_r$ coincide with $q_2, \dots, q_r$ in some order. This shows that in the factorization $m = p_1 p_2 \dots p_r$ the prime factors are uniquely determined up to their order. The prime factors $p_1, p_2, \dots, p_r$ need not be distinct. By collecting equal prime factors into powers we obtain our theorem.

Though the reader is probably very familiar with the Unique Factorization Theorem and has applied it many times for simplifying fractions he should not regard it as self-evident. There are rings in which there exists no decomposition into prime elements at all, and even if a factorization into prime elements exists it need not be unique, as we shall see later (Chapter 4). In fact, rings with unique factorization into prime elements form a rather special though important class of rings.

It is clear that all positive divisors of m are the products

$$\prod_{p|m} p^{\delta_p} \quad \text{with} \quad 0 \leqq \delta_p \leqq \mu_p .$$

It is sometimes convenient to use the formal notation

$$m = \prod_p p^{\mu_p}$$

where p ranges over all prime numbers and $\mu_p = 0$ if p does not divide m. If

$$a = \prod_p p^{\alpha_p}, \qquad b = \prod_p p^{\beta_p},$$

then

$$(a, b) = \prod_p p^{\delta_p}, \qquad \delta_p = \min(\alpha_p, \beta_p),$$

is the greatest common divisor of a and b. The product

$$l = \prod_p p^{\lambda_p}, \qquad \lambda_p = \max(\alpha_p, \beta_p),$$

is called the *least common multiple* of a and b. Clearly, l is divisible by a and b, and every integer that is divisible by a and b is also divisible by l. Similarly the least common multiple of more than two integers can be defined.

2.5. Congruences. Residue classes

Let m be a natural number, $m > 1$. If a and b are integers such that m divides $a - b$, then we write

$$a \equiv b \pmod{m}$$

and say that a is *congruent to b modulo m.*

We shall prove that congruence mod m is an equivalence relation, i.e. that this relation is reflexive, symmetric, and transitive.

Clearly, $a \equiv a \pmod{m}$ for every integer a since $a - a = 0$ is divisible by m. If $a - b$ is divisible by m, then so is $b - a$. Therefore $a \equiv b \pmod{m}$ implies that $b \equiv a \pmod{m}$. Finally, if $a \equiv b \pmod{m}$ and $b \equiv c \pmod{m}$, then $m \mid (a - b)$ and $m \mid (b - c)$. Hence m also divides $(a - b) + (b - c) = a - c$ so that $a \equiv c \pmod{m}$.

This shows that congruence mod m is an equivalence relation. The corresponding equivalence classes are called *residue classes* mod m.

Division of a and b by m gives

$$a = qm + r, \quad 0 \leqq r < m,$$

$$b = q'm + r', \quad 0 \leqq r' < m.$$

The difference

$$a - b = (q - q')\,m + (r - r')$$

is divisible by m if and only if $r - r'$ is divisible by m. Due to the above inequalities we obtain $-m < r - r' < m$. Hence $m \mid (r - r')$ if and only if $r = r'$. In other words, a and b belong to the same residue class mod m if and only if they have the same remainder when they are divided by m. This accounts for the name residue class. The integers $0, 1, \ldots, m - 1$ form a system of representatives for the residue classes mod m and so do, for instance, $1, 2, \ldots, m$.

We now introduce a concept which will play an important role in the study of many algebraic systems.

Let S be any algebraic system in which $x \circ y$ is an operation and $x \sim y$ an equivalence relation. We say that the equivalence relation $\sim$ is *compatible* with the operation $\circ$ if

$$x_1 \sim y_1 \text{ and } x_2 \sim y_2 \text{ imply that } (x_1 \circ x_2) \sim (y_1 \circ y_2).$$

As a first example we shall prove that in $\boldsymbol{Z}$ congruence mod m is compatible with addition and multiplication.

2.5.1. Theorem. *In the ring of the integers congruence* mod m *is compatible with addition and multiplication, in other words, if*

$$a_1 \equiv a_2 \pmod{m} \textit{ and } b_1 \equiv b_2 \pmod{m}, \tag{1}$$

then

$$a_1 + b_1 \equiv a_2 + b_2 \pmod{m} \tag{2}$$

and

$$a_1 b_1 \equiv a_2 b_2 \pmod{m}. \tag{3}$$

Proof. Due to (1),

$$m \mid (a_1 - a_2) \text{ and } m \mid (b_1 - b_2).$$

It follows that $a_1 - a_2 + b_1 - b_2 = (a_1 + b_1) - (a_2 + b_2)$ is divisible by m which proves (2). From $m \mid (a_1 - a_2)$ we conclude that m divides

$$c(a_1 - a_2) = ca_1 - ca_2$$

for any integer c. Hence

$$a_1 \equiv a_2 \pmod{m}$$

implies that

$$ca_1 \equiv ca_2 \pmod{m}.$$

By multiplying the first congruence (1) by b_1 and the second by a_2 we obtain

$$a_1b_1 \equiv a_2b_1 \pmod{m}, \quad a_2b_1 \equiv a_2b_2 \pmod{m}$$

so that by the transitivity

$$a_1b_1 \equiv a_2b_2 \pmod{m}.$$

This completes the proof.

The last proof shows that, like an equation, a congruence remains true if both sides are multiplied by the same number. However, the rule that both sides of an equation may be divided by the same non-zero number does not hold for congruences. For instance $2 \cdot 5 \equiv 2 \cdot 10 \pmod{10}$ and $2 \not\equiv 0 \pmod{10}$ but $5 \not\equiv 10 \pmod{10}$.

To find an additional condition under which division of a congruence is permitted we need a generalization of the lemma in § 2.2.

2.5.2. Lemma. *If $m \mid ab$ and $(a, m) = 1$, then $m \mid b$.*

Proof. The proof is exactly the same as in the previous case. From $(a, m) = 1$ it follows that $au + mv = 1$ with suitable integers u, v. Multiplication by b gives

$$abu + mbv = b.$$

Due to $m \mid ab$ the left-hand side is divisible by m and hence $m \mid b$. This proves the lemma.

We now come to the condition under which both sides of a congruence may be divided by an integer.

2.5.3. *If* $(a, m) = 1$, *the congruence*

$$ax \equiv ay \pmod{m}$$

implies that

$$x \equiv y \pmod{m}.$$

Proof. The congruence $ax \equiv ay \pmod{m}$ means that m divides $ax - ay = a(x - y)$. By the lemma $m \mid a(x - y)$ and $(a, m) = 1$ implies that $m \mid (x - y)$ so that $x \equiv y \pmod{m}$ and the proposition is proved.

The next theorem and its corollary deal with the solubility of a linear congruence.

2.5.4. *The congruence*

$$ax \equiv 1 \pmod{m}$$

has a solution x *if and only if* $(a, m) = 1$. *The residue class of the solution is uniquely determined.*

Proof. If a solution exists, then $m \mid (ax - 1)$ so that $ax - 1 = km$ for some integer k. Hence, $ax - km = 1$ which shows that $(a, m) = 1$. Conversely, let $(a, m) = 1$. Then there exist integers x and k such that $ax - km = 1$ or $ax - 1 = km$ which gives $ax \equiv 1 \pmod{m}$. If x' is another solution, then $ax \equiv ax' \pmod{m}$ and hence, by Theorem 2.5.3, $x \equiv x' \pmod{m}$. This completes the proof.

2.5.5. Corollary. *If* $(a, m) = 1$, *then the congruence*

$$ay \equiv b \pmod{m}$$

has a solution y *for any given integer* b. *The residue class of the solution is uniquely determined.*

Indeed, there is a solution x of $ax \equiv 1 \pmod{m}$. Multiplication by b gives $a(bx) \equiv b \pmod{m}$. The uniqueness of the residue class of the solution follows in the same way as above.

2.6. The residue class ring

We will show that the compatibility of congruence mod m with addition and multiplication enables us to define the sum and the product of two residue classes.

Let A and B be two residue classes mod m. We choose arbitrary elements a_1, b_1 of A and B, respectively. The sum $a_1 + b_1$ is contained in some residue class S. For another choice of elements $a_2 \in A$, $b_2 \in B$ the sum $a_2 + b_2$ also belongs to S for, by Theorem 2.5.1,

$$a_1 + b_1 \equiv a_2 + b_2 \pmod{m}.$$

Therefore the residue class S does not depend on the choice of the particular elements in A and B but is uniquely determined by A and B alone. This suggests to call S the *sum of the residue classes A and B* and to write

$$A + B = S.$$

Similarly we can define the product of two residue classes. For $a_1 \in A$ and $b_1 \in B$ the product $a_1 b_1$ belongs to some residue class P. Again, Theorem 2.5.1 tells us that P does not depend on the choice of the elements a_1 and b_1 but is uniquely determined by A and B. Hence we write

$$AB = P$$

and call P the *product of the residue classes A and B.*

Since addition and multiplication of residue classes is based on the corresponding operations with integers, it is clear that the postulates (R_1) through (R_7) also hold for addition and multiplication of residue classes. Therefore the residue classes mod m form a commutative ring. The residue class I that contains 1 obviously plays the role of the unit element. Thus we can state the following theorem:

2.6.1. THEOREM. *The residue classes* mod m *form a commutative ring with unit element.*

The residue class ring mod m will be denoted by $\mathbf{Z}/(m)$.

We shall now restate the Theorems 2.5.3 and 2.5.4 as propositions about the residue class ring. To do this we must first express the condition $(a, m) = 1$ in terms of residue classes.

2.6.2. LEMMA. *If a_1 and a_2 belong to the same residue class* mod m, *then $(a_1, m) = (a_2, m)$.*

Proof. Since a_1 and a_2 belong to the same residue class mod m, it follows that $a_1 = a_2 + km$ for some integer k. This shows that every common divisor of a_2 and m also divides a_1. By symmetry, every common divisor of a_1 and m also divides a_2. This proves the lemma.

In particular, if an integer in some residue class is relatively prime to m, then all integers in that residue class have the same property. Such residue classes are called *prime residue classes* mod m.

We now restate the propositions 2.5.3 and 2.5.4. Capital Latin letters denote residue classes mod m.

2.6.3. *If A is a prime residue class, then $AX = AY$ implies that $X = Y$.*

2.6.4. *In the residue class ring* mod m *the equation*

$$AX = I$$

has a solution X if and only if A is a prime residue class. The solution is uniquely determined.

2.6.5. COROLLARY. *If A is a prime residue class, then the equation $AY = B$ has a unique solution in the residue class ring for any given residue class B.*

Later we shall use the following obvious lemma:

2.6.6. LEMMA. *If A and B are prime residue classes* mod m *then so is the product AB.*

Proof. We choose arbitrary elements $a \in A$, $b \in B$. Since A and B are prime residue classes mod m, we have $(a, m) = (b, m) = 1$. Hence there exist integers x, y, u, v such that

$$ax + my = 1, \quad bu + mv = 1.$$

Multiplication of these equations gives

$$ab(xu) + m(axv + byu + myv) = 1$$

which shows that $(ab, m) = 1$. Therefore AB is a prime residue class.

2.7. Simultaneous congruences. Euler's function

2.7.1. Chinese Remainder Theorem. *Let $m_1, m_2, \ldots, m_r$ be natural numbers that satisfy the condition*

$$(m_i, m_k) = 1 \quad \textit{for} \quad i \neq k. \tag{1}$$

Put

$$m = m_1 m_2 \ldots m_r.$$

The simultaneous congruences

$$\begin{aligned} x &\equiv a_1 \pmod{m_1}, \\ x &\equiv a_2 \pmod{m_2}, \\ &\cdots\cdots\cdots\cdots\cdots \\ x &\equiv a_r \pmod{m_r} \end{aligned} \tag{2}$$

where $a_1, a_2, \ldots, a_r$ are arbitrary given integers have a solution x. The residue class mod m *of the solution is uniquely determined. Moreover, $(x, m) = 1$ if and only if $(a_i, m_i) = 1$ for $i = 1, 2, \ldots, r$.*

Proof. Let

$$m_i' = m/m_i \qquad (i = 1, 2, \ldots, r).$$

From (1) it follows that

$$(m_1', m_2', \ldots, m_r') = 1.$$

Therefore there exist integers $z_1, z_2, \ldots, z_r$ such that

$$z_1 m_1' + z_2 m_2' + \ldots + z_r m_r' = 1.$$

We write

$$e_i = z_i m_i' \tag{3}$$

so that

$$e_1 + e_2 + \ldots + e_r = 1. \tag{4}$$

(3) implies that

$$e_k \equiv 0 \pmod{m_i} \text{ for } k \neq i, \tag{5}$$

and hence (4) gives

$$e_i \equiv 1 \pmod{m_i} \text{ for } i = 1, 2, \ldots, r. \tag{6}$$

Every integer x that satisfies the condition

$$x \equiv a_1 e_1 + a_2 e_2 + \ldots + a_r e_r \pmod{m} \tag{7}$$

is a solution of (2). Indeed, (7) is all the more satisfied if we replace (mod m) by (mod m_i). By (5) and (6) we thus obtain

$$x \equiv a_i \pmod{m_i} \text{ for } i = 1, 2, \dots, r.$$

If x^* is another solution of (2), then we have

$$x^* \equiv x \pmod{m_i} \text{ for } i = 1, 2, \dots, r.$$

Since $x^* - x$ is divisible by every m_i it follows from (1) that $x^* - x$ is divisible by m.

The condition $(x, m) = 1$ is satisfied if and only if $(x, m_i) = 1$ for $i = 1, 2, \dots, r$. By Lemma 2.6.2 and (2) it follows that $(x, m_i) = (a_i, m_i)$. This completes the proof.

The number of prime residue classes mod m is denoted by $\varphi(m)$. The function φ is called *Euler's function.*

To exclude a trivial case we have so far assumed that $m > 1$. Clearly, the single residue class mod 1 is prime. Therefore we put $\varphi(1) = 1$.

Since $1, 2, \dots, m$ form a system of representatives for the residue classes mod m, and since prime residue classes mod m are represented by integers that are relatively prime to m, $\varphi(m)$ is equal to the number of those integers in the sequence $1, 2, \dots, m$ that are relatively prime to m.

2.7.2. Theorem. *If*

$$m = m_1 m_2 \dots m_r \textit{ where } (m_i, m_k) = 1 \textit{ for } i \neq k,$$

then

$$\varphi(m) = \varphi(m_1)\,\varphi(m_2)\dots\varphi(m_r).$$

Proof. By Theorem 2.7.1., x is contained in a prime residue class mod m if and only if each a_i is contained in a prime residue class mod m_i. Each a_i may be chosen out of $\varphi(m_i)$ prime residue classes mod m_i. Thus there are precisely $\varphi(m_1)\,\varphi(m_2)\dots\varphi(m_r)$ ways to choose $a_1, a_2, \dots, a_r$ such that x belongs to a prime residue class mod m, and different choices of the a_i lead to different residue classes of x. Moreover, each prime residue class mod m of x occurs for some choice of $a_1, a_2, \dots, a_r$. This proves the theorem.

For the decomposition

$$m = \prod_{p|m} p^{\mu_p}$$

of m into powers of distinct prime numbers the theorem gives

$$\varphi(m) = \prod_{p|m} \varphi(p^{\mu_p}). \tag{8}$$

We can therefore find $\varphi(m)$ for every natural number m if the values of Euler's function for powers of prime numbers are known.

To determine $\varphi(p^\mu)$ for a power of a prime number p we proceed as follows. $\varphi(p^\mu)$ is equal to the number of those integers in the sequence

$$1, 2, \ldots, p^\mu \tag{9}$$

that are relatively prime to p^μ or, what is the same, relatively prime to p. To obtain these $\varphi(p^\mu)$ integers we have to remove those integers from the sequence (9) that are divisible by p. These are obviously the integers px with $1 \leqq x \leqq p^{\mu-1}$. The remaining $p^\mu - p^{\mu-1}$ integers of the sequence (9) are relatively prime to p. This gives

$$\varphi(p^\mu) = p^\mu - p^{\mu-1} = p^\mu(1 - 1/p).$$

Hence it follows from (8) that

$$\varphi(m) = \prod_{p|m} p^\mu (1 - 1/p) = m \prod_{p|m} (1 - 1/p).$$

Exercises

1. Prove that for any natural number $n > 1$

$$1/\sqrt{1} + 1/\sqrt{2} + \ldots + 1/\sqrt{n} > \sqrt{n}.$$

2. Let α and β be distinct real or complex numbers. Define

$$\gamma_1 = (\alpha^2 - \beta^2)/(\alpha - \beta), \quad \gamma_2 = (\alpha^3 - \beta^3)/(\alpha - \beta),$$

$$\gamma_k = (\alpha + \beta)\, \gamma_{k-1} - \alpha\beta\gamma_{k-2} \quad (k = 3, 4, \ldots).$$

Show that

$$\gamma_n = (\alpha^{n+1} - \beta^{n+1})/(\alpha - \beta).$$

3. Prove that any natural number $n > 1$ that is not a prime number has at least one prime divisor p such that $p \leqq \sqrt{n}$.
4. Let $n > 1$ be a given natural number. Show that there exist sequences of n consecutive natural numbers none of which is a prime number.

5. Prove that the product of three consecutive integers is divisible by 6 and the product of four consecutive integers by 24.
6. Prove that for any natural number n
   (i) $n^3 - n$ is divisible by 6,
   (ii) $n^5 - n$ is divisible by 30.
7. Find (20384, 10725).
8. Represent (15, 21, 35) in the form $15x + 21y + 35z$ with integers x, y, z.
9. Let $a_1 = 2$, $a_{n+1} = a_n^2 - a_n + 1$ for $n = 1, 2, \ldots$
   (i) Prove that $(a_i, a_k) = 1$ for $i \neq k$.
   (ii) Use (i) to show that there are infinitely many prime numbers.
   (iii) Show that 19 does not divide any a_n.
10. Prove that $(2^{2^m} + 1, 2^{2^n} + 1) = 1$ if $m \neq n$.
11. Prove that $(a, bc) = (a, b)(a, c)$ if $(b, c) = 1$.
12. Let $[a, b]$ denote the least common multiple of a and b.
    (i) Prove that $(a, b)[a, b] = ab$.
    (ii) Prove that $(a, [b, c]) = [(a, b), (a, c)]$.
13. If $2^n - 1$ is a prime number, prove that n is a prime number.
14. Prove that $\sum_{k=1}^{n} 1/k$ is not an integer if $n > 1$.
    (Hint: consider the highest power of 2 dividing n.)
15. Is the equivalence relation $|\alpha| = |\beta|$ in $\boldsymbol{C}$ compatible with addition and multiplication?
16. In $\boldsymbol{Z}$ define the relation $a \sim b$ to mean that a and b are divisible by the same prime numbers apart from the respective powers. Is this equivalence relation compatible with addition and multiplication?
17. Partition the complex numbers into three classes according as the imaginary part is positive, equal to zero, or negative. Is the corresponding equivalence relation compatible with addition and multiplication?
18. For a natural number

$$n = 10^k a_k + 10^{k-1} a_{k-1} + \ldots + 10a_1 + a_0, \quad 0 \leqq a_i \leqq 9$$

put

$$s = a_0 + a_1 + a_2 + \ldots + a_k,$$
$$t = a_0 - a_1 + a_2 - a_3 + \ldots + (-1)^k a_k.$$

Prove that

$$n \equiv s \pmod 9, \quad n \equiv t \pmod{11}.$$

19. Show that there is no natural number n such that $3n^2 - 1$ is a square number.
20. Prove that there exist infinitely many prime numbers p such that $p \equiv 3 \pmod 4$.
21. Prove that $\mathbf{Z}/(p)$ is a field if and only if p is a prime number.
22. Let $\{a\}$ denote the residue class of a mod 7.
    (i) Solve the equations

$$\{5\}\{x\} = \{4\}, \quad \{y\}^2 - \{y\} + \{1\} = \{0\}.$$

    (ii) Find a quadratic equation that has no solution in $\mathbf{Z}/(7)$.
23. Let $(a, m) = d$. Prove that the congruence $ax \equiv b \pmod m$ is soluble if and only if d divides b. How many solutions are there?
24. Let n be a given natural number. Show that there are at most finitely many natural numbers x such that $\varphi(x) = n$.
25. (i) Show that $\varphi(x)$ is even if $x > 2$.
    (ii) Show that there is no natural number x such that $\varphi(x) = 14$.
26. Characterize those natural numbers n for which $\varphi(2n) = \varphi(n)$.
27. Let $\varphi(x) \equiv 2 \pmod 4$. Show that $x = p^\alpha$ or $x = 2p^\alpha$ where p is a prime number and $p \equiv 3 \pmod 4$.
28. Prove that $\varphi(ab)\,\varphi(d) = \varphi(a)\,\varphi(b)\,d$ if $(a, b) = d$.
29. Prove that $\varphi(d) \mid \varphi(n)$ if $d \mid n$.
30. Let a, b, c be given natural numbers and $(a, b) = 1$. Prove that there exists a natural number x such that $(ax + b, c) = 1$.
31. Let $n > 1$. Show that the sum of all natural numbers less than n and relatively prime to n is $1/2\, n\varphi(n)$.

3

Groups

We begin the study of abstract algebraic systems with groups because several concepts and results of group theory will be used later for the study of rings and fields. Group theory is a wide and highly developed branch of mathematics. In this book we have to restrict ourselves to some basic results. For a more comprehensive treatment we refer to the textbooks in the bibliography.

3.1. Semigroups

An algebraic system S with one operation is called a semigroup if the *associative law* holds, i.e.

$$(ab)\,c = a(bc) \text{ for arbitrary elements } a, b, c \text{ of } S.$$

There are two ways of forming products whose factors are three given elements a, b, c taken in this order, namely $(ab)\,c$ and $a(bc)$. By the associative law, these two products coincide. Thus we may omit the brackets and refer without ambiguity to the product abc of the elements a, b, c, taken in this order. (In general, however, the product depends on the order of the elements.)

Now, let n be a natural number, $n \geq 3$, and let $a_1, a_2, \ldots, a_n$ be n elements of a semigroup. There are various ways of forming products whose factors are these n elements taken in the given order. These ways are described by brackets indicating the successive steps in which the multiplication is to be carried out. We shall prove that all these ways lead to the same final result, in other words, *the product of n elements of a semigroup is uniquely determined solely by these elements and their order*. This result permits us to denote the product simply by $a_1a_2 \dots a_n$ without brackets.

For $n = 3$, our proposition is the associative law. We assume that any product of fewer than n elements is uniquely determined by the elements and their order. We consider a certain method of forming the product of $a_1, a_2, \dots, a_n$. At the last step of this method, we have to carry out a multiplication

$$x_1x_2 = (a_1 \dots a_k)\,(a_{k+1} \dots a_n).$$

Owing to the inductive hypothesis, the products

$$x_1 = a_1 \dots a_k, \quad x_2 = a_{k+1} \dots a_n$$

are uniquely determined. The final step in some other method of computing the product of $a_1, a_2, \dots, a_n$ is a multiplication

$$y_1y_2 = (a_1 \dots a_l)\,(a_{l+1} \dots a_n).$$

We have to show that $x_1x_2 = y_1y_2$. Clearly, we may assume that $k < l$. By our inductive assumption, $c = a_{k+1} \dots a_l$ is uniquely determined, and we have $x_1c = y_1$, $cy_2 = x_2$. The associative law gives

$$x_1x_2 = x_1(cy_2) = (x_1c)\,y_2 = y_1y_2,$$

and this completes the proof.

The associative law enables us to define powers a^n for natural numbers n, namely

$$a^n = aa \dots a \quad (n \text{ factors}).$$

Clearly, we have the familiar laws

$$a^ma^n = a^{m+n}, \quad (a^m)^n = a^{mn}. \tag{1}$$

We say that two elements a and b *commute* if $ab = ba$. In this particular case we obtain

$$(ab)^n = a^nb^n. \tag{2}$$

A semigroup S is called *commutative* or *abelian* if any two elements of S commute. In an abelian semigroup any product remains unchanged if we alter the order of its factors in an arbitrary way.

In this case we may also use the product symbol

$$\prod_{i=1}^{n} a_i = a_1 a_2 \dots a_n .$$

The set of all $n \times n$ matrices whose elements belong to $\boldsymbol{Z}$ is a familiar example of a semigroup. $\boldsymbol{N}$ is an abelian semigroup with respect to multiplication and also with respect to addition.

If a semigroup S contains an element e_1, such that

$$e_1 a = a \quad \text{for every} \quad a \in S,$$

then e_1 is called a *left identity*. Similarly e_2 is a *right identity* if

$$a e_2 = a \quad \text{for every} \quad a \in S.$$

If S contains both a left identity e_1 and a right identity e_2, then e_1 and e_2 coincide because

$$e_1 = e_1 e_2 = e_2 .$$

In this case the element $e = e_1 = e_2$ has the property

$$ea = ae = a \text{ for every } a \in S$$

and is called the *identity* of S. The definite article is justified since a semigroup cannot contain more than one identity; indeed, if e' is any identity, then $e' = e'e = e$. The semigroup of all even natural numbers with respect to addition and also the same set with respect to multiplication show that a semigroup need not contain an identity.

A semigroup that consists of the powers of a single element a is called *cyclic*. In case all powers of a are distinct from each other, we speak of an *infinite cyclic semigroup*. A cyclic semigroup with identity is one whose elements other than the identity are powers of a single element a. If we define a^0 to be the identity, then the equations (1) hold for all non-negative integers m, n.

3.2. Groups

Let S be a semigroup with identity e. For a given element a of S, there may exist an element a^{-1} in S such that

$$aa^{-1} = a^{-1}a = e .$$

Such an element a^{-1} is called an *inverse* of a.

The preliminary definition of a group in § 1.5 says that a group is a semigroup with identity in which every element has an inverse. It turns out, however, that a group can be defined by slightly weaker postulates. Thus we define a group as follows:

DEFINITION. *An algebraic system G with a binary operation is called a group if the following conditions are satisfied*:

(A) $(ab)\,c = a(bc)$ *for arbitrary elements* a, b, c *of* G (*associative law*).

(U_r) *G contains at least one* right identity, *i.e. an element e with the property*

$$ae = a \textit{ for every } a \in G.$$

(I_r) *For a fixed right identity e and an arbitrary element a of G, there is at least one element a^{-1} in G such that*

$$aa^{-1} = e.$$

We call a^{-1} a right inverse of a with respect to e.

These postulates imply the existence and uniqueness of the (two-sided) identity and of (two-sided) inverses.

3.2.1. THEOREM. *A group G contains one and only one identity, i.e. an element e such that*

$$ae = ea = a \textit{ for every } a \in G.$$

For every element a of G, there is one and only one inverse a^{-1} in G such that

$$a^{-1}a = aa^{-1} = e.$$

Proof. Let e be a right identity such that every $a \in G$ has at least one right inverse a^{-1} with respect to e. We have $aa^{-1} = e$ and $a^{-1} = a^{-1}e = a^{-1}aa^{-1}$. Multiplying both sides of the last equation on the right by a right inverse $(a^{-1})^{-1}$ of a^{-1} (with respect to e), we obtain

$$e = a^{-1}(a^{-1})^{-1} = a^{-1}aa^{-1}(a^{-1})^{-1} = a^{-1}ae = a^{-1}a.$$

This shows that every a^{-1} with $aa^{-1} = e$ also satisfies the equation $a^{-1}a = e$; in other words, every right inverse is a left inverse. Thus we may speak of inverses without referring to the side. Moreover, it follows that a is an inverse of a^{-1}. From $aa^{-1} = e$ and $a^{-1}a = e$

we obtain

$$ea = aa^{-1}a = ae = a.$$

Consequently, e is not only a right but also a left identity. Since even a semigroup has at most one identity, we conclude that e is the only identity of G. Finally, let a^* be any element such that $aa^* = e$. Multiplying both sides of this equation on the left by a^{-1}, we obtain $a^{-1}aa^* = a^{-1}$ or $a^* = a^{-1}$. Thus, the inverse is unique. This completes the proof.

In a multiplicatively written group the identity is often called the unit element.

For two arbitrary elements a, b of a group, we have $abb^{-1}a^{-1} = e$. Since the inverse is unique, we conclude that

$$(ab)^{-1} = b^{-1}a^{-1}.$$

Similarly, we obtain for more than two factors

$$(a_1a_2 \ldots a_{n-1}a_n)^{-1} = a_n^{-1}a_{n-1}^{-1} \ldots a_2^{-1}a_1^{-1}.$$

For a negative integer n, we define

$$a^n = (a^{-1})^{|n|} \quad (n < 0)$$

and by a^0 we understand the identity for every a. It is easily verified that in view of these definitions the laws of exponents (1) in § 3.1 hold for arbitrary integers m, n. In case a and b commute, (2) in § 3.1 is also satisfied for an arbitrary integer n.

An *abelian* group is one in which any two elements commute.

Of course, in an additively written group, the associative law reads as follows:

$$(a + b) + c = a + (b + c).$$

The identity is often called the null element or the zero element and is denoted by 0. It is defined by

$$a + 0 = 0 + a = a \text{ for every } a.$$

The additive inverse of a is denoted by $-a$, and instead of $a + (-b)$ we simply write $a - b$. Instead of the powers we have the multiples

$$na = a + a + \ldots + a \quad (n \text{ summands}).$$

Analogously to the above definition of powers for arbitrary integral exponents we put[1])

$$0a = 0$$

and for a negative integer n we define

$$na = |n|\,(-a) \quad (n < 0).$$

It is clear that then the additive analogues to (1) in § 3.1, namely

$$ma + na = (m + n)\,a, \quad (mn)\,a = n(ma),$$

hold for arbitrary integers m, n.[2]) In case a and b commute we also have

$$n(a + b) = na + nb.$$

A *finite* group is one that consists of a finite number of elements. The number of elements of a finite group G is called the *order* of G and is denoted by $|G|$. The order of an infinite group is to be understood as its cardinal number.

If a and b are any two elements of a group G, then G contains one and only one element x such that $ax = b$, namely $x = a^{-1}b$; and G also contains one and only one element y such that $ya = b$, namely $y = ba^{-1}$. The next theorem shows that these properties together with the associative law characterize a group.

3.2.2. Theorem. *An algebraic system G with a binary operation is a group if and only if the following conditions are satisfied:*

(A) *The associative law.*

(Q) *For any two elements a, b of G there exists at least one element x such that*

$$ax = b$$

and at least one element y such that

$$ya = b$$

(existence of left and right quotients).

---

[1]) Note that 0 on the left-hand side denotes the integer 0, whereas 0 on the right-hand side means the null element of the group.

[2]) Note that the symbol + has different meanings on both sides of the first equation. On the left-hand side it denotes the group operation whereas, on the right-hand side, it means addition of the integers m and n. There is no danger of confusion because the meaning of + is obvious by the elements to which this operation applies.

Proof. We have to show that the postulates (U_r) and (I_r) in the definition of a group are satisfied. By (Q), the equation $ax = a$ has at least one solution $x = e_a$. Using (Q) again, we see that, for an arbitrary $b \in G$, we can find an element y of G such that $ya = b$. By (A), we obtain

$$be_a = (ya)\, e_a = y(ae_a) = ya = b\,.$$

Thus e_a is a right identity in G. By (Q), for every $c \in G$ the equation $cx = e_a$ has at least one solution x, i.e. every element of G has at least one right inverse with respect to e_a. Thus, (U_r) and (I_r) are satisfied; hence G is a group.

A group G is completely determined when a rule is known by which the product of any two elements can be found. In the case of a finite group, this rule can be expressed by the so-called *Cayley group table.* This is a square table with $|G|$ rows and $|G|$ columns. The rows and columns are labelled by the element of G. At the intersection of the row labelled a and the column labelled b there stands the product ab:

right factor

$\dots b \dots$

	$\vdots$	$\vdots$
left factor	a	$\dots\dots\dots\, ab$
	$\vdots$	

In each row and in each column of a group table every element occurs exactly once. For the row labelled a contains all elements ay, $y \in G$, and for given a and b there is a unique y such that $ay = b$. A similar argument applies to the columns. A square table such that in every row and in every column each element of a certain set occurs exactly once is called a *Latin square.* However, not every Latin square may be regarded as a group table since the associative law need not hold.

In § 1.5 we mentioned two examples of groups, namely, the group of all permutations of a given set and the group GL(n, $\mathbf{K}$). We shall give some more examples.

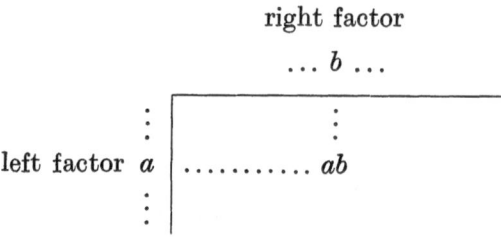

Example 1. Since every group contains the identity, a group of order one consists of the identity alone. The integer 1 under multiplication or the integer 0 under addition are examples of groups of order one.

Example 2. The integers form an abelian group with respect to addition. Indeed, the integers form a ring, and (R_2), (R_3), (R_4) of § 1.5 are precisely the group postulates in additive notation. Moreover, (R_1) means that the group is abelian.

Example 3. Let α be a real number other than 0, 1, —1. Then all powers α^n with integral exponents n are distinct and form an abelian group with respect to multiplication.

Example 4. Let m be a natural number, $m > 1$. For $k = 0, 1, \ldots, m-1$ let $\{k\}$ denote the residue class mod m that contains k. The m residue classes mod m form an abelian group with respect to addition. Indeed, by Theorem 2.6.1, they form a ring, and by (R_1) through (R_4) the elements of a ring form an abelian group under addition. The group operation is given by the following rule.

$$\{k\} + \{l\} = \begin{cases} \{k+l\} & \text{if} \quad k+l < m, \\ \{k+l-m\} & \text{if} \quad k+l \geqq m. \end{cases}$$

Example 5. Let m be a natural number, $m > 1$, and let

$$\varepsilon = \cos(2\pi/m) + i \sin(2\pi/m).$$

The distinct powers of ε are $\varepsilon^0 = 1, \varepsilon, \varepsilon^2, \ldots, \varepsilon^{m-1}$. They form an abelian group under multiplication. Since $\varepsilon^m = 1$, we have for k, $l = 0, 1, \ldots, m-1$

$$\varepsilon^k \varepsilon^l = \begin{cases} \varepsilon^{k+l} & \text{if} \quad k+l < m, \\ \varepsilon^{k+l-m} & \text{if} \quad k+l \geqq m. \end{cases}$$

Example 6. For real numbers x, y with $0 \leqq x < 1$, $0 \leqq y < 1$ we define addition mod 1 as follows:

$$x \dot{+} y = \begin{cases} x+y & \text{if} \quad x+y < 1, \\ x+y-1 & \text{if} \quad x+y \geqq 1. \end{cases}$$

It is easy to see that all real numbers x with $0 \leqq x < 1$ form an abelian group with respect to addition mod 1. The subset of the rational numbers of this interval also forms a group under the same operation. Another important group is obtained by again taking a subset, namely 0 and those rational numbers x of the interval $0 \leqq x < 1$ whose denominators are powers of a given prime number p. This group is called a *group of type* p^∞.

Example 7. The Cayley table

	e	a	b	c
e	e	a	b	c
a	a	e	c	b
b	b	c	e	a
c	c	b	a	e

defines a group of order 4 which is called the *four-group*.

Example 8. The group of all permutations of a set of n elements is called the *symmetric group of degree* n and is denoted by S_n. As we already remarked in § 1.3, the order of S_n is $n!$. It is clear that the rules for the operation in S_n, expressed, for instance, by the Cayley table, only depend on the number n of the permuted elements but not on the nature of these elements or their notation. So, we may regard the group S_3 of order 6 as the group of all permutations of the set $\{1, 2, 3\}$. Using the same notation as in § 1.3 we write

$$a = \begin{pmatrix} 1 & 2 & 3 \\ 2 & 3 & 1 \end{pmatrix}, \quad b = \begin{pmatrix} 1 & 2 & 3 \\ 1 & 3 & 2 \end{pmatrix}.$$

The other elements of S_3 can be expressed by a and b, namely

$$a^2 = \begin{pmatrix} 1 & 2 & 3 \\ 3 & 1 & 2 \end{pmatrix}, \quad ab = \begin{pmatrix} 1 & 2 & 3 \\ 3 & 2 & 1 \end{pmatrix}, \quad a^2b = \begin{pmatrix} 1 & 2 & 3 \\ 2 & 1 & 3 \end{pmatrix}, \quad a^3 = e = \begin{pmatrix} 1 & 2 & 3 \\ 1 & 2 & 3 \end{pmatrix}.$$

One easily verifies that, in addition to $a^3 = e$, the following equations hold:

$$b^2 = e, \quad ba = a^2b.$$

A group is called *cyclic* if it consists of the powers (or the multiples in additive notation) of a single element. Examples 1 and 4 as well as 5 show that for any given natural number there exists a cyclic group of that order.

If a cyclic group C consists of all powers of an element a, then we have to distinguish two cases according as the powers of a are all distinct or not. If all the powers a^k with $k = 0, \pm 1, \pm 2, \ldots$ are distinct, then C is an *infinite cyclic group*. So, the groups in the Examples 2 and 3 are infinite cyclic groups.

If not all the powers of a are distinct, then there holds an equation $a^k = a^l$. We may assume that $k > l$. It follows $a^{k-l} = e$. Hence there exist natural numbers n such that $a^n = e$. Let m be the least natural number for which $a^m = e$. Then the powers

$$a^0 = e, a, a^2, \ldots, a^{m-1} \tag{1}$$

are distinct from one another and every power of a coincides with one of these elements so that C is a *cyclic group of order m*. Indeed, any equation $a^u = a^v$, with $0 \leqq u < v \leqq m - 1$, would imply that $a^{v-u} = e$ where $1 \leqq v - u \leqq m - 1$, which contradicts the choice of m. Further, let t be any integer. Dividing t by m we obtain

$$t = qm + r, \quad 0 \leqq r \leqq m - 1.$$

Due to $a^m = e$ this gives

$$a^t = a^{qm+r} = (a^m)^q\, a^r = a^r.$$

Therefore every power of a coincides with one of the elements (1).

The Cayley table of a finite group certainly contains all information about this group so that, at first sight, it seems to be a suitable device for the investigation of the group. It turns out, however, that in most cases the study of a group by means of its Cayley table is inexpedient and that important properties of a group are not clearly evident from its Cayley table.

We give some examples of a more appropriate method to describe a particular group.

For instance, the Cayley table of a cyclic group of order m would be a very redundant way to describe the rule

$$a^k a^l = \begin{cases} a^{k+l} & \text{if } k + l < m, \\ a^{k+l-m} & \text{if } k + l \geqq m \end{cases} \quad (k, l = 0, 1, \ldots, m - 1).$$

This rule again is a consequence of the single equation $a^m = e$. To give all information about a cyclic group of order m it is therefore sufficient to exhibit the *generating element* a and the *defining relation* $a^m = e$.

The elements of the four-group (Example 7) are e, a, b, ab $(= c)$. Hence, we say that the four group is generated by a and b. These elements satisfy the equations $a^2 = b^2 = e$, $ab = ba$. It is easy to see that these equations enable us to compute the product of any two elements. Therefore the equations are said to be defining relations of the four-group.

In the notation of Example 8, the elements of S_3 are

$$e, a, a^2, b, ab, a^2b.$$

Thus we may say that S_3 is generated by a and b. These two elements satisfy the equations $a^3 = b^2 = e$, $ba = a^2b$. One easily verifies that, by these equations, one can find the product of any two elements of S_3, e.g.

$$(ab)(a^2b) = a(ba)ab = a(a^2b)ab = bab = a^2bb = a^2.$$

3.3. Isomorphisms. Automorphisms

A simple calculation shows that the matrices

$$I = \begin{bmatrix} 1 & 0 \\ 0 & 1 \end{bmatrix}, \quad A = \begin{bmatrix} -1 & 0 \\ 0 & -1 \end{bmatrix}, \quad B = \begin{bmatrix} 1 & 0 \\ 0 & -1 \end{bmatrix}, \quad C = \begin{bmatrix} -1 & 0 \\ 0 & 1 \end{bmatrix}$$

form a group M whose Cayley table is obtained from that of the four-group in Example 7 of § 3.2 by substituting I, A, B, C for e, a, b, c, respectively. As is easily verified, the permutations

$$\varepsilon = \begin{pmatrix} 1 & 2 & 3 & 4 \\ 1 & 2 & 3 & 4 \end{pmatrix}, \quad \alpha = \begin{pmatrix} 1 & 2 & 3 & 4 \\ 2 & 1 & 4 & 3 \end{pmatrix},$$

$$\beta = \begin{pmatrix} 1 & 2 & 3 & 4 \\ 3 & 4 & 1 & 2 \end{pmatrix}, \quad \gamma = \begin{pmatrix} 1 & 2 & 3 & 4 \\ 4 & 3 & 2 & 1 \end{pmatrix}$$

form a group P, and the Cayley table of P is obtained from that in Example 7 if one replaces e, a, b, c, by $\varepsilon, \alpha, \beta, \gamma$, respectively.

The groups M and P differ from one another in the nature of their elements, in the group operation, and in the notation. However, the same formal rules for the group operation hold in both of them. Therefore these two groups are said to be isomorphic. We could generalize this concept to arbitrary finite groups and call two finite groups isomorphic if they have the same Cayley tables except possibly for the notation. But it turns out that isomorphism plays an important role not only in the theory of finite groups but also for infinite groups and for many other algebraic systems. Therefore we prefer to give a definition which does not refer to Cayley tables. It also applies to infinite groups and can easily be extended to other algebraic systems.

Let G_1 and G_2 be two groups and let σ be a bijective mapping of G_1 onto $G_2 = G_1\sigma$ such that

$$(xy)\,\sigma = (x\sigma)\,(y\sigma) \text{ for any two elements } x, y \text{ of } G.$$

Then, σ is called an *isomorphism* of G_1 onto G_2. We also say that G_1 is isomorphic to G_2 and write $G_1 \cong G_2$.

It is easy to see that $\cong$ is an equivalence relation. Groups in one and the same class of isomorphic groups are said to belong to the same type. Isomorphic groups may differ from one another in the nature of their elements (e.g. permutations, matrices, numbers), in the operation (e.g. product of permutations, multiplication of matrices, addition or multiplication of numbers), and in the notation, but all the formal laws that do not refer to the particular nature of the group elements or the operation are the same for all groups of the same type. We therefore speak of the type of the infinite cyclic group, of the four-group, or of the symmetric group of degree n or, briefly, of *the* infinite cyclic group, *the* four-group, and *the* symmetric group of degree n.

The groups in the Examples 2 and 3 of § 3.2 are isomorphic and belong to the type of the infinite cyclic group. The groups in the Examples 4 and 5 are also isomorphic, both are realizations of the type of the cyclic group of order m.

If G_1 and G_2 are isomorphic, then, in general, there are several, possibly infinitely many isomorphisms of G_1 onto G_2.

Every isomorphism σ of G_1 onto G_2 carries the identity e of G_1 into the identity $e\sigma$ of G_2; indeed, $ee = e$ implies that $(e\sigma)\,(e\sigma) = e\sigma$ so that $e\sigma$ is the identity of G_2. Moreover,

$$(x^{-1})\,\sigma = (x\sigma)^{-1} \text{ for every } x \in G_1. \tag{1}$$

For, $xx^{-1} = e$ gives $(x\sigma)\,(x^{-1})\,\sigma = e\sigma$, and hence $(x^{-1})\,\sigma$ is the inverse of $x\sigma$ in G_2.

It is clear that the definition of an isomorphism is not restricted to groups but applies to any algebraic system with one operation, e.g. to semigroups.

An *automorphism* of a group G is an isomorphism of G onto itself. Let ϱ and σ be two automorphisms of G. For any two elements x, y of G we obtain

$$(xy)\,\varrho\sigma = [(xy)\,\varrho]\,\sigma = [(x\varrho)\,(y\varrho)]\,\sigma = [(x\varrho)\,\sigma]\,[(y\varrho)\,\sigma] = (x\varrho\sigma)\,(y\varrho\sigma)$$

which shows that the product of two automorphisms of G is again an automorphism of G. Since the identity mapping of G onto itself is obviously an automorphism of G and since the inverse of an automorphism is also an automorphism it follows that all automorphisms of G form a group, the *automorphism group* $A(G)$ of G.

Let us consider a particular and important kind of automorphisms. We choose an arbitrary element c of the group G and define the mapping τ_c as follows:

$$\tau_c\colon x \to c^{-1}xc \quad (x \in G).$$

It is clear that τ_c is injective, and $x = c^{-1}(cxc^{-1})\,c$ shows that every element of G is an image under τ_c so that τ_c is surjective. Moreover, for any two elements x, y of G, we obtain

$$(xy)\,\tau_c = c^{-1}(xy)\,c = (c^{-1}xc)\,(c^{-1}yc) = (x\tau_c)\,(y\tau_c),$$

hence τ_c is an automorphism of G. We call τ_c the *inner automorphism* generated by c.

Two elements x_1 and x_2 of G are said to be *conjugate* if there is at least one inner automorphism of G that carries x_1 into x_2. If $c^{-1}x_1c = x_2$, then we also say that x_2 is obtained from x_1 by *transformation* by c.

We now form the product of two inner automorphisms τ_c, τ_d:

$$x(\tau_c\tau_d) = (x\tau_c)\,\tau_d = d^{-1}(c^{-1}xc)\,d = (cd)^{-1}\,x(cd) = x\tau_{cd}.$$

Since this holds for every $x \in G$, we have

$$\tau_c\tau_d = \tau_{cd}. \tag{2}$$

This gives, in particular, $\tau_c\tau_{c^{-1}} = \tau_e$, where τ_e is the identity mapping of G onto itself. We infer that the inner automorphisms of G form a group. This group will be denoted by $J(G)$.

All automorphisms other than the inner ones are called *outer* automorphisms.

It is clear that an abelian group has no inner automorphism other than the identity. On the other hand, there are groups all of whose automorphisms are inner ones.

3.4. Embedding of abelian semigroups in groups

Let S be a semigroup. We ask whether there exists a group G that contains S as a subset such that the operations in G and S coincide for the elements of S. If such a group G exists, then we say that S can be *embedded* in G.

Suppose that S is embedded in a group G. Then, an equation $ax = ay$ between elements a, x, y of S implies that $x = y$. Indeed, the equation $ax = ay$ can be regarded as an equation in G. Since G is a group, it contains a^{-1}, and multiplication of our equation on the left by a^{-1} yields $x = y$. Similarly, any equation $xa = ya$ in S implies that $x = y$ provided that S can be embedded in a group.

A semigroup is said to be *cancellative* if either of the equations $ax = ay$ and $xa = ya$ implies that $x = y$. From the above it is clear that a semigroup cannot be embedded in a group unless it is cancellative. It turns out that in the case of abelian semigroups this necessary condition is also sufficient.

3.4.1. THEOREM. *Every cancellative abelian semigroup can be embedded in an abelian group.*

Proof. Let S be a cancellative abelian semigroup. We shall construct an abelian group which contains a subset that is a semigroup isomorphic to S.

Since S is abelian we use additive notation. Then any equation $a + x = a + y$ in S implies that $x = y$.

We form the Cartesian product $P = S \times S$, i.e. the set of all ordered pairs (a, b) with $a, b \in S$. In P we define the relation

$$(a, b) \sim (a_1, b_1) \tag{1}$$

to mean that

$$a + b_1 = a_1 + b.$$

We first prove that this is an equivalence relation. It is clear that our relation is reflexive and symmetric. Suppose that $(a, b) \sim (a_1, b_1)$ and $(a_1, b_1) \sim (a_2, b_2)$. Then, $a + b_1 = a_1 + b$ and $a_1 + b_2 = a_2 + b_1$. Adding these two equations we obtain $a + b_1 + a_1 + b_2 = a_1 + b + a_2 + b_1$. Since S is cancellative it follows that $a + b_2 = a_2 + b$ so that $(a, b) \sim (a_2, b_2)$. Hence, (1) is an equivalence relation. Therefore, we obtain a partition of P into equivalence classes. Note that all pairs (x, x) with $x \in S$ form one equivalence class.

We now define addition in P as follows:

$$(a, b) + (c, d) = (a + c, b + d). \tag{2}$$

It turns out that this operation is compatible with the relation (1) in the sense of § 2.5. Indeed, let

$$(a, b) \sim (a_1, b_1), \quad (c, d) \sim (c_1, d_1).$$

This means $a + b_1 = a_1 + b$, $c + d_1 = c_1 + d$. Addition of these equations gives

$$(a + c) + (b_1 + d_1) = (a_1 + c_1) + (b + d)$$

which shows that

$$(a + c, b + d) \sim (a_1 + c_1, b_1 + d_1).$$

Hence the relation (1) is compatible with the operation (2).

This enables us to define addition of equivalence classes similarly to the addition of residue classes in § 2.6.

Let M and N be two equivalence classes. We choose pairs $(a, b) \in M$ and $(c, d) \in N$. Then the pair

$$(a, b) + (c, d) = (a + c, b + d)$$

is contained in some equivalence class K. If we choose other pairs $(a_1, b_1) \in M$, $(c_1, d_1) \in N$, then, due to the compatibility of our relation with addition of pairs,

$$(a_1 + c_1, b_1 + d_1) \sim (a + c, b + d)$$

so that the pair $(a_1, b_1) + (c_1, d_1)$ belongs to the same class K. Therefore we may write

$$M + N = K$$

where the class K is uniquely determined by the classes M and N alone and does not depend on the choice of particular elements in M and N. Thus addition in the set G of the equivalence classes is defined.

It is clear that addition in G is commutative and associative since, by (2), the commutative and associative laws in P are consequences of the corresponding laws in S.

Let 0 denote the class of all pairs (x, x), $x \in S$. For $(a, b) \in M$ we obtain

$$(a, b) + (x, x) = (a + x, b + x) \sim (a, b)$$

which means that

$$M + 0 = M.$$

Hence 0 is the null element in G. Moreover,

$$(a, b) + (b, a) = (a + b, a + b) \in 0.$$

In other words, the class M' that contains (b, a) satisfies the equation

$$M + M' = 0.$$

Consequently, every element of G has an inverse in G. This shows that G is a group.

For a fixed element a of S all the pairs $(x, a + x)$ with $x \in S$ form an equivalence class. We denote this class by $C(a)$. Since S is cancellative a relation

$$(x, a + x) \sim (y, b + y)$$

implies that $a = b$. In other words, $a \neq b$ implies that $C(a) \neq C(b)$. This shows that the mapping

$$a \to C(a) \quad (a \in S) \tag{3}$$

of S into G is injective. Let S^* denote the subset of G that consists of all classes of the form $C(u)$. Then, (3) defines a bijective mapping of S onto S^*. From

$$(x, a + x) + (y, b + y) = (x + y, a + b + x + y)$$

we finally conclude that

$$C(a) + C(b) = C(a + b).$$

This shows that (3) is an isomorphism of S onto S^*, a subset of G. This completes the proof.

If S is the semigroup of all natural numbers with respect to addition, then our construction yields the additive group of all integers. In this case, the pair (a, b) stands for the difference $b - a$.

In case S is the semigroup of all natural numbers with respect to multiplication, the construction leads to the multiplicative group of all positive rational numbers. The pair (a, b) corresponds to the fraction b/a, and the equivalence relation means that two fractions b/a and b'/a' are regarded as equal if and only if $ab' = a'b$. The set S^* corresponds to all fractions of the form ax/x.

3.5. Subgroups

A non-empty subset of a group is called a *complex*.

A complex U of a group G is called a *subgroup* if U is a group with respect to the operation defined in G.

Every group G contains two *trivial* subgroups, namely G itself and the subgroup that consists of the identity e only. The latter is also denoted by e.

It is clear that a complex U of a group G is a subgroup if and only if the following two conditions hold:

(a) Whenever $x \in U$ and $y \in U$, then $xy \in U$.

(b) Whenever $x \in U$, then $x^{-1} \in U$.

For since U is not empty the two conditions imply that $xx^{-1} = e$ belongs to U. The associative law in U is a consequence of the associative law in G. The conditions (a) and (b) can be replaced by a single condition:

3.5.1. *A complex U of a group is a subgroup if and only if the following condition is satisfied:*

For any two elements x, y of U, the product xy^{-1} belongs to U.

Proof. Taking $x = y$, this condition implies that $xx^{-1} = e$ belongs to U. From $e \in U$ and $x \in U$, we conclude that $ex^{-1} = x^{-1}$ is in U, i.e. condition (b) is satisfied. Finally, let x and y be any two elements of U. It follows that $y^{-1} \in U$ and therefore $x(y^{-1})^{-1} = xy$

belongs to U. Thus, condition (a) is satisfied. It is obvious that, conversely, (a) and (b) imply the condition of our proposition. This completes the proof.

3.5.2. Corollary. *The intersection of an arbitrary set of subgroups of a group G is again a subgroup of G.*

Of course, in additive notation the condition in 3.5.1 reads:

If $x \in U$ and $y \in U$, then $x - y \in U$.

The rational numbers x with $0 \leqq x < 1$ under addition mod 1 form a subgroup of the group of all real numbers of this interval with respect to the same operation (cf. Example 6 in § 3.2). The group of type p^∞ in the same example is a subgroup of the group of the rational numbers. The group of the inner automorphisms of a group G is a subgroup of the automorphism group of G.

Let G be any group. Let $Z = Z(G)$ denote the complex of all those elements of G that commute with every element of G, i.e.

$$Z(G) = \{x: x \in G, xa = ax \text{ for every } a \in G\}.$$

From $x \in Z$ and $y \in Z$ it follows that

$$xya = xay = axy \text{ for every } a \in G$$

so that $xy \in Z$. From $xa = ax$ we conclude that $a^{-1}x^{-1} = x^{-1}a^{-1}$. If a ranges over the whole group G then so does a^{-1}. Therefore, $x \in Z$ implies that $x^{-1} \in Z$. This shows that $Z(G)$ is a subgroup of G. We call $Z(G)$ the *centre* of G.

Every abelian group coincides with its centre. On the other hand there are groups whose centre consists of the identity only, for instance the group S_3 (cf. Example 8 in § 3.2).

If A is any complex of a group G, then there exist subgroups of G that contain all the elements of A, for instance the group G itself. The intersection of all subgroups of G that contain A is denoted by $\langle A \rangle$. We call $\langle A \rangle$ the subgroup *generated* by A. It is clear that every subgroup that contains A also contains $\langle A \rangle$. In this sense, $\langle A \rangle$ is the smallest subgroup of G that contains A. We consider the complex C that consists of the identity and all finite products

$$a_1^{m_1} a_2^{m_2} \ldots a_r^{m_r} \quad (r = 1, 2, \ldots)$$

where the a_i are arbitrary, not necessarily distinct elements of A and the m_i are integers. It is easy to see that C is a subgroup that contains A. On the other hand, every subgroup that contains A also contains all these products. Consequently $C = \langle A \rangle$. Note that distinct products may represent one and the same element of G; for example $ba = a^2b$ in the group S_3 (cf. Example 8 in § 3.2). For several complexes $A_1, A_2, \ldots, A_r$ we define $\langle A_1, A_2, \ldots, A_r \rangle = \langle A_1 \cup A_2 \cup \ldots \cup A_r \rangle$.

In case $\langle A \rangle = G$, the elements of A are said to be a set of *generating elements* of G (cf. the examples at the end of § 3.2).

If A consists of a single element a, then $\langle a \rangle$ is the cyclic subgroup generated by a. It consists of all distinct powers a^k with integral exponents k.

If $\langle a \rangle$ is an infinite cyclic group, then a is said to be an *element of infinite order*. If, however, $\langle a \rangle$ is a finite cyclic group, of order m say, then a is called an *element of order m*. Thus an element $a \neq e$ of G is of order m if $a^m = e$ but $a^k \neq e$ for $1 \leqq k \leqq m - 1$.

It is clear that in any group the identity is the only element of order 1. Of course, in a finite group every element has finite order. In infinite groups, all possibilities concerning the orders of elements actually occur. There are infinite groups in which every element other than the identity is of infinite order. On the other hand, there are infinite groups all of whose elements have finite order. Finally, it is easy to find infinite groups that contain elements of infinite order as well as elements of finite order other than the identity.

By the *product AB* of two *complexes A* and B, we mean the complex whose elements are the distinct products ab with $a \in A$, $b \in B$. Note that an element x of AB may have more than one representation as a product of an element of A by an element of B, namely

$$x = a_1 b_1 = a_2 b_2 = \ldots \qquad (a_1, a_2, \ldots \in A; b_1, b_2, \ldots \in B).$$

Such an element, however, is counted only once as an element of AB. For example, for a subgroup U of G we obtain $UU = U$.

It is clear that the associative law $(AB)\,C = A(BC)$ holds for arbitrary complexes, because it is even satisfied for the single elements.

The complexes AB and BA may be distinct. If we have $AB = BA$, then we say that A and B commute. Note that $AB = BA$ does not imply that every element of A commutes with every element of B. The equation $AB = BA$ only means that every product ab with $a \in A$, $b \in B$ can also be written in the form $b_1 a_1$ with $a_1 \in A$, $b_1 \in B$ and that, conversely, every product $b_1 a_1$ can also be expressed as a product ab.

The complex of all a^{-1} with $a \in A$ is denoted by A^{-1}. For a subgroup U we obviously have $U^{-1} = U$. In view of $(ab)^{-1} = b^{-1}a^{-1}$ we obtain $(AB)^{-1} = B^{-1}A^{-1}$.

Proposition 3.5.1 can be stated as follows:

3.5.1*. *A complex U is a subgroup if and only if $UU^{-1} \subseteq U$.*

If U is a subgroup, then we actually have $UU^{-1} = U$ since U contains the identity.

For two subgroups U and V of a group G, the product UV is not necessarily a subgroup.

3.5.3. *The product UV of two subgroups U and V is a subgroup if and only if $UV = UV$.*

Proof. Let $UV = VU$. In view of $U^{-1} = U$, $V^{-1} = V$ we obtain

$$UV(UV)^{-1} = UVV^{-1}U^{-1} = UVVU = UUVV = UV.$$

Therefore UV is a subgroup by 3.5.1*. Conversely, if UV is a subgroup, then we have

$$UV = (UV)^{-1} = V^{-1}U^{-1} = VU.$$

This completes the proof.

Let U be a subgroup of a group G. To exclude a trivial case we will assume that $U \neq G$. We say that an element $a \in G$ is *right congruent* mod U to $b \in G$ and write

$$a \equiv_r b \pmod{U} \tag{1}$$

if and only if $ab^{-1} \in U$. In other words, (1) holds if and only if $a = ub$ for some $u \in U$.

3.5.4. Lemma. *Right congruence* mod U *is an equivalence relation in G.*

Proof. For any $a \in G$ we have $aa^{-1} \in U$ so that $a \equiv_r a \pmod{U}$. Hence right congruence is reflexive. From $ab^{-1} \in U$, we conclude that $(ab^{-1})^{-1} = ba^{-1}$ belongs to U since U is a group. Hence, $a \equiv_r b \pmod{U}$ implies that $b \equiv_r a \pmod{U}$ so that right congruence is symmetric. Finally, let $a \equiv_r b \pmod{U}$ and $b \equiv_r c \pmod{U}$, i.e. $ab^{-1} \in U$ and $bc^{-1} \in U$. Since U is a group, the product $(ab^{-1})(bc^{-1}) = ac^{-1}$ also belongs to U so that $a \equiv_r c \pmod{U}$. This shows that right congruence is transitive, and the lemma is proved.

The equivalence classes corresponding to the relation (1) are called *right cosets* of U. Since (1) holds if and only if $a = ub$ with $u \in U$, the right coset of U that contains b is the set

$$Ub = \{ub : u \in U\}.$$

By Theorem 1.2.1, we obtain a partition of G into right cosets of U:

$$G = \bigcup_{x \in R} Ux. \tag{2}$$

Here x ranges over a complex R that contains one and only one element of each right coset of U. Such a complex R is called a *right transversal* to U in G. The element x in Ux is said to be a *representative* of the coset Ux. Of course, one of the cosets is U itself.

Similarly, we can define left congruence mod U, namely

$$a \equiv_l b \pmod{U}$$

if and only if $b^{-1}a \in U$. What we proved for right congruence carries over to left congruence. In particular, we obtain a partition

$$G = \bigcup_{y \in L} yU$$

of G into *left cosets* of U. The complex L is said to be *left transversal* to U in G.

Clearly, in an abelian group, we need not distinguish between right and left cosets.

3.5.5. *If R is a right transversal then R^{-1} is a left transversal to U in G.*

Proof. By replacing the elements in (2) by their inverses we obtain

$$G = G^{-1} = \bigcup_{x \in R} x^{-1}U^{-1} = \bigcup_{x \in R} x^{-1}U = \bigcup_{y \in R^{-1}} yU,$$

which proves the proposition.

3.5.6. Corollary. *The number of distinct right cosets of U is equal to the number of distinct left cosets of U.*[1])

The number of distinct right (or left) cosets of U in G is called the *index* of U in G and is denoted by $|G: U|$.

If U is the identity subgroup, then each coset consists of a single element, and we have $|G: e| = |G|$.

Let G be a finite group. Then the index of any subgroup is, of course, finite. Equation (2) is a partition of G into $|G: U|$ cosets each of which contains $|U|$ elements. Counting the number of elements on both sides of (2), we obtain

3.5.7. Lagrange's Theorem. *For any subgroup U of a finite group G*

$$|G| = |G: U|\,|U| = |G: U|\,|U: e|.$$

In particular, the order as well as the index of any subgroup are divisors of $|G|$.

By applying 3.5.6 to the case when U is the cyclic subgroup generated by an element of order m, we obtain the following

3.5.8. Corollary. *In a finite group G, the order of any element is a divisor of $|G|$.*

An immediate consequence of the corollary is

3.5.9. Fermat's Theorem. *Every element x of a finite group of order g satisfies the equation $x^g = e$.*

Proof. The order m of x is a divisor of g, say $g = mq$. This gives

$$x^g = x^{mq} = (x^m)^q = e^q = e,$$

which proves the theorem.

3.6. Cyclic groups

We first determine all subgroups of a cyclic group.

3.6.1. Theorem.

(i) *Every subgroup of a cyclic group is cyclic.*

---

[1]) If there are infinitely many cosets, then number stands for the cardinal number.

(ii) *Every subgroup other than the identity subgroup of an infinite cyclic group is of finite index. Conversely, for every given natural number d, an infinite cyclic group contains one and only one subgroup of index d.*

(iii) *If d is a positive divisor of the order m of a cyclic group* $\langle a \rangle$, *then* $\langle a \rangle$ *contains one and only one subgroup of index d. For any positive divisor f of m,* $\langle a \rangle$ *contains precisely one subgroup of order f.*

Proof. Let $\langle a \rangle$ be any cyclic group. It is obviously sufficient to consider the non-trivial subgroups of $\langle a \rangle$. Let U be a subgroup of $\langle a \rangle$, $U \neq e$, $U \neq \langle a \rangle$. If U contains some power a^n, then a^{-n} also belongs to U. This shows that U contains at least one power of a with a positive exponent. Let d be the least positive exponent such that $a^d \in U$. We call d the minimal exponent of U. Now let a^s be any element of U. We divide s by d to obtain

$$s = qd + r, \quad 0 \leqq r < d.$$

From $a^s \in U$ and $a^d \in U$, we conclude that $a^s\,(a^{-d})^q = a^r$ belongs to U. Since d is the minimal exponent of U, this leads to a contradiction unless $r = 0$. Thus the exponents of all powers of a that belong to U are divisible by d. This means $U = \langle a^d \rangle$; hence (i) is proved.

Note that every subgroup is uniquely determined by its minimal exponent.

For any integer n, there is a uniquely determined pair of integers k, j such that

$$n = kd + j, \quad 0 \leqq j \leqq d - 1.$$

This shows that an arbitrary element $a^n = (a^d)^k a^j$ of $\langle a \rangle$ has a unique representation

$$a^n = ua^j \quad \text{with} \quad u \in \langle a^d \rangle, \quad 0 \leqq j \leqq d - 1.$$

Consequently

$$\langle a \rangle = \langle a^d \rangle \cup \langle a^d \rangle\, a \cup \ldots \cup \langle a^d \rangle\, a^{d-1} \tag{1}$$

is the decomposition of $\langle a \rangle$ into cosets of $U = \langle a^d \rangle$. This gives, in particular,

$$|\langle a \rangle : \langle a^d \rangle| = d.$$

Since every subgroup of $\langle a\rangle$ is uniquely determined by its minimal exponent and since the minimal exponent is equal to its index, it follows that $\langle a\rangle$ contains at most one subgroup of a given index.

If $\langle a\rangle$ is an infinite cyclic group and d any given natural number, then $\langle a\rangle$ always contains a subgroup of index d, namely $\langle a^d\rangle$. This proves (ii).

In case $\langle a\rangle$ is a finite cyclic group of order m, then any possible minimal exponent of a subgroup must divide m since it is the index of this subgroup. If d is a positive divisor of m, then $\langle a^d\rangle$ is a subgroup with the minimal exponent d and hence of index d. Indeed, if $m = df$, then $\langle a^d\rangle$ consists of the elements

$$a^0 = e,\ a^d,\ a^{2d},\ \ldots,\ a^{(f-1)d}. \tag{2}$$

This proves the first part of (iii). The second part follows from the fact that the order and the index of a subgroup of a finite group uniquely determine each other.

This completes the proof.

We shall now determine the order of the subgroup that is generated by an arbitrary element of a finite cyclic group.

3.6.2. *In a cyclic group $\langle a\rangle$ of order m, the element $\langle a^h\rangle$ generates a subgroup of order $m/(h, m)$, where (h, m) denotes the greatest common divisor of h and m.*

Proof. Put $(h, m) = d$. We first show that

$$\langle a^h\rangle = \langle a^d\rangle. \tag{3}$$

Since d divides h, we see that $a^h \in \langle a^d\rangle$; hence $\langle a^h\rangle \subseteq \langle a^d\rangle$. On the other hand, there exist two integers u and v such that $d = mu + hv$. This gives

$$a^d = a^{mu}a^{hv} = a^{hv}$$

so that $a^d \in \langle a^h\rangle$; hence $\langle a^d\rangle \subseteq \langle a^h\rangle$. The two inclusions prove (3). Now, d divides m, say $m = df$. Consequently, $\langle a^d\rangle$ consists of the elements $a^0 = e,\ a^d, \ldots, a^{(f-1)d}$ so that $|\langle a^h\rangle| = |\langle a^d\rangle| = f = m/d = m/(h, m)$. This proves our proposition.

We are now ready to determine the number of elements of a given order in a finite cyclic group.

3.6.3. *Let f be a positive divisor of the order of a finite cyclic group $\langle a \rangle$. Then $\langle a \rangle$ contains $\varphi(f)$ elements of order f, where φ denotes Euler's function.*

Proof. The order m of $\langle a \rangle$ has a factorization $m = df$. An element a^h is of order f if and only if the subgroup $\langle a^h \rangle$ is of order f. By 3.6.2, the necessary and sufficient condition for $|\langle a^h \rangle| = f$ is $(h, m) = d$. Hence, the number of elements of order f is equal to the number of integers h that satisfy the conditions

$$1 \leqq h \leqq m, \quad (h, m) = d.$$

Since all these integers h are divisible by d, we may write $h = dk$ with $1 \leqq k \leqq f$. Thus, the number of elements of order f is equal to the number of integers k that satisfy the conditions

$$1 \leqq k \leqq f, \quad (k, f) = 1.$$

By the definition of Euler's function φ in § 2.7, the number of these integers k is equal to $\varphi(f)$. This completes the proof.

3.6.4. Corollary. *Euler's function satisfies the equation*

$$\sum_{f|m} \varphi(f) = m$$

where f ranges over all positive divisors of m.

Proof. For every positive divisor f of the order m of a cyclic group $\langle a \rangle$, 1 and m included, we collect the $\varphi(f)$ elements of order f in $\langle a \rangle$. In this way, every element of $\langle a \rangle$ is collected precisely once.

3.6.5. Corollary. *A cyclic group $\langle a \rangle$ of order m contains $\varphi(m)$ generating elements, i.e. elements a^r such that $\langle a^r \rangle = \langle a \rangle$. The generating elements are the powers a^r with $(r, m) = 1$.*

Proof. An element a^r generates the whole group if and only if its order is equal to m. Hence there are $\varphi(m)$ generating elements. The second part of our proposition follows from 3.6.2.

3.7. Homomorphisms

We first introduce a concept which will be used for the study of homomorphisms.

In § 3.3 we defined conjugacy for elements of a group. This definition can readily be extended to arbitrary complexes, in particular to subgroups.

Let A and B be two complexes of a group G. We say that A is *conjugate* to B if $A = x^{-1}Bx$ for at least one element x of G.

It is easy to see that conjugacy is an equivalence relation. Indeed, $A = e^{-1}Ae = eAe$ so that conjugacy is reflexive. Further, $A = x^{-1}Bx$ implies that $B = z^{-1}Az$ with $z = x^{-1}$ which means that conjugacy is symmetric. Finally, $A = x^{-1}Bx$ and $B = y^{-1}Cy$ imply that $A = x^{-1}y^{-1}Cyx = (yx)^{-1}C(yx)$; hence, conjugacy is transitive. In particular, due to the symmetry, we may say that A and B are conjugate.

A *normal* or *invariant* complex is one which coincides with all its conjugates. In other words, A is normal if and only if $x^{-1}Ax = A$ for all $x \in G$.

If a complex A satisfies the condition that $x^{-1}Ax \subseteq A$ for all $x \in G$, then A is normal. For $x^{-1}Ax \subset A$ would imply for $z = x^{-1}$ that $A \subset z^{-1}Az$ which contradicts the condition on A.

In particular, a subgroup K of G is normal if and only if $x^{-1}Kx = K$ for every $x \in G$.

To indicate that K is a normal subgroup we write $K \trianglelefteq G$ or $K \triangleleft G$ in case $K \neq G$.

Any group G obviously contains two trivial normal subgroups, namely e and G. A group is said to be *simple* if it contains no normal subgroup other than the two trivial ones.

If $K \trianglelefteq G$, then $Kx = xK$ for every $x \in G$ and vice versa. Thus, for a normal subgroup, right and left cosets coincide. Note that $Kx = xK$ does not mean that every single element of K commutes with x. It only means that every product k_1x, $k_1 \in K$, can also be expressed in the form xk_2 with $k_2 \in K$ but possibly $k_1 \neq k_2$, and vice versa.

If $K \triangleleft G$ and H is a subgroup such that $K \subseteq H \subseteq G$, then K is also normal in H. For $K \trianglelefteq H$ is a consequence of $x^{-1}Kx = K$ for every $x \in H$, and this is even true for all $x \in G$.

In general, however, the relation of being a normal subgroup is not transitive. That is, from $K \triangleleft H$ and $H \triangleleft G$ it does not follow that $K \triangleleft G$.

An immediate consequence of the definition of a normal complex is:

3.7.1. *The intersection of any set of normal subgroups is a normal subgroup.*

Let K be a normal subgroup and U an arbitrary subgroup of G. It follows that $KU = UK$, because $Kx = xK$ even holds for each single element x. Thus, by 3.5.2, KU is a subgroup of G. If, moreover, U is normal, then KU too is a normal subgroup. Indeed, for any $x \in G$ we have $x^{-1}KUx = x^{-1}Kxx^{-1}Ux = KU$. This gives

3.7.2. *If at least one of the subgroups U, V is normal in G, then UV is a subgroup. If both U and V are normal, then so is UV.*

We now come to the definition of a homomorphism.

Let G and G^* be two groups. A mapping

$$\sigma : x \to x\sigma \quad (x \in G)$$

of G onto G^* is called a *homomorphism* of G onto G^* if

$$(xy)\,\sigma = (x\sigma)\,(y\sigma) \text{ for any two elements } x, y \text{ of } G.$$

Homomorphisms play a very important role not only in group theory but also for the study of other algebraic systems. [1])

As an example, we consider the group GL(n, **K**) (cf. § 1.5). If we assign to every matrix x of that group its determinant $|x|$, then the multiplication theorem for determinants gives $|xy| = |x|\,|y|$. Hence the mapping $\sigma : x \to |x|$ is a homomorphism of GL(n, **K**) onto the multiplicative group of all non-zero elements of **K**.

As another example, we define a mapping σ of the group S_3, in the notation of Example 8 in § 3.2, onto 1 or -1, respectively, as follows: $e \to 1$, $a \to 1$, $a^2 \to 1$, $b \to -1$, $ab \to -1$, $a^2b \to -1$. It is easily verified that σ is a homomorphism of S_3 onto the multiplicative group of order 2 consisting of 1 and -1.

Clearly, a homomorphism of G onto G^* that is injective is an isomorphism of G onto G^*.

---

[1]) The term homomorphism is also used in a more general sense, namely to denote any mapping of a group G *into* a group G^* such that the image of a product is always equal to the product of the images. A surjective homomorphism is then called an *epimorphism*, and an injective homomorphism is said to be a *monomorphism*. The set of all homomorphisms of G into G^* is denoted by Hom (G, G^*).

Any homomorphism σ of G onto G^* carries the identity e of G into the identity of $G^* = G\sigma$. For $ee = e$ gives $(e\sigma)(e\sigma) = e\sigma$ so that $e\sigma$ is the identity of G^*. Moreover,

$$(x^{-1})\,\sigma = (x\sigma)^{-1} \text{ for every } x \in G. \tag{1}$$

Indeed, $xx^{-1} = e$ gives $(x\sigma)(x^{-1})\,\sigma = e\sigma$ so that $(x^{-1})\,\sigma$ is the inverse of $x\sigma$.

It is obvious that σ carries any subgroup U of G into a subgroup $U\sigma = U^*$ of $G\sigma = G^*$. Conversely, let U^* be any subgroup of $G^* = G\sigma$ and let U denote the inverse image of U^* in G. Then U is a subgroup of G. Indeed, let x and y be arbitrary elements of U. Then $x\sigma$ and $y\sigma$ belong to U^* and since U^* is a group, we have $(x\sigma)(y\sigma)^{-1} \in U^*$. But by (1), $(x\sigma)(y\sigma)^{-1} = (xy^{-1})\,\sigma$ so that $xy^{-1} \in U$. Hence, U is a subgroup.

We now turn to a more detailed study of homomorphisms. Let σ be an arbitrary homomorphism of the group G onto the group $G^* = G\sigma$. By K we denote the complex of all those elements of G whose image under σ is the identity e^* of G^*. It follows from the previous remark that K is a subgroup of G. Moreover, for an arbitrary element x of G and any $u \in K$ we have

$$(x^{-1}ux)\,\sigma = (x\sigma)^{-1}(u\sigma)(x\sigma) = (x\sigma)^{-1}\,e^*(x\sigma) = e^*.$$

Thus, $u \in K$ implies that $x^{-1}ux \in K$ for every $x \in G$. Hence, K is a normal subgroup of G. K is called the *kernel* of the homomorphism σ.

The homomorphism σ can now be described as follows: Let

$$G = K \cup Ka \cup Kb \cup \ldots$$

be the partition of G into cosets of the kernel K. (Since K is a normal subgroup, we need not distinguish between right and left cosets.) Every element of a coset Ka has the form ua with $u \in K$. Applying σ to this element we find

$$(ua)\,\sigma = (u\sigma)(a\sigma) = e^*(a\sigma) = a\sigma.$$

This shows that σ carries all the elements of a coset of K into one and the same element of G^*. Conversely, if two elements of G have the same image under σ, then they belong to the same coset of K. For $x\sigma = y\sigma$ implies that $(xy^{-1})\,\sigma = (x\sigma)(y\sigma)^{-1} = e^*$, which means that $xy^{-1} \in K$ or $x \in Ky$. Consequently, there is a one-to-one correspondence between the cosets of K and the elements of G^*.

Let us mention two extreme cases. If $K = G$, then we have the trivial homomorphism of G onto the group e^* of order 1. On the other hand, if K consists of the identity only, then σ is injective and hence an isomorphism of G onto G^*.

So far, our starting point was a given homomorphism of G onto some group. We saw that this homomorphism has a normal subgroup of G as its kernel. We now ask whether, conversely, every normal subgroup of G is the kernel of a suitable homomorphism. The answer is in the affirmative.

Let K be a normal subgroup of G. We shall construct a group G^* and a homomorphism of G onto G^* whose kernel is K. First we construct the group G^*. Let

$$G = K \cup Ka \cup Kb \cup \ldots$$

be the partition of G into cosets of K. We prove that the cosets of K form a group with respect to multiplication. Since K is a normal subgroup, we have

$$(Ka)\,(Kb) = KaKb = KKab = Kab = Kc$$

where c is the representative of the coset that contains the element ab. This shows that the product of two cosets of K is a coset of K. The coset K plays the role of the identity, since

$$(Ka)\,K = KaK = KKa = Ka.$$

Finally, if Kd is the coset that contains a^{-1}, then we see that $KaKd = K$; for, as we have proved, $KaKd$ is some coset of K, and since aa^{-1} is contained in that coset it must be the coset K. We conclude that the cosets of a normal subgroup K form a group under multiplication. This group is called the *factor group* of G with respect to K and is denoted by G/K.

A mapping of G onto G/K can be defined in a most obvious way: to every element of G we assign the coset of K in which it is contained. From $x \in Ka$ and $y \in Kb$, it follows that $xy \in KaKb$. This shows that our mapping is a homomorphism of G onto G/K. It is called the *natural homomorphism* of G onto G/K. The identity of

G/K is the coset K, and the natural homomorphism maps an element of G onto K if and only if it is contained in K, in other words, K is the kernel of the natural homomorphism of G onto G/K.

Summarizing our results, we obtain one of the most important theorems of group theory:

3.7.3. Homomorphism Theorem for Groups. *To any homomorphism of a group G onto a group G^* there corresponds, as its kernel, a normal subgroup K of G consisting of all elements of G that are mapped onto the identity of G^*. The group G^* is isomorphic to the factor group G/K. Conversely, if K is any normal subgroup of G, then the natural homomorphism of G onto the factor group G/K has K as its kernel.*

The kernels of the above examples are the normal subgroup of all those matrices whose determinants are equal to the unit element of **K** and the cyclic normal subgroup $\langle a \rangle$, respectively.

The coset decomposition (1) in § 3.6 shows that every factor group of a cyclic group is cyclic; for the factor group $\langle a \rangle / \langle a^d \rangle$ is generated by the coset $\langle a^d \rangle a$.

We now study the behaviour of subgroups under homomorphisms.

3.7.4. I. Isomorphism Theorem. *If K is a normal subgroup and H an arbitrary subgroup of G, then $H \cap K$ is a normal subgroup of H and*

$$HK/K \cong H/(H \cap K).$$

Proof. The natural homomorphism of G onto $G^* = G/K$ maps any subgroup H of G onto some subgroup H^* of G^*. The kernel of the homomorphism of H onto H^* is obviously $H \cap K$. Thus, by Theorem 3.7.3, $H \cap K$ is a normal subgroup of H, and we have

$$H/(H \cap K) \cong H^*. \tag{2}$$

We now determine the complex $\bar{H}$ of *all* those elements of G whose images are contained in H^*. As we observed above, $\bar{H}$ is a subgroup of G; moreover, it is clear that H is contained in $\bar{H}$. In general, however, $\bar{H}$ is larger than H; for if a is an arbitrary element of H, then the whole coset aK belongs to $\bar{H}$. On the other hand, $\bar{H}$ contains only those cosets that can be represented by an element of H.

This shows that $\bar{H} = HK$, and by the definition of $\bar{H}$, we have

$$\bar{H}/K = HK/K = H^*. \tag{3}$$

The theorem now follows from (2) and (3).

Before we proceed, we remark that, under any homomorphism σ of G onto $G\sigma$, two conjugate elements of G have conjugate images in $G\sigma$. Indeed, from $x = c^{-1}yc$ we obtain $x\sigma = (c\sigma)^{-1}(y\sigma)(c\sigma)$.

We now restrict our attention to the subgroups H of G that contain the kernel K. The proof of Theorem 3.7.4 shows that there is a one-to-one correspondence between these subgroups and the subgroups of G/K; for in this case, we have $\bar{H} = HK = H$. The following theorem describes this correspondence in detail.

3.7.5. II. Isomorphism Theorem. *Let K be a normal subgroup of G. There is a one-to-one correspondence between the subgroups H of G that contain K and the subgroups of G/K defined by*

$$H \leftrightarrow H/K \quad (K \subseteq H \subseteq G).$$

Two subgroups H_1 and H_2 are conjugate in G if and only if H_1/K and H_2/K are conjugate in G/K. In particular, H is a normal subgroup of G if and only if H/K is a normal subgroup of G/K. In this case, the respective factor groups are isomorphic, i.e.

$$G/H \cong (G/K)/(H/K). \tag{4}$$

Proof. As we already observed, the proof of Theorem 3.7.4 yields the one-to-one correspondence.

If H_1 and H_2 are conjugate in G, then their images H_1/K and H_2/K are conjugate in G/K, because conjugacy is preserved under every homomorphism.

The subgroup H_1/K of G/K consists of cosets of K. Taking a representative of each one of these cosets, we obtain a complex R_1 such that $H = KR_1$. A subgoup of G/K that is conjugate to H_1/K has the form

$$H_2/K = (Ka)^{-1}(H_1/K)(Ka).$$

Thus, we have

$$H_2 = (Ka)^{-1}KR_1(Ka) = a^{-1}KR_1a = a^{-1}H_1a.$$

This shows that conjugate subgroups of G (containing K) correspond to conjugate subgroups of G/K and vice versa.

Since a subgroup is normal if and only if it coincides with all its conjugates, it follows that H/K is a normal subgroup of G/K if and only if H is a normal subgroup of G.

Finally, suppose that H/K is a normal subgroup of G/K. We first perform the natural homomorphism of G onto G/K and then the natural homomorphism of G/K onto $(G/K)/(H/K)$. This yields a homomorphism of G onto $(G/K)/(H/K)$. It is easily checked that H is the kernel of this homomorphism so that (4) holds. This completes the proof.

As an application of the Isomorphism Theorems we prove a theorem that will be useful later.

3.7.6. Zassenhaus' Lemma. *Let U and V be subgroups of the group G, and let $U_0 \trianglelefteq U$, $V_0 \trianglelefteq V$. Then*

$$U_0(U \cap V_0) \trianglelefteq U_0(U \cap V), \quad V_0(V \cap U_0) \trianglelefteq V_0(V \cap U)$$

and

$$U_0(U \cap V)/U_0(U \cap V_0) \cong V_0(V \cap U)/V_0(V \cap U_0).$$

Proof. Since U_0 and V_0 are normal in U and V, respectively, it follows that the products that occur in our proposition are groups.

By Theorem 3.7.4, we obtain

$$U_0(U \cap V)/U_0 \cong (U \cap V)/(U_0 \cap U \cap V) = (U \cap V)/(V \cap U_0). \quad (5)$$

By symmetry, $(V \cap U_0) \trianglelefteq (U \cap V)$ implies that $(U \cap V_0) \trianglelefteq (U \cap V)$, and this gives

$$(V \cap U_0)(U \cap V_0) \trianglelefteq (U \cap V).$$

By Theorem 3.7.5, to this normal subgroup of $U \cap V$ there corresponds a unique normal subgroup of $(U \cap V)/(V \cap U_0)$, namely

$$(V \cap U_0)(U \cap V_0)/(V \cap U_0). \quad (6)$$

Under the isomorphism (5), the normal subgroup (6) of the right-hand side and the normal subgroup

$$U_0(V \cap U_0)(U \cap V_0)/U_0 = U_0(U \cap V_0)/U_0$$

of the left-hand side correspond to each other. Since the factor groups with respect to corresponding normal subgroups on both sides of (5) are isomorphic, we have

$$U_0(U \cap V)/U_0(U \cap V_0) \cong (U \cap V)/(V \cap U_0)\,(U \cap V_0). \qquad (7)$$

Since the right-hand side of (7) is symmetric with respect to U_0, U and V_0, V we conclude that

$$V_0(V \cap U)/V_0(V \cap U_0) \cong (V \cap U)/(U \cap V_0)\,(V \cap U_0). \qquad (8)$$

Combining (7) and (8), we obtain the theorem.

3.8. Subnormal series

A finite sequence

$$G = G_0 \supseteq G_1 \supseteq G_2 \supseteq \ldots \supseteq G_{l-1} \supseteq G_l = e \qquad (1)$$

of subgroups of a group G, beginning with G and ending with the identity is called a *subnormal series* of G if each G_i is a normal subgroup in G_{i-1}. The number l is said to be the *length* of the subnormal series, and the factor groups G_{i-1}/G_i are called its *factors*.

We speak of a subnormal series *without repetitions* if all its terms are distinct from one another.

A *refinement* of (1) is (1) itself and every subnormal series

$$G = G_0^* \supseteq G_1^* \supseteq G_2^* \supseteq \ldots \supseteq G_{m-1}^* \supseteq G_m^* = e \qquad (2)$$

such that each G_i occurs among the G_j^*. In other words, (2) is obtained from (1) by inserting new terms. A *proper* refinement of (1) is one that contains at least one term that does not occur in (1).

Two subnormal series of G are said to be *isomorphic* if they have the same length and if it is possible to establish a one-to-one correspondence between their factors, taken in a suitable order, such that corresponding factors are isomorphic.

The cyclic group $\langle a \rangle$ of order 6, for example, has the subnormal series

$$\langle a \rangle \supset \langle a^2 \rangle \supset e \quad \text{and} \quad \langle a \rangle \supset \langle a^3 \rangle \supset e.$$

They are isomorphic since the factors in both series are cyclic of orders 2 and 3, respectively.

Two isomorphic subnormal series with repetitions obviously remain isomorphic if we eliminate the repetitions by cancelling repeated terms.

3.8.1. Schreier's Refinement Theorem. *Any two subnormal series of a given group have isomorphic refinements.*

Proof. Let

$$G = G_0 \supseteq G_1 \supseteq G_2 \supseteq \ldots \supseteq G_{r-1} \supseteq G_r = e$$

and

$$G = H_0 \supseteq H_1 \supseteq H_2 \supseteq \ldots \supseteq H_{s-1} \supseteq H_s = e$$

be subnormal series of the group G. We refine these series as follows. Between G_{i-1} and G_i we insert the $s - 1$ groups

$$G_{i,k} = G_i(G_{i-1} \cap H_k) \quad (k = 1, \ldots, s-1)$$

and between H_{k-1} and H_k we insert the $r - 1$ groups

$$H_{i,k} = H_k(H_{k-1} \cap G_i) \quad (i = 1, \ldots, r-1).$$

The groups $G_{i,0} = G_{i-1}$, $G_{i,s} = G_i$, $H_{0,k} = H_{k-1}$, $H_{r,k} = H_k$ are those that occur in the given series. By inserting all these groups we obtain subnormal series since, by 3.7.6, $G_{i,k} \trianglelefteq G_{i,k-1}$ and $H_{i,k} \trianglelefteq H_{i-1,k}$; note that this also holds for $k = 0$, $k = s$, $i = 0$, and $i = r$. Further, 3.7.6 gives

$$G_i(G_{i-1} \cap H_{k-1})/G_i(G_{i-1} \cap H_k) \cong H_k(H_{k-1} \cap G_{i-1})/H_k(H_{k-1} \cap G_i),$$

i.e.

$$G_{i,k-1}/G_{i,k} \cong H_{i-1,k}/H_{i,k} \quad (i = 1, \ldots, r;\ k = 1, \ldots, s).$$

Hence, the refined series are isomorphic. This completes the proof.

If the refined series contain repeated terms, then we can remove the repetitions without destroying the isomorphism.

A *composition series* is a subnormal series without repetitions that admits no proper refinement. An infinite group need not have a composition series, but for any finite group there exists at least one; for, every subnormal series of a finite group G, for example $G \supset e$, admits only a finite number of proper refinements.

By Theorem 3.7.5, the factors of every composition series are simple groups, for there is a one-to-one correspondence between the non-trivial normal subgroups H/G_i of G_{i-1}/G_i and the normal subgroups H of G_{i-1} that satisfy $G_{i-1} \supset H \supset G_i$. For the same reason, every subnormal series without repetitions whose factors are simple groups is a composition series.

3.8.2. Jordan-Hölder's Theorem. *If the group G possesses a composition series, then any two composition series of G are isomorphic.*

Proof. By Theorem 3.8.1, any two composition series have isomorphic refinements, while, on the other hand, they admit no proper refinement so that they are isomorphic without any insertion. This completes the proof.

In the definition of isomorphism between subnormal series, we disregarded the order in which the factors occur. The above example of the cyclic group of order 6 shows that this is necessary for the validity of Theorem 3.8.2.

A group is called *soluble*[1]) if it has a subnormal series all of whose factors are abelian.

Clearly, any refinement of a subnormal series with abelian factors has again abelian factors. It is easy to see that the only simple abelian groups are the cyclic groups whose order is a prime number. Thus, we obtain the following characterization of finite soluble groups:

3.8.3. Theorem. *A finite group is soluble if and only if the orders of all factors in a composition series are prime numbers.*

By the *commutator* of the elements x, y of a group G we understand the element $[x, y] = x^{-1}x^{-1}xy$.

We have $yx[x, y] = xy$ so that x and y commute if and only if $[x, y] = e$. Let C denote the complex of all commutators of any two elements of G. The complex C itself need not be a subgroup. The subgroup $\langle C \rangle$ generated by C is called the *commutator subgroup* or the *derived group* of G. We shall denote it by G' or $G^{(1)}$. It is clear that G is abelian if and only if $G' = e$.

For any automorphism σ of G, (1) in § 3.3 gives

$$[x, y]\,\sigma = (x^{-1}y^{-1}xy)\,\sigma = (x\sigma)^{-1}\,(y\sigma)^{-1}\,(x\sigma)\,(y\sigma) = [x\sigma, y\sigma].$$

Thus every automorphism of G carries any commutator into a commutator. This holds, in particular, for every inner automorphism of G, in other words, the complex C is normal. Therefore the commutator subgroup $G' = \langle C \rangle$ is a normal subgroup of G.

[1]) The reason for this name will appear in § 7.2.

3.8.4. Theorem. *The factor group G/G' is abelian. Conversely, if N is a normal subgroup such that G/N is abelian, then G' is contained in N.*

Proof. Since every commutator belongs to G', we obtain

$$xG'\,yG' = xyG' = yx[x, y]\,G' = yxG' = yG'\,xG'$$

which shows that G/G' is abelian. Now let G/N be abelian. Then we have

$$N = Nx^{-1}Ny^{-1}NxNy = Nx^{-1}y^{-1}xy = N[x, y]$$

so that $[x, y] \in N$. This gives $C \subseteq N$ and hence $G' \subseteq N$, and the theorem is proved.

The so-called *higher commutator subgroups* of G are defined as follows:

$$G^{(0)} = G,$$

$$G^{(i)} = \text{commutator subgroup of } G^{(i-1)} \quad (i = 1, 2, \ldots).$$

For a subgroup H of G, we obtain by induction $H^{(i)} \subseteq G^{(i)}$ for $i = 1, 2, \ldots$

3.8.5. Theorem. *A group G is soluble if and only if $G^{(s)} = e$ for some natural number s.*

Proof. Suppose that $G^{(s)} = e$. Then

$$G \supset G^{(1)} \supset G^{(2)} \supset \ldots \supset G^{(s-1)} \supset G^{(s)} = e$$

is a subnormal series whose factors are abelian so that G is soluble. Conversely, let

$$G \supset G_1 \supset G_2 \supset \ldots \supset G_{r-1} \supset G_r = e$$

be a subnormal series with abelian factors. We shall prove that

$$G^{(i)} \subseteq G_i \quad (i = 1, 2, \ldots, r). \tag{3}$$

This implies that $G_{(}^{s)} = e$ for some $s \leqq r$. Since G/G_1 is abelian, it follows from Theorem 3.8.4 that $G^{(1)} \subseteq G_1$. Thus, (3) is true for $i = 1$. Suppose that $G^{(i-1)} \subseteq G_{i-1}$. Since G_{i-1}/G_i is abelian we can apply Theorem 3.8.4 to obtain $G'_{i-1} \subseteq G_i$ where G'_{i-1} denotes the commutator subgroup of G_{i-1}. From $G^{(i-1)} \subseteq G_{i-1}$ it follows that $G^{(i)} \subseteq G'_{i-1}$. Thus we obtain

$$G^{(i)} \subseteq G'_{i-1} \subseteq G_i$$

which proves (3). This completes the proof of the theorem.

3.8.6. THEOREM. *If G is soluble, then every subgroup of G as well as every factor group G/K is soluble.*

Proof. If H is a subgroup of G, then we have $H^{(i)} \subseteq G^{(i)}$. Hence, $G^{(s)} = e$ implies that $H^{(s)} = e$. Consequently, H is soluble.

The commutator subgroup G/K is generated by the cosets

$$x^{-1}Ky^{-1}KxKyK = x^{-1}y^{-1}xyK.$$

This gives

$$(G/K)' = G'K/K$$

and by induction

$$(G/K)^{(i)} = G^{(i)}K/K \quad (i = 1, 2, \ldots).$$

From $G^{(s)} = e$, it follows that $(G/K)^{(s)} = K/K$. This proves the theorem.

3.9. Direct products

If two complexes A and B of a group have the property that $ab = ba$ for any two elements $a \in A$, $b \in B$ then we say that A and B commute elementwise.

3.9.1. LEMMA. *Two normal subgroups A and B of a group G for which $A \cap B = e$ commute elementwise.*

Proof. Let a and b be any two elements of A and B, respectively. The commutator $[a, b]$ can be written as follows:

$$[a, b] = a^{-1}(b^{-1}ab) = (a^{-1}b^{-1}a)\, b.$$

Since A and B are normal, this gives $[a, b] \in A$ as well as $[a, b] \in B$ so that $[a, b] = e$, which proves the lemma.

Let $G_1, G_2, \ldots, G_n$ be non-trivial subgroups of a group G. We say that G is the *direct product* of $G_1, G_2, \ldots, G_n$ and write

$$G = G_1 \times G_2 \times \ldots \times G_n$$

if the following conditions are satisfied:

(i) *Any element x of G has a unique representation*

$$x = x_1 x_2 \dots x_n \quad \textit{with} \quad x_i \in G_i \qquad (i = 1, 2, \dots, n). \tag{1}$$

(ii) *Any two distinct subgroups G_i, G_j commute elementwise.*
The subgroups G_i are called *direct factors*. A group is said to be *directly decomposable* if it can be represented as a direct product of non-trivial subgroups.

By (ii), the direct factors may be arranged in an arbitrary order. The element x_i in the representation (1) is called the *G_i-component* of x.

If the representation (1) of $y \in G$ is

$$y = y_1 y_2 \cdots y_n,$$

then, by (ii),

$$xy = (x_1 y_1)(x_2 y_2) \dots (x_n y_n)$$

where the brackets indicate the G_i-components of xy. To put it briefly, multiplication is carried out componentwise. It follows that

$$y^{-1} = (y_1^{-1})(y_2^{-1}) \dots (y_n^{-1})$$

and

$$y^{-1} x y = (y_1^{-1} x_1 y_1)(y_2^{-1} x_2 y_2) \dots (y_n^{-1} x_n y_n). \tag{2}$$

If x is an element of G_i so that $x = x_i$, $x_j = e$ for $j \neq i$, then (2) gives

$$y^{-1} x_i y = y_i^{-1} x_i y_i$$

which shows that $y^{-1} x_i y$ belongs to G_i. So we have

(iii) *Every direct factor is a normal subgroup.*

It is readily seen that

$$G/G_i \cong G_1 \times \dots \times G_{i-1} \times G_{i+1} \times \dots \times G_n.$$

As another easy consequence, we obtain the following property of a direct product

(iv) $G_i \cap \langle G_1, \dots, G_{i-1}, G_{i+1}, \dots, G_n \rangle = e \quad (i = 1, 2, \dots, n).$

For, if x_i is an element of this intersection, then (ii) gives

$$x_i = x_1 \dots x_{i-1} x_{i+1} \dots x_n \text{ where } x_j \in G_j,$$

and using (i) we conclude that $x_j = e$ for $j = 1, 2, \dots, n$.

Conversely, (iii) and (iv) imply that G is the direct product of the G_i. We may even replace (iv) by a weaker condition.

3.9.2. *Let* $G_1, G_2, \ldots, G_n$ *be normal subgroups of* G *such that* $G = \langle G_1, G_2, \ldots, G_n\rangle$ *and*

$$G_i \cap \langle G_1, \ldots, G_{i-1}\rangle = e \quad (i = 2, \ldots, n). \tag{3}$$

Then, G *is the direct product of* $G_1, G_2, \ldots, G_n$.

Proof. By (3), we have $G_i \cap G_j = e$ for $i \neq j$. From 3.9.1 we conclude that G_i and G_j commute elementwise. Thus, (ii) is satisfied.

Since $G_1, G_2, \ldots, G_n$ are normal subgroups, we obtain

$$G = \langle G_1, G_2, \ldots, G_n\rangle = G_1G_2 \ldots G_n.$$

Hence, every $x \in G$ has at least one representation (1). Suppose that

$$x = x_1x_2 \ldots x_n = x_1'x_2' \ldots x_n'$$

are two such representations. Then, there is some integer i, $1 \leqq i \leqq n$, such that $x_i \neq x_i'$, but $x_{i+1} = x_{i+1}', \ldots, x_n = x_n'$. This gives

$$x_ix_i'^{-1} = (x_1^{-1}x_1')\,(x_2^{-1}x_2') \ldots (x_{i-1}^{-1}x_{i-1}'),$$

which shows that $x_ix_i'^{-1} \neq e$ is contained in G_i as well as in $\langle G_1, \ldots, G_{i-1}\rangle$. Since this contradicts (3), we infer that the representation (1) is unique. Thus, (i) is satisfied, and our proof is complete.

The concept of a direct product can also be used to construct new groups from given ones. Let A and B be two groups whose unit elements are denoted by e_A and e_B, respectively. We take all pairs (a, b), $a \in A$, $b \in B$, and define multiplication as follows

$$(a_1, b_1)\,(a_2, b_2) = (a_1a_2, b_1b_2).$$

It is easily verified that these pairs form a group G with respect to this operation. The unit element is (e_A, e_B). Moreover, the elements (a, e_B) form a subgroup A^*, that is obviously isomorphic to A, and the pairs (e_A, b) form a subgroup B^* isomorphic to B. It is evident that $G = A^* \times B^*$. In view of the isomorphisms $A \cong A^*$ and $B \cong B^*$, the group we have constructed is called the direct product of A and B. In the same way one can construct the direct product of any finite number of given groups.

A group is called *completely reducible* if it is the direct product of simple groups. The following theorem states a frequently used property of such groups.

3.9.3. THEOREM. *Every non-trivial normal subgroup H of a completely reducible group G is a direct factor of G, i.e. there exists a direct decomposition $G = H \times B$.*

Proof. Let

$$G = G_1 \times G_2 \times \ldots \times G_n$$

be a direct decomposition into simple factors G_i. Clearly,

$$G = HG = HG_1G_2 \ldots G_n. \tag{4}$$

We shall show that suitable factors G_i on the right-hand side can be cancelled such that the product of the remaining factors is direct. Since G_1 is simple and H is normal in G, the intersection $G_1 \cap H$ coincides with G_1 or is e. In the first case, the factor G_1 on the right-hand side of (4) is redundant and can be cancelled. In the second case, we have $HG_1 = H \times G_1$. In the same way, we can deal with all the factors G_i. Suppose that among $G_1, \ldots, G_{i-1}$ precisely $G_{k_1}, \ldots, G_{k_r}$ have not been cancelled so that we have

$$HG_1 \ldots G_{i-1} = H \times G_{k_1} \times \ldots \times G_{k_r}.$$

The intersection $G_i \cap HG_1 \ldots G_{i-1}$ is G_i or e. In the first case, G_i is cancelled, in the second case we have

$$HG_1 \ldots G_i = H \times G_{k_1} \times \ldots \times G_{k_r} \times G_i.$$

Eventually, we obtain

$$G = HG_1 \ldots G_n = H \times G_{k_1} \times \ldots \times G_{k_t} = H \times B.$$

Incidentally, we have proved that the complementary direct factor B can be chosen as the direct product of suitable G_i.

If we use additive notation, then we speak of the *direct sum* instead of the direct product and write

$$G = G_1 \oplus G_2 \oplus \ldots \oplus G_n.$$

The decomposition of a group into the direct product of subgroups can also be defined in the case of infinitely many factors.

The group G is said to be the direct product of its subgroups G_λ, where λ ranges over an index set Λ, if the following conditions are satisfied:

(i) G is generated by the subgroups G_λ.

(ii) Any subgroup of G that is generated by a finite number of the subgroups G_λ is the direct product of these subgroups. We write

$$G = \prod_{\lambda \in \Lambda}^{\times} G_\lambda .$$

In the case of a finite number of subgroups, this definition obviously coincides with our previous definition.

From (i) and (ii), it follows that every element x of G has a unique representation as a product of finitely many factors belonging to distinct subgroups G_λ, i.e. $x = x_{\lambda_1} x_{\lambda_2} \ldots x_{\lambda_r}$ where $x_{\lambda_i} \in G_{\lambda_i}$. We simplify the notation by writing

$$x = \prod_{\lambda \in \Lambda} x_\lambda$$

and call x_λ the G_λ-component of x, where only finitely many components are distinct from the identity. The product of x and

$$y = \prod_{\lambda \in \Lambda} y_\lambda$$

is

$$xy = \prod_{\lambda \in \Lambda} x_\lambda y_\lambda .$$

In the same way as in the case of finitely many factors, we find that every G_λ is a normal subgroup of G and that, for every index λ,

$$G_\lambda \cap \langle G_\mu, \mu \neq \lambda \rangle = e .$$

If an arbitrary set of groups G_λ, $\lambda \in \Lambda$, is given then we can use the process described above to construct a group G that contains, for every index λ, a subgroup G_λ^* isomorphic to G_λ and is the direct product of its subgroups G_λ^*. Let G be the set of the formal expressions of the form

$$x = \prod_{\lambda \in \Lambda} x_\lambda \quad (x_\lambda \in G_\lambda) \tag{5}$$

where only finitely many of the elements x_λ are distinct from the identity e_λ of the corresponding G_λ. In G we define a binary operation as follows

$$xy = \Big(\prod_{\lambda \in \Lambda} x_\lambda\Big)\Big(\prod_{\lambda \in \Lambda} y_\lambda\Big) = \prod_{\lambda \in \Lambda} (x_\lambda y_\lambda) .$$

It is readily seen that G is a group with respect to this operation. For a fixed $\mu \in \Lambda$, the expressions (5) in which $x_\lambda = e_\lambda$ for all $\lambda \neq \mu$ obviously form a subgroup G_μ^* of G which is isomorphic to G_μ. We leave it as an exercise for the reader to show that G is the direct product of the subgroups G_λ^*. As in the case of finitely many factors we call G the direct product of the G_λ.

We finally observe that, in a similar way, one can define the direct product of a given set of semigroups with identity. It is obvious that this construction yields a semigroup.

3.10. Permutation groups

A permutation group is a group whose elements are permutations of some set Ω and whose operation is the multiplication of permutations as defined in § 1.3.

Let G be a group of permutations of the set Ω. We shall refer to the elements of Ω as *symbols*. The image of a symbol α under the permutation $x \in G$ is denoted by αx. For $y \in G$ we have

$$\alpha(xy) = (\alpha x)\, y.$$

If α and β are two symbols, then we say that α is *connected* to β under G and write $\alpha \sim \beta$ if there is at least one $x \in G$ such that $\alpha x = \beta$. We will show that $\sim$ is an equivalence relation. For the identity e of G we have $\alpha e = \alpha$ for every $\alpha \in \Omega$ so that $\sim$ is reflexive. If $\alpha \sim \beta$, i.e. $\alpha x = \beta$ for some $x \in G$, then $\beta x^{-1} = \alpha$ and therefore $\beta \sim \alpha$. Hence, $\sim$ is symmetric. Finally, let $\alpha \sim \beta$ and $\beta \sim \gamma$ which means that $\alpha x = \beta$ and $\beta y = \gamma$ for suitable $x, y \in G$. This gives $\alpha(xy) = (\alpha x)\, y = \beta y = \gamma$, i.e. $\alpha \sim \gamma$. Thus, $\sim$ is transitive.

Consequently, G gives rise to a partition of Ω into equivalence classes. These classes are called the *orbits* of G. Thus, we obtain the following proposition:

3.10.1. *Every group G of permutations of the set Ω defines a partition of Ω into mutually disjoint subsets, the orbits of G. Two symbols α and β of Ω belong to the same orbit if and only if there is a permutation x in G such that $\alpha x = \beta$.*

We now restrict ourselves to the case that Ω and therefore G are finite.

Let α be any symbol, and denote by G_α the complex of all elements u of G such that $\alpha u = \alpha$. G_α is not empty since at least $e \in G_\alpha$. It is obvious that G_α is a subgroup of G. We call G_α the *stabilizer subgroup* of α.

3.10.2. Theorem. *Let G be a group of permutations of a finite set Ω and let Λ be an orbit of G. Then the number of symbols in Λ is equal to the index $|G : G_\alpha|$ where G_α is the stabilizer subgroup of an arbitrary symbol α of Λ. In particular, the number of symbols in any orbit divides the order $|G|$.*

Proof. Let

$$G = G_\alpha \cup G_\alpha a \cup G_\alpha b \cup \dots \tag{1}$$

be the decomposition of G into right cosets of G_α. All permutations of a coset $G_\alpha c$ carry α into one and the same symbol; for, if u is an element of G_α, then $\alpha(uc) = (\alpha u)\, c = \alpha c$. Conversely, if x and y carry α into the same symbol, then x and y belong to the same coset of G_α. Indeed, $\alpha x = \alpha y$ implies that $\alpha(xy^{-1}) = \alpha$ and hence that $xy^{-1} \in G_\alpha$, i.e. $x \in G_\alpha y$. Thus, the number of symbols in the orbit of α is equal to the number of cosets in the decomposition (1). This proves the theorem.

A permutation group is called *transitive* on Ω if there is only one single orbit. This means that for any two symbols α, β there is at least one $x \in G$ such that $\alpha x = \beta$. For G to be transitive it is obviously sufficient that for a fixed symbol α and any other symbol β there exists at least one $x \in G$ such that $\alpha x = \beta$. A permutation group that is not transitive is called *intransitive*.

3.10.3. Theorem. *Every group is isomorphic to a transitive permutation group.*

Proof. Let G be an arbitrary group. We shall construct a permutation group $R(G)$ that is isomorphic to G. The group $R(G)$ will be a group of permutations of the elements of G.

To any element a of G we assign the mapping

$$\varrho(a) : x \to xa \quad (x \in G).$$

If $x \neq y$, then $x\varrho(a) = xa \neq ya = y\varrho(a)$; thus, $\varrho(a)$ is injective. For an arbitrary element y of G we have $(ya^{-1})\,\varrho(a) = ya^{-1}a = y$ so that every element of G is an image under $\varrho(a)$. This shows that $\varrho(a)$ is a permutation of the elements of G.

For $a, b \in G$ we obtain for any $x \in G$

$$x\big(\varrho(a)\,\varrho(b)\big) = \big(x\varrho(a)\big)\,\varrho(b) = (xa)\,b = x(ab) = x\varrho(ab).$$

This gives

$$\varrho(a)\,\varrho(b) = \varrho(ab).$$

Moreover, for $a \neq e$, $\varrho(a)$ is distinct from the identity permutation. Therefore, the mapping

$$a \to \varrho(a) \quad (a \in G)$$

is an isomorphism of G onto the group $R(G)$ of all permutations $\varrho(a)$. It is clear that $R(G)$ is transitive. This completes the proof.

For a finite group G, the group $R(G)$ is called the *regular representation* of G.

A permutation of a set Ω that contains n symbols is said to be of *degree* n. Let $\Omega = \{\xi_1, \xi_2, \ldots, \xi_n\}$. As in § 1.3, a permutation x of Ω can be denoted by

$$x = \begin{pmatrix} \xi_1 & \xi_2 & \ldots & \xi_n \\ \xi_1 x & \xi_2 x & \ldots & \xi_n x \end{pmatrix} \tag{2}$$

or, briefly, by

$$x = \begin{pmatrix} \xi \\ \xi x \end{pmatrix} \qquad (\xi \in \Omega).$$

We say that a leaves ξ fixed if $\xi a = \xi$.

It is often more expedient to use another notation. Let $\alpha_1, \alpha_2, \ldots, \alpha_l$ be distinct symbols. By a cycle

$$(\alpha_1, \alpha_2, \ldots, \alpha_l)$$

of *length* l we mean the permutation that carries α_1 into α_2, α_2 into $\alpha_3, \ldots, \alpha_{l-1}$ into α_l and α_l into α_1, while all symbols other than $\alpha_1, \alpha_2, \ldots, \alpha_l$, if there are any, remain fixed. Of course

$$(\alpha_1, \alpha_2, \ldots, \alpha_l) = (\alpha_2, \alpha_3, \ldots, \alpha_l, \alpha_1) = \ldots = (\alpha_l, \alpha_1, \ldots, \alpha_{l-1}).$$

It is easy to see that

$$(\alpha_1, \alpha_2, \ldots, \alpha_{l-1}, \alpha_l)^{-1} = (\alpha_l, \alpha_{l-1}, \ldots, \alpha_2, \alpha_1), \quad (\alpha_1, \ldots, \alpha_l)^l = e$$

and l is precisely the order of a cycle of length l. Obviously, two cycles commute if they are disjoint, i.e. if there is no symbol that occurs in both of them. A cycle of length 2 is called a *transposition*.

It is evident that every permutation of a finite set can be written as a product of disjoint cycles; for instance

$$\begin{pmatrix} 1 & 2 & 3 & 4 & 5 & 6 & 7 & 8 & 9 & 10 & 11 & 12 \\ 8 & 4 & 9 & 5 & 2 & 6 & 3 & 1 & 10 & 7 & 12 & 11 \end{pmatrix} = (2, 4, 5)\,(9, 10, 7, 3)\,(1, 8)\,(11, 12).$$

Clearly this decomposition is unique except for the order in which the cycles are written. Occasionally, it is convenient to include also the symbols that remain fixed. For this end, we use cycles of length one to denote the fixed symbols. Thus, the above permutation can also be written as follows

$$(6)\,(1, 8)\,(11, 12)\,(2, 4, 5)\,(9, 10, 7, 3).$$

Suppose that a permutation x of degree n is expressed as a product of s disjoint cycles including the cycles of length one if there are symbols that remain fixed, and let $l_1, l_2, \ldots, l_s$ be the lengths of the cycles. With a suitable numbering we may assume that $l_1 \leqq l_2 \leqq \ldots \leqq l_s$. Evidently, $l_1 + l_2 + \ldots + l_s = n$. The sequence $(l_1, l_2, \ldots, l_s)$ is called the *type* of the permutation x. The above permutation is of type (1, 2, 2, 3, 4). Since disjoint cycles commute and a cycle of length l has order l, we conclude that the order of a permutation of type $(l_1, l_2, \ldots, l_s)$ is equal to the least common multiple of $l_1, l_2, \ldots, l_s$.

For two permutations

$$x = \begin{pmatrix} \xi \\ \xi x \end{pmatrix}, \quad a = \begin{pmatrix} \xi \\ \xi a \end{pmatrix}$$

we get

$$a^{-1}xa = \begin{pmatrix} \xi a \\ \xi \end{pmatrix}\begin{pmatrix} \xi \\ \xi x \end{pmatrix}\begin{pmatrix} \xi \\ \xi a \end{pmatrix} = \begin{pmatrix} \xi a \\ \xi \end{pmatrix}\begin{pmatrix} \xi \\ \xi x \end{pmatrix}\begin{pmatrix} \xi x \\ (\xi x)\,a \end{pmatrix} = \begin{pmatrix} \xi a \\ (\xi x)\,a \end{pmatrix}.$$

This gives the following rule: $a^{-1}xa$ is obtained by replacing each ξ by ξa in both rows of (2). If

$$x = (\alpha_1, \ldots, \alpha_k)\,(\beta_1, \ldots, \beta_l) \ldots (\gamma_1, \ldots, \gamma_m) \tag{3}$$

is the decomposition of x into disjoint cycles, then $a^{-1}xa$ has the decomposition

$$a^{-1}xa = (\alpha_1 a, \ldots, \alpha_k a)\,(\beta_1 a, \ldots, \beta_l a) \ldots (\gamma_1 a, \ldots, \gamma_m a).$$

The latter is obtained by replacing each symbol in (3) by its image under a. This shows, in particular, that x and $a^{-1}xa$ are of the same type. Conversely, if x and y are two permutations of the same type, then one can easily find a permutation a such that $y = a^{-1}xa$. Indeed, write

$$x = (\alpha_1, \ldots, \alpha_k)\,(\beta_1, \ldots, \beta_l)\,\ldots\,(\gamma_1, \ldots, \gamma_m),$$

$$y = (\alpha_1', \ldots, \alpha_k')\,(\beta_1', \ldots, \beta_l')\,\ldots\,(\gamma_1', \ldots, \gamma_m'),$$

then

$$a = \begin{pmatrix} \alpha_1, \ldots, \alpha_k, \beta_1, \ldots, \beta_l, \ldots, \gamma_1, \ldots, \gamma_m \\ \alpha_1', \ldots, \alpha_k', \beta_1', \ldots, \beta_l', \ldots, \gamma_1', \ldots, \gamma_m' \end{pmatrix}$$

has the required property. This gives the following result:

3.10.4. Theorem. *Two permutations of the symmetric group S_n are conjugate if and only if they are of the same type.*

Every cycle can be written as a product of transpositions, viz.

$$(\alpha_1, \alpha_2, \ldots, \alpha_l) = (\alpha_1, \alpha_2)\,(\alpha_1, \alpha_3)\,\ldots\,(\alpha_1, \alpha_l). \tag{4}$$

Since every permutation is a product of cycles, it can be represented as a product of transpositions. In general, however, the transpositions are not disjoint so that they need not commute. Moreover, the representation of a given permutation as a product of transpositions is not unique but, as we shall see later, the number of transpositions that occur as factors is always even or always odd.

From now on, it will be convenient to take for the set Ω whose permutations are studied the set of the natural numbers $\{1, 2, \ldots, n\}$.

As we just observed, all transpositions generate S_n, but only a small part of them is really needed. For example, the $n - 1$ transpositions

$$(1, 2), (1, 3), \ldots, (1, n)$$

are sufficient to generate S_n. For, we have $(1, \alpha)\,(1, \beta)\,(1, \alpha) = (\alpha, \beta)$.

By applying the permutation

$$a = \begin{pmatrix} 1 & 2 & \ldots & n \\ \alpha_1 & \alpha_2 & \ldots & \alpha_n \end{pmatrix}$$

to the product

$$\Delta = \prod_{\substack{i,k=1\\ i>k}}^{n} (i-k),$$

we obtain

$$\Delta^a = \prod_{\substack{i,k=1\\ i>k}}^{n} (\alpha_i - \alpha_k).$$

It is clear that $\Delta^a = \Delta$ or $\Delta^a = -\Delta$, because except for the sign the factors in both products are all the differences between distinct members in the sequence $1, \ldots, n$. We put

$$\Delta^a = \chi(a)\,\Delta$$

where $\chi(a) = 1$ or -1. There exist permutations a for which $\chi(a) = -1$, for example $a = (n-1, n)$. The number $\chi(a)$ is called the *character* of the permutation a. If b is another permutation, we have

$$\Delta^{ab} = (\Delta^a)^b = \chi(a)\,\Delta^b = \chi(a)\,\chi(b)\,\Delta$$

and on the other hand

$$\Delta^{ab} = \chi(ab)\,\Delta\,.$$

Therefore

$$\chi(a)\,\chi(b) = \chi(ab)\,.$$

Thus, the mapping $a \to \chi(a)$ is a homomorphism of S_n onto the cyclic group of order 2 consisting of 1 and -1. The kernel A_n of this homomorphism is called the *alternating group* of degree n. Clearly, A_n is a normal subgroup of index 2 in S_n.

From $1 = \chi(e) = \chi(b^{-1}b) = \chi(b^{-1})\,\chi(b)$, i.e. $\chi(b^{-1}) = \chi(b)$, and

$$\chi(b^{-1}ab) = \chi(b^{-1})\,\chi(a)\,\chi(b) = \chi(a)$$

it follows that two permutations have the same character if they are conjugate in S_n.

As we observed above, the character of the transposition $(n-1, n)$ is equal to -1. Since, by Theorem 3.10.4, all transpositions are conjugate in S_n, it follows that the character of every transposition is equal to -1. If a permutation a can be represented as a product of s transpositions, we obtain therefore $\chi(a) = (-1)^s$. The representation of a as a product of transpositions is not unique. If a can also be written as a product of t transpositions, then we also have

$\chi(a) = (-1)^t$. Hence, s and t are both even or both odd. This shows that in every representation of a permutation a as a product of transpositions the number of transpositions is always either even or odd according as $\chi(a) = 1$ or $\chi(a) = -1$. Therefore, the permutations a for which $\chi(a) = 1$ are called *even* permutations while the permutations a with $\chi(a) = -1$ are called *odd* permutations.

From (4) we conclude that a cycle of length l is even or odd according as l is odd or even.

We now study the solubility of the symmetric groups. The results will be used later.

The group S_2 is cyclic of order 2 and therefore soluble. The alternating group A_3 is the cyclic group generated by (1, 2, 3), and

$$S_3 \supset A_3 \supset e$$

is a composition series whose factors are of order 2 and 3, respectively. Hence, S_3 is soluble.

To prove that S_4 is soluble we construct a composition series in which the orders of the factors are prime numbers. If we take A_4 as the first term after S_4, then the first factor is of order 2. Next, we have to find a normal subgroup of A_4 whose index in A_4 is a prime number. In § 3.3 we observed that the permutations

$$e, \ a = (1, 2)\,(3, 4), \ b = (1, 3)\,(2, 4), \ c = (1, 4)\,(2, 3)$$

form a group V which is isomorphic to the four-group. Since a, b, c are products of two transpositions, V is a subgroup of A_4. Moreover, a, b, c form a class of conjugate elements in S_4. Hence V is a normal subgroup of S_4 and therefore also a normal subgroup of A_4. Since A_4 is of order 12, we have $|A_4 : V| = 3$.

The group V contains three subgroups of order 2, namely

$$C_1 = \langle a \rangle, \quad C_2 = \langle b \rangle, \quad C_3 = \langle c \rangle .$$

Since V is abelian, every C_i is a normal subgroup of V (but not of A_4 or S_4). Consequently,

$$S_4 \supset A_4 \supset V \supset C_1 \supset e$$

is a composition series the orders of whose factors are 2, 3, 2, 2, respectively. This shows that S_4 is soluble. In the above composition series the group C_1 may be replaced by C_2 or C_3.

Later, three subgroups of order 8 of S_4 will also play a role. It is easily verified, that the elements of V together with

$$(1, 2), (3, 4), (1, 3, 2, 4), (1, 4, 2, 3)$$

form a subgroup U_1 of order 8. Two other subgroups of order 8 of S_4 are

$$U_2 = \{V, (1, 3), (2, 4), (1, 4, 3, 2), (1, 2, 3, 4)\},$$

$$U_3 = \{V, (1, 4), (2, 3), (1, 2, 4, 3), (1, 3, 4, 2)\}.$$

The three subgroups U_1, U_2, U_3 are conjugate, namely

$$(2, 3, 4)^{-1}\, U_1(2, 3, 4) = U_2,$$

$$(2, 4, 3)^{-1}\, U_1(2, 4, 3) = U_3.$$

Moreover, $U_1 \cap U_2 \cap U_3 = V$.

So far we saw that S_2, S_3, and S_4 are soluble. The symmetric groups S_n with $n > 4$, however, are not soluble. Since the orders of the alternating group A_n for $n > 4$ (even for $n > 3$) are not prime numbers, this is a consequence of the following theorem:

3.10.5. **Theorem.** *The alternating groups A_n with $n > 4$ are simple.*

Proof. In the proof, distinct lower case Greek letters stand for distinct integers of the sequence $1, 2, \ldots, n$.

Let N be a normal subgroup of A_n for $n > 4$ and $N \neq e$. We shall prove that $N = A_n$. For $x \in N$ and $a \in A_n$ we have

$$xax^{-1}a^{-1} = x(ax^{-1}a^{-1}) \in N.$$

We first show that N contains a permutation of the form $(\alpha, \beta)\,(\gamma, \delta)$. In any case, N contains a permutation x of one of the forms listed below, where the dots may, but need not, stand for further elements or cycles. In each of the four cases we choose a suitable permutation a of A_n and form the product $xax^{-1}a^{-1}$.

	x	a	$xax^{-1}a^{-1}$
(i)	$(\alpha, \beta, \gamma, \delta, \ldots) \ldots$	(β, γ, δ)	(α, δ, γ)
(ii)	$(\alpha, \beta, \gamma)(\delta, \varepsilon, \ldots) \ldots$	$(\beta, \gamma, \varepsilon)$	$(\alpha, \varepsilon, \gamma, \beta, \delta)$
(iii)	(α, β, γ)	(β, γ, δ)	$(\alpha, \delta)(\beta, \gamma)$
(iv)	$(\alpha, \beta)(\gamma, \delta) \ldots$	(α, β, γ)	$(\alpha, \delta)(\beta, \gamma)$

In the cases (iii) and (iv) we immediately obtain a permutation of the required type in N. Case (ii) is reduced to case (i), and (i) is reduced to (iii).

Next, we show that N contains all permutations of the form $(\alpha, \beta)(\gamma, \delta)$. Without loss of generality, we may assume that $(1, 2)(3, 4) \in N$. Precisely one of the permutations

$$c = \begin{pmatrix} 1\,2\,3\,4 \ldots \\ \alpha\,\beta\,\gamma\,\delta \ldots \end{pmatrix} \quad \text{or} \quad c = \begin{pmatrix} 1\,2\,3\,4 \ldots \\ \beta\,\alpha\,\gamma\,\delta \ldots \end{pmatrix}$$

belongs to A_n since they only differ by a transposition. In both cases we obtain

$$c^{-1}(1, 2)(3, 4)\,c = (\alpha, \beta)(\gamma, \delta).$$

Since N is a normal subgroup of A_n, we have $(\alpha, \beta)(\gamma, \delta) \in N$.

Finally, we prove that, for $n > 4$, the permutations $(\alpha, \beta)(\gamma, \delta)$ generate A_n. In the representation of the elements of A_n as products of transpositions, the transpositions occur in pairs since their number is even. Such pair is either of the form $(\alpha, \beta)(\gamma, \delta)$ or of the form $(\alpha, \beta)(\alpha, \delta)$. In view of $n > 4$, we have in the second case

$$(\alpha, \beta)(\alpha, \delta) = (\alpha, \beta)(\xi, \eta)(\xi, \eta)(\alpha, \delta)$$

with ξ and η distinct from α, β, and δ. It follows that $N = A_n$. This completes the proof.

3.11. Sylow subgroups and p-groups

We first prove a generalization of Theorem 3.10.2 which we shall need later.

A homomorphism of a group G onto a permutation group P of degree m is called a *permutational representation* of degree m of G.

We have

$$G/N \cong P \tag{1}$$

where N denotes the kernel of the homomorphism or, as we also say, the kernel of the representation. We shall always assume that the degree m is finite.

Let Λ be an orbit of P and $\alpha \in \Lambda$. By P_α we denote the stabilizer subgroup of α in P. There is a unique subgroup G_α of G such that G_α/N corresponds to P_α under the isomorphism (1). We also refer to G_α as the stabilizer subgroup of α because G_α consists precisely of those elements of G whose images in P leave α fixed. Due to (1), we have $|G: G_\alpha| = |P: P_\alpha|$, By Theorem 3.10.2, the number of elements in Λ is equal to $|P: P_\alpha|$. So, we obtain the following result:

3.11.1. THEOREM. *Let Λ be an orbit of a permutational representation of the group G. Then the number of elements in Λ is equal to the index $|G: G_\alpha|$ where G_α is the stabilizer subgroup of an arbitrary element α of Λ.*

The converse of Lagrange's Theorem 3.5.6 is not true, i.e. if d is a given divisor of the order of the finite group G, then G does not necessarily contain a subgroup of order d. A basic result of the theory of finite groups is the following theorem, which asserts that the converse of Lagrange's Theorem holds if we confine ourselves to subgroups of prime power order.

3.11.2. THEOREM. *If p^t is any prime power that divides the order of the finite group G, then G contains at least one subgroup of order p^t.*

Proof. We denote the order of G by g and put $g = p^t r$. (Note that r may be divisible by p). Let $K_1, K_2, \ldots, K_m$ be all the complexes of G containing p^t elements each. Their number m is given by

$$m = \binom{g}{p^t} = \frac{p^t r(p^t r - 1) \ldots (p^t r - k) \ldots (p^t r - p^t + 1)}{p^t \cdot 1 \ldots k \ldots (p^t - 1)}.$$

Let p^s be the highest power of p that divides r, where the case $s = 0$ is not excluded. It is easily seen that m is not divisible by p^{s+1}, for in the representation above of m as a fraction each pair of factors $(p^t r - k)$ and k in the numerator and denominator, respectively, is divisible by the same power of p.

To any element x of G we assign the permutation

$$\begin{pmatrix} K_1 & K_2 & \dots & K_m \\ K_1 x & K_2 x & \dots & K_m x \end{pmatrix}$$

of the complexes $K_1, K_2, \dots, K_m$. Since $K_i(xy) = (K_i x)\, y$, we obtain a permutational representation of degree m of G. We consider its orbits. Since m is not divisible by p^{s+1}, there is at least one orbit Λ such that the number of complexes contained in Λ is not divisible by p^{s+1}. Let l denote the number of complexes in Λ. By renumbering the complexes, if necessary, we may assume that K_1 belongs to Λ. Let H be the stabilizer subgroup of K_1 and put $|H| = h$. By Theorem 3.11.1, we have $g = hl$. The divisor l of g is not divisible by p^{s+1}, which implies that l divides r; hence, in particular, $l \leqq r$. If a is any element of K_1, then $K_1 = K_1 H$ shows that K_1 contains at least h distinct elements au, where u ranges over all the elements of H. This shows that $h \leqq p^t$. So, we obtain the inequalities

$$g = hl \leqq hr \leqq p^t r = g$$

which imply that $h = p^t$. Consequently, H is a subgroup of order p^t. This completes the proof.

As a special case of Theorem 3.11.2 we obtain:

3.11.3. Cauchy's Theorem. *If p is a prime divisor of the order of G, then G contains at least one element of order p.*

In particular, if p^n is the highest power of p that divides the order of G, then G contains at least one subgroup of order p^n. Every subgroup of order p^n of G is called a *Sylow p-subgroup*. The main facts on Sylow p-subgroups are summarized in:

3.11.4. Sylow's Theorem.

(a) *Any two Sylow p-subgroups of G are conjugate.*

(b) *Every subgroup of G whose order is a power of p is contained in a Sylow p-subgroup.*

(c) *If r denotes the number of Sylow p-subgroups in G, then* $r \equiv 1 \pmod p$.

Proof. Let $P_1, P_2, \ldots, P_r$ be all the Sylow p-subgroups of G. To any element x of P_1 we assign the permutation

$$\begin{pmatrix} P_1 & P_2 & \ldots & P_r \\ x^{-1}P_1x & x^{-1}P_2x & \ldots & x^{-1}P_rx \end{pmatrix}$$

to obtain a permutational representation of degree r of P_1. All these permutations leave P_1 fixed. Moreover, P_1 is the only Sylow p-subgroup that remains fixed under all these permutations. For, suppose that we have $x^{-1}P_kx = P_k$ for all $x \in P_1$ and some $k \neq 1$; then P_1P_k would be a group and $P_k \lhd P_1P_k$. By Theorem 3.7.4, the order of P_1P_k would be a power of p and $|P_1P_k| > |P_k|$, but this is impossible since $|P_k|$ is the highest power of p that divides $|G|$. We now consider the orbits of our permutational representation of P_1. One of these orbits consists of the single Sylow p-subgroup P_1, whereas all the other orbits contain more than one P_i. By Theorem 3.11.1, the numbers of Sylow p-subgroups in the latter orbits are divisors of $|P_1|$, so they are powers of p with positive exponents. So, we have $r \equiv 1 \pmod{p}$, which proves (c).

Suppose that $P_1, \ldots, P_s$ are conjugate in G. For the proof of (a), we have to show that $s < r$ leads to a contradiction. Transformation by elements of P_1 permutes $P_1, \ldots, P_s$. As above, we conclude that $s \equiv 1 \pmod{p}$. In case $s < r$, there is another system, $P_{s+1}, \ldots, P_{s+t}$ say, of Sylow p-subgroups that are conjugate in G. Transformation by elements of P_1 yields a permutational representation of P_1 of degree t. From our argument above, we conclude that the numbers of Sylow p-subgroups in the orbits of this representation are powers of p with positive exponents. This gives $t \equiv 0 \pmod{p}$. On the other hand, we can transform $P_{s+1}, \ldots, P_{s+t}$ by elements of P_{s+1}. In the same way as above, we then obtain $t \equiv 1 \pmod{p}$. Thus, the assumption $s < r$ leads to a contradiction. This proves (a).

Finally, let U be a subgroup of G whose order is a power of p. If we transform $P_1, \ldots, P_r$ by the elements of U, then we obtain a permutational representation of U of degree r. Since $|U|$ is a power of p, the numbers of Sylow p-subgroups in the orbits of this representation are powers of p. From $r \equiv 1 \pmod{p}$, we conclude that at least one orbit consists of a single Sylow p-subgroup, P_k

say. Then we have $u^{-1}P_k u = P_k$ for every $u \in U$. This shows that UP is a subgroup of G whose order is a power of p. But, since $|P_k|$ is the highest power of p dividing $|G|$, we have $|UP_k| = |P_k|$ and hence $U \subseteq P_k$. This completes the proof of (b).

The groups U_1, U_2, U_3 considered in § 3.10 are the Sylow 2-subgroups of S_4. One immediately verifies (a), (b), (c) of the last theorem for this case.

Every automorphism of a group G can be regarded as a permutation of the elements of G. If we assign to any element a of G the permutation

$$\begin{pmatrix} x \\ a^{-1}xa \end{pmatrix} \qquad (x \in G)$$

which corresponds to the inner automorphism generated by a, then we obtain a permutational representation of G. The orbits of this permutation group are called the *conjugacy classes* of G (cf. §§ 3.3 and 3.7).

Let G be a finite group. The stabilizer subgroup of a given element y of G in our permutational representation consists of all elements $u \in G$ such that $u^{-1}yu = y$. This subgroup is called the *normalizer* of y. Applying Theorem 3.11.1 we obtain the following result:

3.11.5. THEOREM. *The number of elements in a conjugacy class of a finite group G is equal to the index $|G\colon N(y)|$ where $N(y)$ is the normalizer of an arbitrary element y of that class.*

As we defined in § 3.5, the centre of a group G is the subgroup of all $x \in G$ that commute with every element of G. This means that $a^{-1}xa = x$ for every $a \in G$. Hence, an element of the centre is the only element in its conjugacy class. The converse is also obvious: if a conjugacy class consists of a single element x, then x belongs to the centre. If the centre of G consists of the identity only, then G is said to be a group with trivial centre.

We now prove an important property of finite p-groups, i.e. of groups whose order is a power of the prime number p.

3.11.6. THEOREM. *Every finite p-group has a non-trivial centre.*

Proof. Let G be a group of order p^n. Denote by $K_1, K_2, \ldots, K_t$ the conjugacy classes of G. By 3.11.5, the number of elements in K_i is a power p^{n_i} with $n_i \geqq 0$ ($i = 1, \ldots, t$). One of the classes, K_1 say, consists of the identity only so that $n_1 = 0$. Counting the elements in each class, we obtain

$$p^n = 1 + p^{n_2} + \ldots + p^{n_t}.$$

Since the right-hand side is divisible by p, we conclude that not all the exponents $n_2, \ldots, n_t$ are positive. Thus, K_1 is not the only class that consists of a single element and this means that e is not the only element in the centre. This completes the proof.

As an easy consequence of the previous theorem we obtain

3.11.7. THEOREM. *Every finite p-group is soluble.*

Proof. We proceed by induction on the order. Our theorem is obviously true for groups of order p. We assume that all groups of order p^k with $k < n$ are soluble. Let G be of order p^n. By Theorem 3.11.6, the order of the centre Z of G is greater than 1. Hence, both G/Z and Z are soluble by the induction assumption. Therefore there exist composition series

$$G/Z \supset G_1/Z \supset \ldots \supset G_{r-1}/Z \supset Z/Z$$

and

$$Z \supset Z_1 \supset \ldots \supset Z_{s-1} \supset e$$

in which all the indices are equal to p. Consequently the series

$$G \supset G_1 \supset \ldots \supset G_{r-1} \supset Z \supset Z_1 \supset \ldots \supset Z_{s-1} \supset e$$

is a composition series of G all of whose indices are equal to p so that G is soluble as was to be proved.

3.12. Endomorphisms and operators

An *endomorphism* of a group G is an isomorphism or a homomorphism of G into itself. A bijective endomorphism is an automorphism but, in general, an endomorphism is neither injective nor surjective. So, we can also say that an endomorphism of G is a homomorphism (in particular, an isomorphism) of G onto G or onto a subgroup of G distinct from G. The image of an element x of G under an endomorphism α will be denoted by x^α. Since α is a homomorphism, we have for $x, y \in G$

$$(xy)^\alpha = x^\alpha y^\alpha .$$

For example, to any integer k we can assign an endomorphism $\alpha(k)$ of the infinite cyclic group $\langle a \rangle$, namely $a^{\alpha(k)} = a^k$, and hence $(a^l)^{\alpha(k)} = a^{lk}$. It is clear that the $\alpha(k)$ are all the endomorphisms of $\langle a \rangle$.

As in the case of automorphisms, it follows that the product of two endomorphisms of G is again an endomorphism of G. Therefore the endomorphisms of G form a semigroup under multiplication. The endomorphism semigroup of an infinite cyclic group, for example, is isomorphic to the multiplicative semigroup of the integers.

In the case of an abelian group we can also define the sum of two endomorphisms. Let α and β be endomorphisms of the abelian group G. Then, we define the sum $\alpha + \beta$ as follows:

$$x^{\alpha+\beta} = x^\alpha x^\beta \quad (x \in G).$$

The equation

$$(xy)^{\alpha+\beta} = (xy)^\alpha (xy)^\beta = x^\alpha y^\alpha x^\beta y^\beta = x^\alpha x^\beta y^\alpha y^\beta = x^{\alpha+\beta} y^{\alpha+\beta}$$

shows that $\alpha + \beta$ is an endomorphism. It is easily verified that addition of endomorphisms is commutative and associative. We also define the zero endomorphism 0, namely

$$x^0 = e \text{ for every } x \in G,$$

and the additive inverse $-\alpha$ of the endomorphism α, namely

$$x^{-\alpha} = (x^\alpha)^{-1} = (x^{-1})^\alpha \quad (x \in G).$$

It is clear that $-\alpha$ is an endomorphism and $\alpha + (-\alpha) = 0$. It is easy to prove that in view of these definitions the endomorphisms of an abelian group form a ring, the so-called *endomorphism ring*. As an example, we prove one of the distributive laws:

$$x^{\alpha(\beta+\gamma)} = (x^\alpha)^{\beta+\gamma} = (x^\alpha)^\beta (x^\alpha)^\gamma = x^{\alpha\beta} x^{\alpha\gamma} = x^{\alpha\beta+\alpha\gamma}.$$

In general, the endomorphism ring of an abelian group is not commutative.

The endomorphism ring of an infinite cyclic group, for example, is isomorphic to the ring of the integers. Non-commutative endomorphism rings will occur in the next section.

It turns out that it is expedient to generalize the concept of an endomorphism to that of an operator.

By an *operator domain* Ω of a group G we mean a set of symbols $\tau, \omega, \ldots$, with the following properties:

(i) To every $\omega \in \Omega$ and every $x \in G$, there corresponds a unique element $x\omega$ of G.

(ii) For arbitrary elements x, y of G and ω of Ω, we have

$$(xy)\,\omega = (x\omega)\,(y\omega).$$

In view of (ii), every operator $\omega \in \Omega$ induces an endomorphism of G. But we distinguish between an operator and the corresponding endomorphism and, in particular, we do not exclude the possibility that distinct operators induce the same endomorphism of G. This distinction is one of the reasons for the great importance of groups with operators. For instance, when we are given two groups with the same operator domain, then two distinct operators may induce distinct endomorphisms in one of the groups, but the same endomorphism in the other. Or, the operator domain may consist of the elements of some given algebraic system (e.g. a ring) so that equality of operators is defined qua elements of that system; but this does not imply that distinct operators induce distinct endomorphisms.

A subgroup U of G is called *Ω-admissible* if every operator maps U into itself, i.e.

$$U\omega \subseteq U \text{ for every } \omega \in \Omega.$$

Here, of course, $U\omega$ means the group of all $u\omega$ with $u \in U$. If it is clear from the context to which operator domain we refer, then we simply speak of admissible subgroups.

Let G be a group and Ω an operator domain of G. An endomorphism α of G is called an *Ω-endomorphism* if

$$(x\omega)^\alpha = (x^\alpha)\,\omega \text{ for every } x \in G \text{ and every } \omega \in \Omega.$$

This equation means that the mapping α commutes with every mapping induced by an operator.

An analogous concept arises in connection with homomorphisms of one group onto another. Let G and G^* be two groups with the same operator domain Ω. A homomorphism σ of G onto G^* is called an Ω-homomorphism if

$$(x\omega)\,\sigma = (x\sigma)\,\omega \text{ for every } x \in G \text{ and every } \omega \in \Omega.$$

We shall now prove that the Homomorphism Theorem and the two Isomorphism Theorems in § 3.7 remain valid if we consider only admissible subgroups and admissible normal subgroups and if the term homomorphism (isomorphism) is replaced by Ω-homomorphism (Ω-isomorphism). Instead of repeating our previous proofs, we shall only supplement them with a few remarks.

Suppose that the groups G and G^* have the same operator domain Ω and that σ is an Ω-homomorphism of G onto G^*. Let K be the kernel of σ. We prove that K is Ω-admissible. For, K consists of all elements u of G such that $u\sigma$ is the identity e^* of G^*; hence we obtain

$$(u\omega)\,\sigma = (u\sigma)\,\omega = e^*\omega = e^*$$

so that $u\omega \in K$.

By Theorem 3.7.3, the factor group G/K is isomorphic to G^* in the ordinary sense (i.e. disregarding Ω). It remains to show that Ω can be regarded as an operator domain of G/K and that G/K and G^* are even Ω-isomorphic.

First, we define how the operators act on the cosets of K. All the elements of the coset Ka have the form ua with $u \in K$. By applying an operator ω, we obtain

$$(ua)\,\omega = (u\omega)\,(a\omega) = u_1(a\omega)$$

where u_1 belongs to K, since K is Ω-admissible. Thus, ω carries all the elements of Ka into $K(a\omega)$. Therefore we define

$$(Ka)\,\omega = K(a\omega). \tag{1}$$

Then we obtain

$$(KaKb)\,\omega = K(ab)\,\omega = K(a\omega)\,(b\omega)$$

$$= K(a\omega)\,K(b\omega) = [(Ka)\,\omega]\,[(Kb)\,\omega].$$

Consequently, Ω is an operator domain of G/K.

It is now easy to prove that G/K and G^* are Ω-isomorphic. Under the ordinary isomorphism between G/K and G^*, the cosets Kx and the elements $x\sigma$ of G^* correspond to each other. An operator ω carries Kx into $K(x\omega)$ and $x\sigma$ into $(x\sigma)\,\omega = (x\omega)\,\sigma$. But under the ordinary isomorphism between G/K and G^*, also $K(x\omega)$ and $(x\omega)\,\sigma$ correspond to each other. This shows that G/K and G^* are Ω-isomorphic.

This completes the proof of our generalized Homomorphism Theorem, which can be stated as follows:

3.12.1. THEOREM. *To any Ω-homomorphism of a group G onto a group G^* there belongs as its kernel an Ω-admissible normal subgroup K of G. The kernel consists of all elements of G that are mapped onto the identity of G^*. The group G^* is Ω-isomorphic to the factor group G/K. Conversely, if K is an Ω-admissible normal subgroup of G then the natural homomorphism of G onto G/K is an Ω-homomorphism.*

The proof of the Isomorphism Theorems of § 3.7 may now be left as an exercise for the reader. As is readily seen, Zassenhaus' Lemma 3.7.6 can also be extended to groups with operators.

3.13. Vector spaces. Modules

One of the most important examples of groups with operators are the vector spaces.

We recall the definition of a vector space in § 1.5. Let V be an algebraic system with one operation, written as addition. Moreover, let $\mathbf{F}$ be a field and assume that, for any $\alpha \in \mathbf{F}$ and any $v \in V$, there is defined a unique element αv of V. Then V is called a *vector space over* $\mathbf{F}$ if the following axioms are satisfied for arbitrary elements x, y, z of V and α, β of $\mathbf{F}$:

(V_1) $x + y = y + x$.

(V_2) $(x + y) + z = x + (y + z)$.

(V_3) V contains an element 0 such that

$$x + 0 = x \text{ for every } x \in V.$$

(V_4) For every $x \in V$ there exists an element $-x$ in V such that $x + (-x) = 0$.

(V_5) $\alpha(x + y) = \alpha x + \alpha y$.

(V_6) $(\alpha + \beta)\, x = \alpha x + \beta x$.

(V_7) $(\alpha\beta)\, x = \alpha(\beta x)$.

(V_8) $1x = x$ for the unit element 1 of $\mathbf{F}$.

(V_1) through (V_4) mean that V is an abelian group with respect to addition. (V_5) shows that $\mathbf{F}$ is an operator domain of the group V. Finally, (V_6) through (V_8) can be interpreted as follows: Addition and multiplication of operators in the sense of addition and multiplication of endomorphisms of V coincide with addition and multiplication in $\mathbf{F}$, and the unit element of $\mathbf{F}$ acts as the identity operator on V.

We shall refer to the elements of V as vectors.

Any $\mathbf{F}$-admissible subgroup of V is called a *subspace*. For U to be a subspace it is, of course, necessary but also sufficient that the following two conditions are satisfied:

(i) If $u \in U$, then $\alpha u \in U$ for every $\alpha \in \mathbf{F}$.

(ii) If $u \in U$ and $v \in U$, then $u + v \in U$.

For, if $u \in U$ and $v \in U$, then, by (i), $(-1)\, v = -v$ belongs to U, and hence, by (ii), $u - v \in U$. In view of 3.5.1, this shows that U is a subgroup. Moreover, (i) means that U is admissible.

Let $v_1, \ldots, v_r$ be vectors in V. Any expression

$$\alpha_1 v_1 + \ldots + \alpha_r v_r \quad (\alpha_i \in \mathbf{F})$$

is called a *linear combination* of $v_1, \ldots, v_r$ with *coefficients* $\alpha_1, \ldots, \alpha_r$. Note that only finite linear combinations are defined.

Let S be an arbitrary, possibly infinite, set of vectors of V. It is easy to see that all finite linear combinations of vectors of S form a subspace of V. This subspace is said to be *spanned* by S and is denoted by $\langle S \rangle$. This notation agrees with that in § 3.5, for $\langle S \rangle$ is the $\mathbf{F}$-admissible subgroup generated by S. If S is a finite set, $S = \{v_1, \ldots, v_r\}$ say, then we write $\langle S \rangle = \langle v_1, \ldots, v_r \rangle$, and this subspace consists of all linear combinations of $v_1, \ldots, v_r$.

A set $\{w_1, \ldots, w_s\}$ of vectors is said to be *linearly independent* if any equation

$$\alpha_1 w_1 + \ldots + \alpha_s w_s = 0$$

implies that $\alpha_1 = \ldots = \alpha_s = 0$. Otherwise, $w_1, \ldots, w_s$ are called *linearly dependent*. Thus $w_1, \ldots, w_s$ are linearly dependent if the zero vector can be obtained as a linear combination of $w_1, \ldots, w_s$ in which not all the coefficients are equal to zero.

If $w_1, \ldots, w_s$ are linearly independent, then $w_i \neq 0$ for $i = 1, \ldots, s$. Indeed, suppose, for example, that $w_1 = 0$; then we have $1w_1 + 0w_2 + \ldots + 0w_s = 0$ which contradicts the linear independence. Every non-empty subset of a set of linearly independent vectors is also linearly independent. For, if $w_1, \ldots, w_r$ with $r < s$ are linearly dependent, then

$$\alpha_1 w_1 + \ldots + \alpha_r w_r = 0$$

with at least one $\alpha_i \neq 0$, and therefore

$$\alpha_1 w_1 + \ldots + \alpha_r w_r + 0w_{r+1} + \ldots + 0w_s = 0$$

so that $w_1, \ldots, w_s$ are linearly dependent.

An infinite set of vectors is said to be linearly independent if every finite subset is linearly independent.

A set B of vectors is called a *basis* of the vector space V if B is linearly independent and $\langle B \rangle = V$.

3.13.1. *If B is a basis of V then any vector $a \in V$ has a unique representation*

$$a = \alpha_1 b_1 + \ldots + \alpha_n b_n$$

where $b_i \in B$ for $i, \ldots, n$. (The number n may depend on a).

Proof. Since $\langle B \rangle = V$, any vector a has at least one representation as a linear combination of vectors of B. Suppose that a has two distinct representations. By subtracting one representation from the other, we obtain a representation of the zero vector as a linear combination of vectors of B in which not all the coefficients are equal to zero. This contradicts the linear independence of B, and our proposition is proved.

3.13.2. Theorem. *Every vector space has a basis.*

Proof. Let V be any vector space and S a set of vectors such that $\langle S \rangle = V$. Such sets exist, e.g. $S = V$. Let $\mathfrak{T}$ denote the totality of all linearly independent subsets of S. Then $\mathfrak{T}$ is partially ordered with respect to inclusion. Let $\mathfrak{C}$ be a chain in $\mathfrak{T}$ (cf. § 1.2.), i.e.

$$\mathfrak{C} = \{B_\lambda, \lambda \in \Lambda\}$$

where Λ is some index set, and the B_λ are linearly independent sets of vectors such that $B_\lambda \subseteq B_\mu$ or $B_\mu \subseteq B_\lambda$ for each pair $\lambda, \mu \in \Lambda$. We form the union $U = \bigcup_{\lambda \in \Lambda} B_\lambda$. We first show that U is linearly independent. Suppose that, on the contrary, U contains a set of linearly dependent vectors, $v_1, \ldots, v_r$ say. Each v_i is contained in some B_λ, say $v_i \in B_{\lambda_i}$ $(i = 1, \ldots, r)$. Since the B_λ are completely ordered, there is some index k, $1 \leqq k \leqq r$, such that $B_{\lambda_i} \subseteq B_{\lambda_k}$ for $i = 1, \ldots, r$. This gives $v_i \in B_{\lambda_k}$ for $i = 1, \ldots, r$. But this is a contradiction because B_{λ_k} is linearly independent so that B_{λ_k} cannot contain a set of linearly dependent vectors $v_1, \ldots, v_r$. By Zorn's Lemma (1.2.2), there exists a maximal linearly independent set B in S. It is now easy to see that B is a basis of V. We have to show that any $a \in V$ is a linear combination of vectors of B. If a belongs to B, then there is nothing to prove. If a does not belong to B, then the set $B \cup \{a\}$ is linearly dependent since B is a maximal linearly independent set. Therefore we obtain

$$\alpha a + \beta_1 b_1 + \ldots + \beta_n b_n = 0$$

where $b_i \in B$ and not all the coefficients $\alpha, \beta_1, \ldots, \beta_n$ are equal to zero. Since $b_1, \ldots, b_n$ are linearly independent we conclude that $\alpha \neq 0$. This gives

$$a = -\alpha^{-1}\beta_1 b_1 - \ldots - \alpha^{-1}\beta_n b_n$$

which completes the proof.

A vector space is said to be *finite dimensional* if it has a basis of finitely many vectors. We now confine ourselves to finite dimensional vector spaces.

3.13.3. Exchange Lemma. *Let $a_1, a_2, \ldots, a_r$ be vectors in V and $b = \alpha_1 a_1 + \alpha_2 a_2 + \ldots + \alpha_r a_r$ where $\alpha_1 \neq 0$. Then*

$$\langle a_1, a_2, \ldots, a_r \rangle = \langle b, a_2, \ldots, a_r \rangle.$$

Proof. Obviously, $\langle b, a_2, \ldots, a_r \rangle \subseteq \langle a_1, a_2, \ldots, a_r \rangle$. Since $\alpha_1 \neq 0$, we obtain

$$a_1 = \alpha_1^{-1} b - \alpha_1^{-1}\alpha_2 a_2 - \ldots - \alpha_1^{-1}\alpha_r a_r$$

so that every linear combination of $a_1, a_2, \ldots, a_r$ is contained in $\langle b, a_2, \ldots, a_r \rangle$, i.e.

$$\langle a_1, a_2, \ldots, a_r \rangle \subseteq \langle b, a_2, \ldots, a_r \rangle.$$

Both inclusions prove the lemma.

3.13.4. THEOREM. *A vector space that is spanned by r vectors cannot contain more than r linearly independent vectors.*

Proof. Suppose that $V = \langle a_1, \ldots, a_r \rangle$ and let $b_1, \ldots, b_s$ be linearly independent vectors in V. We have to prove that $s \leqq r$. We shall show that the assumption $s > r$ leads to a contradiction.

We have $b_1 = \alpha_1 a_1 + \ldots + \alpha_r a_r$. Since $b_1, \ldots, b_s$ are linearly independent it follows, in particular, that $b_1 \neq 0$. Therefore at least one of the coefficients α_i is distinct from zero. By renumbering the a_i, if necessary, we may assume that $\alpha_1 \neq 0$. Then it follows from 3.13.3 that

$$\langle b_1, a_2, \ldots, a_r \rangle = \langle a_1, a_2, \ldots, a_r \rangle .$$

Therefore, b_2 has a representation

$$b_2 = \beta_1 b_1 + \beta_2 a_2 + \ldots + \beta_r a_r .$$

Since b_1 and b_2 are linearly independent we conclude that at least one of the coefficients $\beta_2, \ldots, \beta_r$ is distinct from zero. By numbering $a_2, \ldots, a_r$ in a suitable way we may assume that $\beta_2 \neq 0$. Then, 3.13.3 gives

$$\langle b_1, b_2, a_3, \ldots, a_r \rangle = \langle b_1, a_2, a_3, \ldots, a_r \rangle = \langle a_1, \ldots, a_r \rangle .$$

Next, b_3 has a representation

$$b_3 = \gamma_1 b_1 + \gamma_2 b_2 + \gamma_3 a_3 + \ldots + \gamma_r a_r .$$

Because b_1, b_2, b_3 are linearly independent, at least one of the coefficients $\gamma_3, \ldots, \gamma_r$ is distinct from zero. We may assume that $\gamma_3 \neq 0$. By 3.13.3, we then obtain

$$\langle b_1, b_2, b_3, a_4, \ldots, a_r \rangle = \langle b_1, b_2, a_3, a_4, \ldots, a_r \rangle = \langle a_1, \ldots, a_r \rangle .$$

Proceeding in this way we arrive at

$$\langle b_1, \ldots, b_r \rangle = \langle a_1, \ldots, a_r \rangle .$$

We assumed that $s > r$. Hence $b_{r+1} \in \langle b_1, \ldots, b_r \rangle$ so that we obtain

$$b_{r+1} = \delta_1 b_1 + \ldots + \delta_r b_r$$

which contradicts the hypothesis that $b_1, \ldots, b_s$ are linearly independent. This completes the proof.

3.13.5. Corollary. *A system of r linear homogeneous equations for s unknowns $x_1, x_2, \ldots, x_s$*

$$a_{11}x_1 + a_{12}x_2 + \ldots + a_{1s}x_s = 0,$$
$$a_{21}x_1 + a_{22}x_2 + \ldots + a_{2s}x_s = 0,$$
$$\ldots\ldots\ldots\ldots\ldots\ldots\ldots\ldots\ldots\ldots$$
$$a_{r1}x_1 + a_{r2}x_2 + \ldots + a_{rs}x_s = 0$$

with coefficients a_{ik} in a field F has at least one non-trivial solution (i.e. a solution other than $x_1 = x_2 = \ldots = x_s = 0$) in F if $s > r$.

Proof. The vector space of all $r \times 1$ matrices with elements in F is obviously spanned by the r matrices

$$\begin{bmatrix} 1 \\ 0 \\ \vdots \\ 0 \end{bmatrix}, \begin{bmatrix} 0 \\ 1 \\ \vdots \\ 0 \end{bmatrix}, \ldots, \begin{bmatrix} 0 \\ 0 \\ \vdots \\ 1 \end{bmatrix}.$$

Since $s > r$ it follows from the previous theorem that the s matrices

$$\begin{bmatrix} a_{11} \\ a_{21} \\ \vdots \\ a_{r1} \end{bmatrix}, \begin{bmatrix} a_{12} \\ a_{22} \\ \vdots \\ a_{r2} \end{bmatrix}, \ldots, \begin{bmatrix} a_{1s} \\ a_{2s} \\ \vdots \\ a_{rs} \end{bmatrix}$$

are linearly dependent. This means that the above system of linear equations has a solution such that not all the x_i are equal to 0, as was to be proved.

3.13.6. Theorem. *All bases of a finite dimensional vector space consist of the same number of vectors.*

Proof. Let $v_1, \ldots, v_m$ and $w_1, \ldots, w_n$ be two bases of the finite dimensional vector space V. Since $w_1, \ldots, w_n$ are linearly independent it follows from Theorem 3.13.4 that $n \leqq m$. Interchanging the roles of the two bases we similarly obtain $m \leqq n$ and the theorem follows.

The number of vectors in any basis of V is called the *dimension* of V and is denoted by $\dim V$. To the zero space we assign the dimension 0.

3.13.7. *Let V be a vector space of dimension n. Then any set of n linearly independent vectors of V forms a basis of V.*

Proof. Let $a_1, \ldots, a_n$ be linearly independent. We have to show that any vector $b \in V$ can be expressed as a linear combination of $a_1, \ldots, a_n$. Since the dimension of V is n, it follows from Theorem 3.13.4 that $b, a_1, \ldots, a_n$ are linearly dependent, i.e.

$$\beta b + \alpha_1 a_1 + \ldots + \alpha_n a_n = 0$$

where not all the coefficients are equal to zero. Since $a_1, \ldots, a_n$ are linearly independent we conclude that $\beta \neq 0$. Therefore the last equation yields a representation of b as a linear combination of $a_1, \ldots, a_n$. This completes the proof.

3.13.8. *If $V = \langle a_1, \ldots, a_r \rangle$, then a suitable subset of $a_1, \ldots, a_r$ is a basis of V.*

Proof. We proceed by induction on r. For $r = 1$ there is nothing to prove. We assume that the proposition is true for the vector space $V_1 = \langle a_1, \ldots, a_{r-1} \rangle$. Therefore we can choose suitable vectors $a_{k_1}, \ldots, a_{k_m}$ out of $a_1, \ldots, a_{r-1}$ such that $a_{k_1}, \ldots, a_{k_m}$ form a basis of V_1. We now consider a_r. If a_r is a linear combination of $a_{k_1}, \ldots, a_{k_m}$, then $a_r \in V_1$ and hence $V = V_1$. In this case, $a_{k_1}, \ldots, a_{k_m}$ form a basis of V. If a_r is not a linear combination of $a_{k_1}, \ldots, a_{k_m}$, then $a_{k_1}, \ldots, a_{k_m}, a_r$ form a basis of V. Indeed, every vector $v \in V$ can be represented as a linear combination

$$v = (\alpha_1 a_1 + \ldots + \alpha_{r-1} a_{r-1}) + \alpha_r a_r .$$

The sum in the brackets belongs to V_1 so that it can be expressed as a linear combination of $a_{k_1}, \ldots, a_{k_m}$. This shows that

$$\langle a_{k_1}, \ldots, a_{k_m}, a_r \rangle = V .$$

It remains to show that $a_{k_1}, \ldots, a_{k_m}, a_r$ are linearly independent. Suppose that

$$\beta_1 a_{k_1} + \ldots + \beta_m a_{k_m} + \beta a_r = 0 . \tag{1}$$

Then, $\beta = 0$, for otherwise (1) would yield a representation of a_r as a linear combination of $a_{k_1}, \ldots, a_{k_m}$. But since $a_{k_1}, \ldots, a_{k_m}$ are linearly independent, it now follows that also $\beta_1 = \ldots = \beta_m = 0$. This proves our proposition.

3.13.9. *If V is of dimension n and U a subspace of dimension m, then $m \leqq n$. If $m = n$, then $U = V$.*

Proof. It is obvious that $m \leqq n$ since U cannot contain more linearly independent vectors than V. If $m = n$, then U contains n linearly independent vectors and hence, by 3.13.7, a basis of V. Therefore $U = V$ and the proposition is proved.

We now prove an analogue to Theorem 3.9.3 for vector spaces.

3.13.10. THEOREM. *Let $v_1, \ldots, v_n$ be a basis of a vector space V and $b_1, \ldots, b_m$ a basis of a subspace U. Then one can choose suitable $t = n - m$ vectors $v_{k_1}, \ldots, v_{k_t}$ out of $v_1, \ldots, v_n$ such that*

$$b_1, \ldots, b_m, v_{k_1}, \ldots, v_{k_t} \tag{2}$$

form a basis of V. Briefly, every basis of a subspace U of V can be extended to a basis of V.

Proof. We choose $v_{k_1}, \ldots, v_{k_t}$ according to the following rule. v_{k_1} is the first vector in the sequence $v_1, \ldots, v_n$ that is not contained in $\langle b_1, \ldots, b_m \rangle$. Suppose that v_{k_i} has been chosen. Then $v_{k_{i+1}}$ is the first vector in the sequence $v_{k_i+1}, v_{k_i+2}, \ldots, v_n$ that is not contained in $\langle b_1, \ldots, b_m, v_{k_1}, \ldots, v_{k_i} \rangle$. After a finite number of steps we arrive at the vectors (2). By the rule for choosing the v_{k_i} it is clear that every v_s that does not occur in (2) can be expressed as a linear combination of the vectors (2). Therefore the vectors (2) span V. It remains to show that they are linearly independent. Suppose that

$$\beta_1 b_1 + \ldots + \beta_m b_m + \gamma_1 v_{k_1} + \ldots + \gamma_t v_{k_t} = 0. \tag{3}$$

We first observe that $\gamma_t = 0$; for otherwise v_{k_t} could be expressed as a linear combination of $b_1, \ldots, b_m, v_{k_1}, \ldots, v_{k_{t-1}}$ which contradicts the rule for choosing the v_{k_i}. Next, we see that $\gamma_{t-1} = 0$. For if $\gamma_t = 0$ and $\gamma_{t-1} \neq 0$, then (3) would yield a representation of $v_{k_{t-1}}$ as a linear combination of $b_1, \ldots, b_m, v_{k_1}, \ldots, v_{k_{t-2}}$ which, again, contradicts the choice of the v_{k_i}. Proceeding in this way we finally obtain $\gamma_t = \gamma_{t-1} = \ldots = \gamma_1 = 0$. But then (3) implies that also $\beta_1 = \ldots = \beta_m = 0$ since $b_1, \ldots, b_m$ are linearly independent as a basis of U. It is clear that $t = n - m$ because (2) is a basis of the n-dimensional vector space V. This completes the proof.

Let V be an n-dimensional vector space over $\mathbf{F}$ and $v_1, \ldots, v_n$ a basis of V. For any vector

$$a = \alpha_1 v_1 + \ldots + \alpha_n v_n$$

of V, the coefficients $\alpha_1, \ldots, \alpha_n$ are called the *coordinates* of a with respect to the basis $v_1, \ldots, v_n$. We arrange the coordinates in a column and write

$$\boldsymbol{\alpha} = \begin{bmatrix} \alpha_1 \\ \vdots \\ \alpha_n \end{bmatrix} = [\alpha_i].$$

If $\boldsymbol{\beta} = [\beta_i]$ are the coordinates of a vector b, then $a + b$ has obviously the coordinates $\boldsymbol{\alpha} + \boldsymbol{\beta} = [\alpha_i + \beta_i]$. Moreover, the vector λa with $\lambda \in \mathbf{F}$ has the coordinates $\lambda\boldsymbol{\alpha} = [\lambda\alpha_i]$. This shows that the mapping

$$a \to \boldsymbol{\alpha} \quad (a \in V)$$

is an $\mathbf{F}$-isomorphism (cf. § 3.12) of V onto the vector space $\mathbf{F}^n$ defined in § 1.5. Thus, we obtain:

3.13.11. *Every n-dimensional vector space over a field $\mathbf{F}$ is $\mathbf{F}$-isomorphic to $\mathbf{F}^n$, the vector space of all $n \times 1$ matrices with elements of $\mathbf{F}$.*

By a *linear mapping*

$$\varrho : v \to v^\varrho \quad (v \in V)$$

of a vector space V into itself we mean an $\mathbf{F}$-endomorphism of the additive group V in the sense of § 3.12. Thus, we have

$$(v + w)^\varrho = v^\varrho + w^\varrho, \quad (\alpha v)^\varrho = \alpha(v^\varrho) \tag{4}$$

for any two vectors $v, w \in V$ and any $\alpha \in \mathbf{F}$.

If $v_1, \ldots, v_n$ is a basis of V and

$$a = \alpha_1 v_1 + \ldots + \alpha_n v_n,$$

then (4) gives

$$a^\varrho = (\alpha_1 v_1 + \dots + \alpha_n v_n)^\varrho = (\alpha_1 v_1)^\varrho + \dots + (\alpha_n v_n)^\varrho$$
$$= \alpha_1 (v_1^\varrho) + \dots + \alpha_n (v_n^\varrho).$$

This shows that ϱ is uniquely determined by the images $v_1^\varrho, \dots, v_n^\varrho$ of a basis. Each v_i^ϱ is a linear combination of $v_1, \dots, v_n$,

$$v_i^\varrho = \sum_{k=1}^{n} \varrho_{ik} v_k \quad (i = 1, \dots, n). \tag{5}$$

With the notations

$$\boldsymbol{v} = \begin{bmatrix} v_1 \\ \vdots \\ v_n \end{bmatrix}, \quad \boldsymbol{v}^\varrho = \begin{bmatrix} v_1^\varrho \\ \vdots \\ v_n^\varrho \end{bmatrix}, \quad R = \begin{bmatrix} \varrho_{11} & \cdots & \varrho_{1n} \\ \cdots & \cdots & \cdots \\ \varrho_{n1} & \cdots & \varrho_{nn} \end{bmatrix}$$

(5) can be written as follows

$$\boldsymbol{v}^\varrho = R\boldsymbol{v}. \tag{6}$$

Conversely, any $n \times n$ matrix R with elements of F thus defines a linear mapping ϱ of V into itself. As is well known, ϱ is bijective if and only if R is a non-singular matrix. If σ is another linear mapping and S its matrix, then we can form the sum $\varrho + \sigma$ and the product $\varrho\sigma$ in the sense of § 3.12. We obtain

$$\boldsymbol{v}^{\varrho+\sigma} = \boldsymbol{v}^\varrho + \boldsymbol{v}^\sigma = R\boldsymbol{v} + S\boldsymbol{v} = (R + S)\,\boldsymbol{v},$$
$$\boldsymbol{v}^{\varrho\sigma} = (\boldsymbol{v}^\varrho)^\sigma = (R\boldsymbol{v})^\sigma = R\boldsymbol{v}^\sigma = RS\boldsymbol{v}.$$

This shows that to $\varrho + \sigma$ and $\varrho\sigma$ there correspond the matrices $R + S$ and RS, respectively.

As we observed in § 3.12, the linear mappings of V into itself form a ring. Generalizing the concept of an isomorphism of groups, we define an isomorphism $x \to x^*$ of a ring P onto a ring P^* as a bijective mapping of P onto P^* such that $(x + y)^* = x^* + y^*$ and $(xy)^* = x^* y^*$. Using these concepts we can state our results as follows:

3.13.12. *The ring of all linear mappings of an n-dimensional vector space over a field* F *into itself is isomorphic to the ring of all* $n \times n$ *matrices with elements in* F.

One easily verifies that for the proofs of the theorems on vector spaces we did not use the commutative law of multiplication in the field **F**. It was sufficient that in **F** every non-zero element α has a multiplicative inverse, i.e. an element α^{-1} such that $\alpha^{-1}\alpha = \alpha\alpha^{-1} = 1$. As we shall see in the next chapter, there exist so-called skew fields, i.e. fields that satisfy all the postulates for a field except the commutative law of multiplication. Therefore the theorems on vector spaces which we proved so far remain valid for vector spaces over skew fields.

We now consider another generalization of vector spaces over a field.

The postulates (V_1) through (V_8) only refer to addition, multiplication and the existence of a unit element in **F** but not to the commutativity of multiplication in **F** or the existence of multiplicative inverses. Therefore these postulates can also be used to define an algebraic system in case **F** is an arbitrary ring with unit element.

Let V be an additively written abelian group and **R** a ring with unit element 1 such that **R** is an operator domain of V. Thus, for each $v \in V$ and each $\alpha \in \mathbf{R}$ there is a unique element αv in V. If the postulates (V_1) through (V_8) are satisfied, then V is called an **R**-*module*. If we wish to emphasize that the operators are written on the left we speak of a *left* **R**-*module*. (Similarly, one can define right **R**-modules.)

The theorems on vector spaces do not necessarily hold for **R**-modules over an arbitrary ring **R**, e.g. an **R**-module need not have a basis.

If $v_1, \ldots, v_r$ are elements of an **R**-module V, then any expression

$$\alpha_1 v_1 + \ldots + \alpha_r v_r \quad (\alpha_i \in \mathbf{R})$$

is again called a linear combination of $v_1, \ldots, v_r$ with coefficients $\alpha_1, \ldots, \alpha_r$. Note that only finite linear combinations are defined. The definition of linear dependence is exactly the same as for vector spaces.

As set B of elements of an **R**-module V is called a *basis* of V if B is linearly independent and if every element of V can be represented as a (finite) linear combination of elements of B with coefficients in **R**. The proof of 3.13.1 remains valid for an arbitrary ring

R so that the representation of any element of V as a linear combination of elements of B is unique.

An **R**-module V with a basis B is called a *free* **R**-module. The elements of B are said to form a system of *free generators*.

If V has a finite basis, consisting of n elements, say, then V is called a free **R**-module of *rank* n. But this does not exclude the possibility that there exist other bases the number of elements in which is distinct from n.

Exercises

1. Let S be a semigroup with identity e. An element $u \in S$ is called a unit if there are elements $x, y \in S$ such that $xu = uy = e$. Prove that $x = y$. Let U denote the set of all units in S. Prove that U is a group and that every group in S containing e is a subgroup of U. Give an example to show that S may contain groups that are not subgroups of U.
2. Prove that a finite cancellative semigroup is a group.
3. Prove that a group G is abelian if $x^2 = e$ for every $x \in G$.
4. Prove that the mapping $x \to x^{-1}$ for every element x of a group G is an automorphism if and only if G is abelian.
5. Show that the symmetric group S_3 has only inner automorphisms and is isomorphic to its automorphism group.
6. Determine the automorphism group of the four-group.
7. Let $f(x)$ and $g(x)$ be rational functions of a variable x. Define the operation $f \circ g$ to mean $f \circ g(x) = f(g(x))$. Show that under this operation the functions $f(x) = 1/x$ and $g(x) = 1/(1 - x)$ generate a group isomorphic to S_3. Find all isomorphisms of this group onto S_3.
8. Show that the multiplicative group of all complex numbers with modulus 1 is isomorphic to the group of all real numbers x, $0 \leqq x < 1$, under addition mod 1.
9. Let U be a complex of a finite group such that $xy \in U$ whenever $x \in U$ and $y \in U$. Prove that U is a subgroup. Give an example to show that the finiteness condition is not redundant.
10. Determine all groups of order 4 and of order 6.
11. Let U, V be subgroups of a group. Prove that the number of

distinct elements in the complex UV is equal to $|U||V||U \cap V|^{-1}$ (Note that UV need not be a subgroup. Hint: Decompose U into left cosets and V into right cosets of $U \cap V$.)

12. Let n be a natural number, relatively prime to the order of a finite group G. Let $x, y \in G$. Prove that $x^n = y^n$ implies that $x = y$. Show that for a given $a \in G$ there exists one and only one $x \in G$ such that $x^n = a$.
13. Prove that in every abelian group the elements of finite order form a subgroup. Do the elements of infinite order together with the identity also form a subgroup?
14. Give examples of infinite groups G_1, G_2, G_3 with the following properties. All elements of G_1 other than the identity have infinite order. All elements of G_2 have finite order. G_3 contains both elements of infinite order and of finite order other than the identity.
15. Let A be the group of all real numbers x, $0 \leqq x < 1$, under addition mod 1. Find the elements of finite order in A. (Cf. Example 6 in § 3.2.)
16. Determine the centre of $\mathrm{GL}(n, \mathbf{K})$.
17. Let a and m be integers with $(a, m) = 1$. Prove that $a^{\varphi(m)} \equiv 1 \pmod{m}$.
18. Let G be a cyclic group of order m and let $m = hk$ with $h > 1$, $k > 1$. Show that the unique subgroup H of order h can be characterized in either of the following ways: (a) H consists of all elements x^k, $x \in G$. (b) H consists of all $x \in G$ such that $x^h = e$.
19. Show that every subgroup of index 2 in an arbitrary group is normal.
20. Let U be a complex of a group G. Define a relation $\sim$ in G as follows: $x \sim y$ if and only if $xy^{-1} \in U$. Prove that this relation is an equivalence relation compatible with the group operation if and only if $U \trianglelefteq G$.
21. Consider the symmetric group S_4 to show that the property of being a normal subgroup is not transitive.
22. Prove that any subgroup of a group G is normal if it contains the commutator subgroup G'.
23. Show that $\mathrm{GL}(n, \mathbf{K}) = UD$ where U is the normal subgroup of all matrices with determinant 1 and D is the subgroup of all

diagonal matrices. Verify that $\mathrm{GL}(n, \mathbf{K})/U \cong D/(U \cap D) \cong \mathbf{K}^*$ where $\mathbf{K}^*$ denotes the multiplicative group of all non-zero elements of the field $\mathbf{K}$.

24. Let C denote the multiplicative group of all non-zero complex numbers. Find endomorphisms of C whose kernels are: (i) the subgroup of all complex numbers with modulus 1, (ii) the subgroup of all positive numbers. Determine the subgroup of all elements of finite order in C.
25. Let $J(G)$ be the group of all inner automorphisms of a group G. Prove that $J(G) \cong G/Z$ where Z denotes the centre of G.
26. Show that a cyclic group of prime power order or of infinite order is directly indecomposable.
27. Represent the multiplicative group of all non-zero complex numbers as a direct product. (Cf. Exercise 24.)
28. Find all direct decompositions of the four group.
29. Show that the transpositions $(1, 2), (2, 3), \ldots, (n-1, n)$ generate the symmetric group S_n. Show that S_n can also be generated by $(1, 2)$ and $(1, 2, \ldots, n)$.
30. Show that $(1, 2, 3), (1, 2, 4), \ldots, (1, 2, n)$ generate the alternating group A_n.
31. Prove that the commutator subgroup of the symmetric group S_n is the alternating group A_n. (Hint: Show that every cycle of length 3 is a commutator.)
32. Let a be a permutation of degree n which is a product of r mutually disjoint cycles, including cycles of length 1. Prove that $\chi(a) = (-1)^{n-r}$.
33. Show that the alternating group A_4 has no subgroup of order 6.
34. Prove that every group of order p^2, where p is a prime number, is abelian.
35. Prove that every finite abelian group is the direct product of its Sylow subgroups.
36. Let G be a group of order pq, where p and q are prime numbers and $p < q$. Prove that G contains a single, and hence normal Sylow q-subgroup.
37. Determine the automorphism group and the endomorphism ring of a cyclic group of order m.
38. Let the group G possess an endomorphism α, distinct from the zero endomorphism and the identity mapping, and such

that $\alpha^2 = \alpha$. Prove that G contains a normal subgroup N and a subgroup H such that $G = NH$ and $N \cap H = e$. Give an example.

39. Prove that the mapping $x \to x^2$ for every element x of a group G is an endomorphism of G if and only if G is abelian. Show that this mapping is an automorphism of G if G is abelian and of odd order.
40. Let the operator domain Ω of the four-group consist of all automorphisms. Show that there is no Ω-automorphism of the four-group other than the identity.
41. Let e, a, b, ab be the elements of the four-group and let the operator domain Ω consist of the single operator ω defined as follows: $a\omega = b$, $b\omega = ab$. Determine the Ω-automorphisms.
42. Let A and B be two subspaces of a vector space. By $A \cup B$ we understand the subspace of all sums $a + b$ with $a \in A$, $b \in B$. Prove that

$$\dim A + \dim B = \dim (A \cup B) + \dim (A \cap B).$$

(Hint: Use 3.13.10.)

4

Rings. Integral domains

In this chapter we first derive some basic properties of arbitrary rings. Then we study rings with properties which are more or less similar to those of the ring of the integers and consider the arithmetic of such rings, i.e. questions related to divisibility, prime elements etc. Most of the results in this chapter will be used later.

4.1. Definitions and examples

Definition. An algebraic system $\mathbf{R}$ with two operations, addition and multiplication, is called a *ring* if the following postulates are satisfied for arbitrary elements of $\mathbf{R}$:

(R_1) $a + b = b + a$ (*commutative law of addition*).

(R_2) $(a + b) + c = a + (b + c)$ (*associative law of addition*).

(R_3) $\mathbf{R}$ *contains an element* 0 *such that*

$$a + 0 = a \text{ for every } a \in \mathbf{R}.$$

0 *is called the* zero element *or, briefly, the* zero *of* $\mathbf{R}$.

(R_4) *For every* $a \in \mathbf{R}$, *there exists an element* $-a$ *in* $\mathbf{R}$ *such that*

$$a + (-a) = 0.$$

$(-a)$ *is called the* additive inverse *of* a.

(R_5) $(ab)\,c = a(bc)$ (*associative law of multiplication*).

(R_6) $(a + b)\,c = ac + bc$, $c(a + b) = ca + cb$ (*distributive laws*).

A ring is said to be *commutative* if

$ab = ba$ *for any two elements* $a, b \in \mathbf{R}$ (*commutative law of multiplication*).

In a commutative ring each distributive law implies the other. The postulates (R_1) through (R_4) mean that the elements of $\mathbf{R}$ form an abelian group with respect to addition. We refer to this group as the *additive group* of $\mathbf{R}$ and denote it by $\mathbf{R}^+$.

Every additively written abelian group A can occur as the additive group of some ring as the following trivial example shows. Define multiplication in A as follows:

$$ab = 0 \text{ for any two elements of } A.$$

It is obvious that, under this definition, A becomes a ring. Such a ring is called a *zero ring*.

The elements of a ring form a semigroup with respect to multiplication.

Due to the associative laws, sums and products may be written without brackets.

For any element $a \in \mathbf{R}$, the first distributive law gives $a0 + a0 = a(0 + 0) = a0$ and hence $a0 = 0$. From the other distributive law we obtain $0a = 0$ for every $a \in \mathbf{R}$.

Instead of $b + (-a)$ we simply write $b - a$. As regards multiplication, the additive inverse satisfies, in an arbitrary ring, the same rules as in any ring of numbers, namely

$$a(-b) = (-a)\,b = -(ab), \quad (-a)\,(-b) = ab.$$

This follows easily from the distributive laws. We have

$$ab + (-a)\,b = (a - a)\,b = 0b = 0$$

and hence $(-a)\,b = -(ab)$. Similarly we obtain $a(-b) = -(ab)$. Moreover,

$$(-a)\,(-b) = -((a)\,(-b)) = -(-(ab)) = ab.$$

The distributive laws can also be interpreted in group theoretical terms. Take a fixed element a and assign to each $x \in \mathbf{R}$ the product xa. Then, the distributive law $(x + y)\,a = xa + ya$ means that the mapping $x \to xa$ is an endomorphism of $\mathbf{R}^+$. Thus, any $a \in \mathbf{R}$

can be regarded as an operator of $\mathbf{R}^+$ so that $\mathbf{R}$ becomes an operator domain of $\mathbf{R}^+$. The equations $x(a+b) = xa + xb$ and $x(ab) = (xa)b$ show that addition and multiplication of the endomorphisms induced by the elements of $\mathbf{R}$ as right operators correspond to addition and multiplication in $\mathbf{R}$. Similarly, the other distributive law, $a(x+y) = ax + ay$, shows that $\mathbf{R}$ is also an operator domain of $\mathbf{R}^+$ if the action of an operator a is defined as multiplication by a on the left. These interpretations of the distributive laws will turn out to be very expedient.

If a ring $\mathbf{R}$ contains an element e such that

$$ae = ea = a \text{ for every } a \in \mathbf{R},$$

then e is called the *unit element* of $\mathbf{R}$. Since, by a result in § 3.1, the semigroup of the elements of $\mathbf{R}$ with respect to multiplication cannot contain more than one identity, it follows that a ring cannot contain more than one unit element. The even integers form an example of a ring without unit element. The unit element of an arbitrary ring is frequently denoted by 1. The elements 1 and 0 coïncide if and only if the ring consists of the single element 0.

A ring $\mathbf{R}$ is said to be a *division ring* or a *skew field* if the non-zero elements of $\mathbf{R}$ form a group with respect to multiplication. In other words, $\mathbf{R}$ is a division ring if and only if $\mathbf{R}$ contains a unit element $1 \neq 0$ and if for every $a \in \mathbf{R}$, $a \neq 0$, there exists an element a^{-1} in $\mathbf{R}$ such that

$$aa^{-1} = a^{-1}a = 1.$$

A commutative division ring is called a *field*.

If a and b are given elements of a division ring and $a \neq 0$, then there are unique elements x and y such that $ax = b$ and $ya = b$, namely $x = a^{-1}b$ and $y = ba^{-1}$. Of course, in a field x and y coincide. In this case we also write $x = y = b/a$.

Example 1. The single element 0 with $0 + 0 = 0$, $00 = 0$.

Example 2. The integers.

Example 3. The rational numbers. They form a field.

Example 4. The so-called Gaussian integers $a + bi$ where a and b are integers.

Example 5. The complex numbers $x + yi$ where x and y are rational numbers. They form a field.

Example 6. The numbers $a + b\sqrt{-5}$ where a and b are integers.

Example 7. The residue class ring mod m. (Cf. Theorem 2.6.1.)

Example 8. The matrices

$$\begin{bmatrix} a & b \\ -b & a \end{bmatrix} \quad \text{where } a \text{ and } b \text{ are integers.}$$

Example 9. Let $\mathbf{R}$ be a commutative ring with unit element. An *algebra* $\mathbf{A}$ over $\mathbf{R}$ is a ring with the following properties:

(i) $\mathbf{A}$ is a free $\mathbf{R}$-module (cf. § 3.13).

(ii) For every $\alpha \in \mathbf{R}$ and any two elements $a, b \in \mathbf{A}$

$$\alpha(ab) = a(\alpha b) = (\alpha a)\, b.$$

Let us first consider the case that $\mathbf{A}$ is a free $\mathbf{R}$-module of finite rank n and let $v_1, \ldots, v_n$ be a basis of $\mathbf{A}$. Then any element a of $\mathbf{A}$ has a unique representation

$$a = \alpha_1 v_1 + \ldots + \alpha_n v_n \quad (\alpha_i \in \mathbf{R}).$$

Since $\mathbf{A}$ is a ring, the products $v_i v_j$ are defined, namely

$$v_i v_j = \sum_{k=1}^{n} \gamma_{ijk} v_k \quad (\gamma_{ijk} \in \mathbf{R}). \tag{1}$$

Due to the distributive laws in $\mathbf{A}$ and (ii), multiplication of arbitrary elements of $\mathbf{A}$ is uniquely determined by (1). We obtain

$$\left(\sum_{i=1}^{n} \alpha_i v_i\right)\left(\sum_{j=1}^{n} \beta_j v_j\right) = \sum_{i=1}^{n}\sum_{j=1}^{n} \alpha_i \beta_j v_i v_j = \sum_{k=1}^{n}\left(\sum_{i=1}^{n}\sum_{j=1}^{n} \alpha_i \beta_j \gamma_{ijk}\right) v_k.$$

It is clear that multiplication in $\mathbf{A}$ is associative if and only if $(v_i v_j)\, v_k = v_i(v_j v_k)$ for $i, j, k = 1, \ldots, n$. A simple calculation shows that the associative law is satisfied if and only if the coefficients in (1) are subject to the following conditions:

$$\sum_{l=1}^{n} \gamma_{ijl}\gamma_{lkm} = \sum_{l=1}^{n} \gamma_{jkl}\gamma_{ilm} \quad (i, j, k, m = 1, \ldots, n). \tag{2}$$

The complex numbers in Example 5 can be regarded as an algebra of rank 2 over the field of the rational numbers. The rings in

Examples 4 and 6 are algebras of rank 2 over the ring of the integers; the bases are 1, i and 1, $\sqrt{-5}$, respectively.

In case **R** is a field, then the rank of an algebra **A** as an **R**-module is also called its dimension.

It is evident that we need not confine ourselves to algebras of finite rank. For, if **A** is a free **R**-module with an infinite basis, then the products of any two elements of the basis as well as all elements of **A** are finite linear combinations of elements of the basis.

We give three other examples of algebras.

Example 10. Let **R** be a commutative ring with unit element 1 and let E_{ik} denote the $n \times n$ matrix whose entries are 1 in the intersection of the i-th row and the k-th column and 0 elsewhere. The well known rules for matrix multiplication give

$$E_{ik}E_{lm} = \begin{cases} E_{im} & \text{for } k = l, \\ 0 & \text{for } k \neq l, \end{cases}$$

where 0 stands for the $n \times n$ zero matrix. Any matrix

$$A = \begin{bmatrix} \alpha_{11} & \cdots & \alpha_{1n} \\ \cdots & \cdots & \cdots \\ \alpha_{n1} & \cdots & \alpha_{nn} \end{bmatrix}$$

with elements of **R** can be represented as follows

$$A = \sum_{i,k=1}^{n} \alpha_{ik} E_{ik} .$$

It is well known, that multiplication of matrices is associative, but the conditions (2) can easily be verified in our case. Therefore, the ring of all $n \times n$ matrices with elements in **R** is an algebra of rank n over **R** with E_{ik} $(i, k = 1, \ldots, n)$ as a basis. It is called the *complete matrix algebra* of degree n over **R**.

Example 11. Let S be a semigroup and **R** a commutative ring with unit element. We shall construct the so-called *semigroup algebra* of S over **R**. Let M be a free **R**-module with a basis whose elements are labelled by the elements of S so that to each $x \in S$ there corresponds precisely one basis element b_x of M. Thus the elements of M are the linear combinations

$$\sum_{x \in S} \alpha_x b_x \quad (\alpha_x \in \mathbf{R}) \tag{3}$$

where at most finitely many α_x are distinct from zero. To obtain an algebra we define multiplication of the b_x as follows:

$$b_x b_y = b_z \quad \text{if} \quad xy = z \quad \text{in} \quad S. \tag{4}$$

Since multiplication in S is associative, the associative law is also satisfied for the products of the b_x. Therefore, by (4), M becomes an algebra over $\mathbf{R}$. In view of (4), the b_x form a semigroup isomorphic to S. Therefore we shall use the same notation for both semigroups and write x instead of b_x. In other words, we form the formal finite linear combinations of elements of S with coefficients in $\mathbf{R}$. Thus we write

$$\sum_{x \in S} \alpha_x x$$

instead of (3).

If G is a finite group, then multiplication in the *group algebra* of G over $\mathbf{R}$ can explicitely be described as follows:

$$\Big(\sum_{x \in G} \alpha_x x\Big)\Big(\sum_{y \in G} \beta_y y\Big) = \sum_{x \in G}\sum_{y \in G} \alpha_x \beta_y xy = \sum_{z \in G}\Big(\sum_{xy=z} \alpha_x \beta_y\Big) z$$

where in the last sum x and y range over all pairs of elements of G for which $xy = z$.

Before we consider the last example of an algebra we will insert a comment on the notation.

Suppose that an algebra $\mathbf{A}$ over $\mathbf{R}$ has a unit element e. We consider the set $\mathbf{R}e$ of all elements of $\mathbf{A}$ of the form αe with $\alpha \in \mathbf{R}$. We find that $\alpha e = \alpha' e$ implies that $\alpha = \alpha'$, moreover $(\alpha + \beta)\, e = \alpha e + \beta e$, $(\alpha\beta)\, e = (\alpha e)\,(\beta e)$. This shows that the mapping $\alpha \to \alpha e$ is a bijective mapping of $\mathbf{R}$ onto $\mathbf{R}e$ which preserves addition and multiplication. Generalizing the group theoretical concept in an obvious way, we say that $\mathbf{R}$ and $\mathbf{R}e$ are isomorphic. (We shall deal with isomorphism of rings in the next section.) In other words, $\mathbf{R}$ and $\mathbf{R}e$ differ only in the notation. This suggests to write α instead of αe and, in particular, to identify the unit element 1 of $\mathbf{R}$ with e. We also say that we have embedded $\mathbf{R}$ in the algebra $\mathbf{A}$.

Example 12. Let $\mathbf{R} = \boldsymbol{R}$, the field of the real numbers. We define an algebra $\mathbf{Q}$ over $\boldsymbol{R}$ which is a vector space of dimension 4 over $\boldsymbol{R}$. Let $1, i, j, k$ be a basis of $\mathbf{Q}$. Thus each element a of $\mathbf{Q}$ has a unique representation

$$a = \alpha_0 + \alpha_1 i + \alpha_2 j + \alpha_3 k \quad (\alpha_0, \alpha_1, \alpha_2, \alpha_3 \in \boldsymbol{R}).$$

Multiplication is defined as follows:

$$i^2 = j^2 = k^2 = -1,\ ij = -ji = k,\ jk = -kj = i,\ ki = -ik = j.$$

For the product of two arbitrary elements we thus obtain

$$(\alpha_0 + \alpha_1 i + \alpha_2 j + \alpha_3 k)(\beta_0 + \beta_1 i + \beta_2 j + \beta_3 k)$$
$$= (\alpha_0\beta_0 - \alpha_1\beta_1 - \alpha_2\beta_2 - \alpha_3\beta_3) + (\alpha_1\beta_0 + \alpha_0\beta_1 - \alpha_3\beta_2 + \alpha_2\beta_3)\, i$$
$$+ (\alpha_2\beta_0 + \alpha_3\beta_1 + \alpha_0\beta_2 - \alpha_1\beta_3)\, j + (\alpha_3\beta_0 - \alpha_2\beta_1 + \alpha_1\beta_2 + \alpha_0\beta_3)\, k.$$

It is easily verified that the associative law holds. This algebra is called the *quaternion algebra*. A simple calculation shows that the quaternion algebra is a *division algebra*, i.e. an algebra that is a division ring. The condition $a \neq 0$ means that $\alpha_0^2 + \alpha_1^2 + \alpha_2^2 + \alpha_3^2 > 0$. To determine b such that $ab = 1$ requires to find real numbers that satisfy the following linear equations:

$$\alpha_0\beta_0 - \alpha_1\beta_1 - \alpha_2\beta_2 - \alpha_3\beta_3 = 1,$$
$$\alpha_1\beta_0 + \alpha_0\beta_1 - \alpha_3\beta_2 + \alpha_2\beta_3 = 0,$$
$$\alpha_2\beta_0 + \alpha_3\beta_1 + \alpha_0\beta_2 - \alpha_1\beta_3 = 0,$$
$$\alpha_3\beta_0 - \alpha_2\beta_1 + \alpha_1\beta_2 + \alpha_0\beta_3 = 0.$$

The determinant of this system is equal to $(\alpha_0^2 + \alpha_1^2 + \alpha_2^2 + \alpha_3^2)^2$. Since the determinant is distinct from zero, there exists a unique solution.

An element a of a ring $\mathbf{R}$ is called a *divisor of zero* if there is an element $b \neq 0$ in $\mathbf{R}$ such that $ab = 0$ or $ba = 0$. Thus 0 is always a divisor of zero, but a ring may contain divisors of zero other than 0, e.g. the E_{ik} in Example 10. Disregarding the trivial divisor of zero, namely 0, we speak of a ring *without divisors of zero* if it does not contain any divisor of zero other than 0. Hence, in a ring without divisors of zero, any equation $ab = 0$ implies that $a = 0$ or $b = 0$ (or both). It is obvious, that any division ring is a ring without divisors of zero; for, if $ab = 0$ and $a \neq 0$, then there exists a^{-1} and we obtain $0 = a^{-1}ab = 1b = b$.

A subset of a ring $\mathbf{R}$ is called a *subring* if it is a ring with respect to addition and multiplication as defined in $\mathbf{R}$. Every ring $\mathbf{R}$ contains two trivial subrings, namely 0 and $\mathbf{R}$ itself.

In any ring $\mathbf{R}$ let $\mathbf{Z}$ be the subset of all those elements z that satisfy the condition

$$za = az \text{ for every } a \in \mathbf{R}.$$

$\mathbf{Z}$ is not empty, because at least $0 \in \mathbf{Z}$. If z_1 and z_2 belong to $\mathbf{Z}$, then

$$(z_1 + z_2)\, a = z_1 a + z_2 a = a z_1 + a z_2 = a(z_1 + z_2),$$

$$(-z_1)\, a = -(z_1 a) = -(a z_1) = a(-z_1),$$

$$z_1 z_2 a = z_1 a z_2 = a z_1 z_2 .$$

This shows that $z_1 + z_2$, $-z_1$, and $z_1 z_2$ also belong to $\mathbf{Z}$ so that $\mathbf{Z}$ is a subring; it is called the *centre* of $\mathbf{R}$.

Clearly, a commutative ring coincides with its centre. If $\mathbf{R}$ contains a unit element 1, then, for any natural number n, the elements

$$n \cdot 1 = 1 + 1 + \ldots + 1 \ (n \text{ summands})$$

and their additive inverses belong to the centre.

4.2. Homomorphisms

Let $\mathbf{R}_1$ and $\mathbf{R}_2$ be two rings and let us suppose that there exists a bijective mapping σ of $\mathbf{R}_1$ onto $\mathbf{R}_2 = \mathbf{R}_1\sigma$ such that

$$(x + y)\,\sigma = x\sigma + y\sigma, \quad (xy)\,\sigma = (x\sigma)\,(y\sigma) \tag{1}$$

for any two elements x, y of $\mathbf{R}$. Then σ is called an *isomorphism* of $\mathbf{R}_1$ onto $\mathbf{R}_2$, and $\mathbf{R}_1$ is said to be isomorphic to $\mathbf{R}_2$. We write $\mathbf{R}_1 \cong \mathbf{R}_2$. It is evident that isomorphism is an equivalence relation.

The mapping

$$\sigma: a + bi \to \begin{bmatrix} a & b \\ -b & a \end{bmatrix}, \tag{2}$$

for instance, is an isomorphism of the ring of the Gaussian integers (Example 4 in § 4.1) onto the matrix ring in Example 8. It is obvious that σ has the property of an isomorphism with respect to addition; as to multiplication, it follows from a straightforward calculation, namely

$$(a + bi)(c + di) = (ac - bd) + (ad + bc)\, i,$$

$$\begin{bmatrix} a & b \\ -b & a \end{bmatrix} \begin{bmatrix} c & d \\ -d & c \end{bmatrix} = \begin{bmatrix} ac - bd & ad + bc \\ -(ad + bc) & ac - bd \end{bmatrix}. \tag{3}$$

Applying σ to the equation $a + 0 = a$ we obtain $a\sigma + 0\sigma = a\sigma$. This shows that the image 0σ of the zero of $\mathbf{R}_1$ is the zero of $\mathbf{R}_2$.

If $\mathbf{R}_1$ has a unit element 1, then 1σ is the unit element of $\mathbf{R}_2$. Indeed, if $a1 = 1a = a$ for every $a \in \mathbf{R}_1$, then $(a\sigma)(1\sigma) = (1\sigma)(a\sigma) = a\sigma$ for every element $a\sigma$ of $\mathbf{R}_1\sigma = \mathbf{R}_2$.

It is evident that any particular property of $\mathbf{R}_1$, e.g. commutativity or being a division ring, is preserved under any isomorphism.

By an *automorphism* of a ring $\mathbf{R}$ we mean an isomorphism of $\mathbf{R}$ onto itself.

If $\alpha = a + bi$ is a complex number, then we denote the conjugate complex number by $\bar{\alpha}$, i.e. $\bar{\alpha} = a - bi$. As is well known, we have

$$\overline{\alpha + \beta} = \bar{\alpha} + \bar{\beta}, \quad \overline{(\alpha\beta)} = \bar{\alpha}\bar{\beta}. \tag{4}$$

This shows that the mapping $\alpha \to \bar{\alpha}$ is an automorphism of the Gaussian integers as well as of the field in Example 5 (§ 4.1) or the field of all complex numbers.

We now introduce the concept of a homomorphism. Its role in the theory of rings is not less important than in group theory.

Let $\mathbf{R}$ and $\mathbf{R}^*$ be two rings and let σ be a mapping of $\mathbf{R}$ onto $\mathbf{R}^* = \mathbf{R}\sigma$ such that

$$(x + y)\,\sigma = x\sigma + y\sigma, \quad (xy)\,\sigma = (x\sigma)(y\sigma) \tag{5}$$

for any two elements x, y of $\mathbf{R}$. Such a mapping is called a *homomorphism* of $\mathbf{R}$ onto $\mathbf{R}^*$.

Clearly, if σ is injective, then it is an isomorphism. But we are mainly interested in homomorphisms which are not injective.

The first condition (5) shows that, in particular, σ is a homomorphism of the additive group $\mathbf{R}^+$ onto the additive group $\mathbf{R}\sigma^+$. Therefore, σ carries the identity 0 of $\mathbf{R}^+$ into the identity of $\mathbf{R}\sigma^+$, in other words, 0σ is the zero of $\mathbf{R}\sigma$.

If $\mathbf{R}$ is a ring with unit element, then so is $\mathbf{R}\sigma$. For, $1a = a1 = a$ for every $a \in \mathbf{R}$ implies that $(1\sigma)(a\sigma) = (a\sigma)(1\sigma) = a\sigma$ so that 1σ is the unit element of $\mathbf{R}\sigma$.

For a more detailed study of the homomorphism σ we first disregard multiplication and consider σ as a homomorphism of $\mathbf{R}^+$ onto $\mathbf{R}\sigma^+$. We can now apply the results of § 3.7 on group homomorphisms. The kernel of σ is a normal subgroup J of $\mathbf{R}^+$ consisting of all elements u of $\mathbf{R}^+$ such that $u\sigma = 0\sigma$. It is redundant to emphasize that J is a normal subgroup because $\mathbf{R}^+$ is abelian. Moreover, all elements of a coset $J + a$ of J have one and the same image in $\mathbf{R}\sigma^+$, namely $a\sigma$, and elements of distinct cosets have distinct images. Finally, the factor group of $\mathbf{R}^+$ with respect to J is isomorphic to $\mathbf{R}\sigma^+$.

We now consider multiplication. For any $u \in J$ and an arbitrary element a of $\mathbf{R}$ we obtain

$$(ua)\,\sigma = (u\sigma)\,(a\sigma) = (0\sigma)\,(a\sigma) = 0\sigma,$$
$$(au)\,\sigma = (a\sigma)\,(u\sigma) = (a\sigma)\,(0\sigma) = 0\sigma$$

since 0σ is the zero of $\mathbf{R}\sigma$. Consequently, $u \in J$ implies that $ua \in J$ and $au \in J$ for every $a \in \mathbf{R}$. We write $Ja = \{ua : u \in J\}$, $aJ = \{au : u \in J\}$. Then, the particular property of J can be expressed as follows:

$$Ja \subseteq J,\ aJ \subseteq J \text{ for every } a \in \mathbf{R}. \tag{6}$$

Such a subgroup J of $\mathbf{R}^+$ is called an *ideal* of $\mathbf{R}$. As in group theory, J is said to be the *kernel* of the homomorphism σ.

Since (6) holds, in particular, for any element a of J, we conclude that J contains the product of any two of its elements so that J is a subring of $\mathbf{R}$. But condition (6) is stronger for it requires not only that the product of two elements of J belong to J but that even the product of an element of J by an arbitrary element of $\mathbf{R}$ is contained in J. In the ring of the Gaussian integers, for example, the real elements, i.e. the integers, form a subring but not an ideal. In § 4.1 we observed that $\mathbf{R}$ can in two ways be considered as an operator domain of $\mathbf{R}^+$ where the action of an element a of $\mathbf{R}$ is defined as multiplication on the right or on the left, respectively. In this context, condition (6) means that J is a subgroup of $\mathbf{R}^+$ which is admissible with respect to $\mathbf{R}$ as a right and a left operator domain in the sense of § 3.12.

We observe that J is an ideal in $\mathbf{R}$ if and only if $u \in J$ and $v \in J$ imply that $u - v \in J$, $ua \in J$, and $au \in J$ for an arbitrary element a of $\mathbf{R}$.

We now prove the ring theoretical analogue to the fact that every normal subgroup of any group occurs as the kernel of a suitable homomorphism.

Let a ring **R** and an ideal J of **R** be given. We shall construct a ring $\mathbf{R}^*$ and a homomorphism σ of **R** onto $\mathbf{R}^*$ such that J is the kernel of σ.

The ideal J is a normal subgroup of the abelian group $\mathbf{R}^+$. The cosets of J have the form $J + a$. They are called the *residue classes modulo* J. From the results of § 3.7 we conclude that the cosets of J form an additive group $\mathbf{R}^+/J$ and that the natural homomorphism of $\mathbf{R}^+$ onto $\mathbf{R}^+/J$ has the kernel J in the group theoretical sense. In particular, the identity of $\mathbf{R}^+/J$ is the coset J.

So far we have only used the fact that J is a subgroup of $\mathbf{R}^+$. That J is even an ideal will now be crucial when we consider multiplication.

For any two non-empty subsets X and Y of **R** we define the product as follows:

$$XY = \{xy : x \in X, y \in Y\}.$$

Forming the product of two cosets of J we thus obtain

$$(J + a)(J + b) = JJ + aJ + Ja + ab. \tag{7}$$

Due to the conditions (6), we have $JJ + aJ + Ja \subseteq J$ so that the right-hand side of (7) is contained in the coset $J + ab$. In other words, if a_1 is an arbitrary element of $J + a$ and b_1 an arbitrary element of $J + b$, then the product $a_1 b_1$ is contained in $J + ab$. Therefore we define

$$(J + a)(J + b) = J + ab.$$

It is easy to see that the cosets of J form a ring with respect to addition in the group theoretical sense and multiplication as just defined. Indeed, addition and multiplication of cosets is based upon the corresponding operations with elements so that the ring postulates remain valid. The ring which we have constructed is called the *residue class ring* of **R** modulo J and is denoted by $\mathbf{R}/J$.

We now consider the mapping σ of **R** onto $\mathbf{R}/J$ which assigns to any element of **R** the coset of J to which it belongs. It is clear that σ is a homomorphism of **R** onto $\mathbf{R}/J$. The kernel of σ consists of all elements of $\mathbf{R}$ whose image is the zero of $\mathbf{R}/J$, namely J. So, J is the kernel of σ. We call σ the *natural homomorphism* of $\mathbf{R}$ onto $\mathbf{R}/J$.

We summarize our results as follows:

4.2.1. HOMOMORPHISM THEOREM FOR RINGS. *To any homomorphism σ of a ring $\mathbf{R}$ onto a ring $\mathbf{R}^*$ there corresponds, as its kernel, an ideal J of $\mathbf{R}$, consisting of all elements of $\mathbf{R}$ that are mapped onto the zero of $\mathbf{R}^*$. The ring $\mathbf{R}^*$ is isomorphic to the residue class ring $\mathbf{R}/J$. Conversely, if J is any ideal in $\mathbf{R}$, then the natural homomorphism of $\mathbf{R}$ onto $\mathbf{R}/J$ has J as its kernel.*

As an example, we consider the ring $\boldsymbol{Z}$ of the integers and the residue class ring mod m as defined in § 2.6. The corresponding ideal consists of all multiples of m and is denoted by (m) so that we write $\boldsymbol{Z}/(m)$ for the residue class ring mod m. A coset of (m) in the additive group $\boldsymbol{Z}^+$ has the form $(m) + a$, i.e. it consists of all integers of the form $km + a$ where k ranges over all integers. This shows that a_1 belongs to $(m) + a$ if and only if $a_1 \equiv a \pmod{m}$, in other words, the residue classes mod m of the integers are residue classes in the above sense.

Our approach in §§ 2.5 and 2.6 was slightly different from our arguments above since in Chapter 2 the group theoretical concepts were not yet available. We shall now show that, also for an arbitrary ring $\mathbf{R}$ and an ideal of $\mathbf{R}$, the residue class ring can be introduced in a way which is completely analogous to that in §§ 2.5 and 2.6.

Let $\mathbf{R}$ be an arbitrary ring and J an ideal in $\mathbf{R}$. We first define

$$a \equiv b \pmod{J}$$

to mean that $a - b \in J$. We say that a is congruent to b modulo J.

In case $\mathbf{R} = \boldsymbol{Z}$ and $J = (m)$, congruence mod J coincides with congruence mod m, in other words, in the ring $\boldsymbol{Z}$ we simply write $a \equiv b \pmod{m}$ instead of $a \equiv b \pmod{(m)}$.

We now show that congruence mod J is an equivalence relation. We have $a \equiv a \pmod{J}$ since $a - a = 0$ is contained in J. From $a \equiv b \pmod{J}$, i.e. $a - b \in J$, it follows that $b - a \in J$ since J, as an additive group, contains the additive inverse of each element; hence, $b \equiv a \pmod{J}$. Finally, $a \equiv b \pmod{J}$ and $b \equiv c \pmod{J}$, i.e. $a - b \in J$ and $b - c \in J$, imply that $(a - b) + (b - c) =$

$a - c$ belongs to J since J is an additive group. This gives $a \equiv c \pmod{J}$.

(The reader will observe that we did not use that J is an ideal but only the fact that J is a subgroup of $\mathbf{R}^+$. This is not surprising, since, if we disregard multiplication, congruence mod J means the same as right congruence (or left congruence since $\mathbf{R}^+$ is abelian) in the sense of § 3.5. So we only repeated the proof of Lemma 3.5.4 in additive notation.)

The equivalence classes with respect to congruence mod J are the residue classes mod J as defined above, namely the cosets $J + a$ of J in $\mathbf{R}^+$. Indeed, x belongs to the residue class of a if and only if $x - a = u$ belongs to J, i.e. $x \in J + a$.

4.2.2. *Congruence* mod J *is compatible with addition and multiplication, in other words, if*

$$a_1 \equiv a_2 \pmod{J} \quad \textit{and} \quad b_1 \equiv b_2 \pmod{J}, \tag{8}$$

then

$$a_1 + b_1 \equiv a_2 + b_2 \pmod{J} \tag{9}$$

and

$$a_1 b_1 \equiv a_2 b_2 \pmod{J}. \tag{10}$$

Proof. From (8) we obtain $a_1 - a_2 \in J$ and $b_1 - b_2 \in J$. Since J is a subgroup of $\mathbf{R}^+$ it follows that $(a_1 - a_2) + (b_1 - b_2) = (a_1 + b_1) - (a_2 + b_2)$ belongs to J. This proves (9). Since J is an ideal, $a_1 - a_2 \in J$ implies that $c(a_1 - a_2) \in J$ and $(a_1 - a_2)\, c \in J$ for every element c of $\mathbf{R}$. Therefore it follows from the first congruence (8) that

$$ca_1 \equiv ca_2 \pmod{J} \text{ and } a_1 c \equiv a_2 c \pmod{J}.$$

Multiplying the first congruence (8) on the right by b_1 and the second on the left by a_2 we obtain

$$a_1 b_1 \equiv a_2 b_1 \pmod{J},\ a_2 b_1 \equiv a_2 b_2 \pmod{J},$$

which gives (10) since congruence is transitive. This completes the proof.

By 4.2.2, we are able to define the sum and the product of residue classes. Let A and B be residue classes mod J. Choose $a_1 \in A$ and $b_1 \in B$ and define the sum $A + B$ to be the residue class that contains

$a_1 + b_1$. From 4.2.2 we conclude that the sum is uniquely defined by A and B alone and does not depend on the choice of a_1 and b_1. Indeed, if we choose $a_2 \in A$ and $b_2 \in B$, then (8) holds so that, by (9), $a_2 + b_2$ belongs to the same residue class as $a_1 + b_1$. The product AB is defined to be the residue class that contains $a_1 b_1$. Again, 4.2.2 shows that the product does not depend on the choice of a_1 and b_1. It is now clear that, under these definitions of addition and multiplication, the residue classes mod J form a ring and that this ring is the same as the residue class ring mod J defined above.

Any ring $\mathbf{R}$ has two trivial ideals, namely the entire ring $\mathbf{R}$ and the zero ideal (0) which consists of the single element 0. A ring is called *simple* if it consists of more than one element and contains no ideals other than the two trivial ones.

If $\mathbf{R}$ is a ring with unit element 1 and J any ideal in $\mathbf{R}$, then $J = \mathbf{R}$ if and only if $1 \in J$. For, if $1 \in J$, then $1a = a$ belongs to J for every $a \in \mathbf{R}$, i.e. $J = \mathbf{R}$; the converse is obvious.

Every division ring is simple. Indeed, let J be an ideal of the division ring $\mathbf{R}$ and assume that $J \neq (0)$. Then, J contains an element $a \neq 0$. Since $\mathbf{R}$ is a division ring there exists a^{-1}, and therefore $aa^{-1} = 1$ belongs to J, i.e. $J = \mathbf{R}$.

An ideal M of a ring $\mathbf{R}$ is said to be *maximal* if $M \neq \mathbf{R}$ and if there is no ideal H of $\mathbf{R}$ such that $M \subset H \subset \mathbf{R}$ with proper inclusions.

4.2.3. *An ideal M of $\mathbf{R}$ is maximal if and only if $\mathbf{R}/M$ is simple.*

Proof. We consider M as a subgroup of $\mathbf{R}^+$ which is admissible with respect to multiplication by elements of $\mathbf{R}$ on the right and on the left. By the generalization of Theorem 3.7.5 to groups with an operator domain, there is a one-to-one correspondence between the admissible subgroups H of $\mathbf{R}^+$ with $M \subseteq H \subseteq \mathbf{R}^+$ and the admissible subgroups of $\mathbf{R}^+/M$. In other words, there is a one-to-one correspondence between the ideals H of $\mathbf{R}$ such that $M \subseteq H \subseteq \mathbf{R}$ and the ideals of the residue class ring $\mathbf{R}/M$. This proves the proposition.

We now prove a theorem on the existence of maximal ideals which will be used later.

4.2.4. THEOREM. *If $\mathbf{R}$ is a ring with unit element* 1 *and if J is* *any ideal of* **R** *such that* $J \neq \mathbf{R}$, *then there exists a maximal ideal* M *of* **R** *that contains* J.

Proof. Let $\mathfrak{T}$ be the set of all ideals K of **R** such that $J \subseteq K$ and $1 \notin K$. $\mathfrak{T}$ is not empty because $1 \notin J$ so that $J \in \mathfrak{T}$. The set $\mathfrak{T}$ is partially ordered by inclusion. Let $\mathfrak{C}$ be any chain in $\mathfrak{T}$ and let U denote the union of all ideals in $\mathfrak{C}$. We show that U is an ideal of **R**. If x and y are any two elements of U, then x and y are contained in ideals of $\mathfrak{C}$, say X and Y, respectively. Since $\mathfrak{C}$ is a chain, we have $X \subseteq Y$ or $Y \subseteq X$. Under suitable notation we may assume that $X \subseteq Y$. Then both x and y are contained in Y. Since Y is an ideal, we conclude that $x - y \in Y$ and also $xa \in Y$ and $ax \in Y$ for any $a \in \mathbf{R}$. This gives $x - y \in U$, $xa \in U$, and $ax \in U$, so that U is an ideal. Moreover, $J \subseteq U$ by the definition of $\mathfrak{T}$ and, finally, 1 does not belong to U since no ideal of $\mathfrak{T}$, and hence of $\mathfrak{C}$, contains 1. Consequently, $U \in \mathfrak{T}$. Thus, $\mathfrak{T}$ satisfies the condition of Zorn's Lemma so that there exists a maximal element M in $\mathfrak{T}$. It is easy to see that M is a maximal ideal of **R**. Since 1 is not contained in M, we have $M \neq \mathbf{R}$. Let H be an ideal such that $M \subset H$. Since M is maximal in $\mathfrak{T}$ we conclude that H does not belong to $\mathfrak{T}$. Since H obviously contains J, it follows that $1 \in H$. This gives $H = \mathbf{R}$, which proves the theorem.

4.3. Commutative rings. Integral domains

In this and the next two sections we shall, step by step, introduce additional conditions on rings until we finally arrive at rings which, to some extent, have properties similar to those of the ring of the integers.

To begin with, we restrict our attention to commutative rings with unit element.

Let a be an element of a commutative ring **R** with unit element. It is evident that the set

$$(a) = \{ax : x \in \mathbf{R}\}$$

is an ideal in **R**. We call it the *principal ideal* generated by a.

4.3.1. *Any simple commutative ring* **R** *with unit element is a field.*

Proof. Let $a \in \mathbf{R}$ and $a \neq 0$. Then $(a) \neq (0)$, and since $\mathbf{R}$ is simple we conclude that $(a) = \mathbf{R}$. This gives $1 \in (a)$. Hence there exists an element $x \in \mathbf{R}$ such that $ax = 1$. This shows that every non-zero element of $\mathbf{R}$ has a multiplicative inverse, i.e. $\mathbf{R}$ is a field.

4.3.2. *Let $\mathbf{R}$ be a commutative ring with unit element. An ideal M of $\mathbf{R}$ is maximal if and only if $\mathbf{R}/M$ is a field.*

Proof. By 4.2.3, M is maximal if and only if $\mathbf{R}/M$ is simple. If $\mathbf{R}$ is a commutative ring with unit element, then so is $\mathbf{R}/M$. Thus our proposition follows from 4.3.1.

Let $\mathbf{R}$ be a commutative ring. An ideal P of $\mathbf{R}$ is called a *prime ideal* if the residue class ring $\mathbf{R}/P$ has no divisors of zero. To put it differently, P is a prime ideal if and only if

$$ab \equiv 0 \pmod{P}$$

implies that $a \equiv 0 \pmod{P}$ or $b \equiv 0 \pmod{P}$ or both.

Since a field contains no divisors of zero, we obtain as a consequence of 4.3.2:

4.3.3. *In a commutative ring with unit element every maximal ideal is a prime ideal.*

The converse is not true, i.e. a prime ideal need not be maximal (cf. Exercise 13 at the end of Chapter 5). But in the next section we shall deal with a class of rings in which every prime ideal is maximal.

When we talk of the arithmetical properties of a ring, we have in mind those properties that are related to divisibility, the greatest common divisor, prime elements and factorization into prime elements. Our aim is to single out a class of rings whose arithmetical properties are similar to those of the integers. This suggests to exclude rings with divisors of zero. Accordingly we make the following definition:

A commutative ring with unit element which consists of more than one element and contains no divisors of zero is called an *integral domain.*

Let a, b elements of an integral domain $\mathbf{R}$ and $a \neq 0$. Then, a is called a *divisor* of b if $\mathbf{R}$ contains an element x such that $ax = b$. We write $a \mid b$ and say that a divides b or that b is a multiple of a.

If $a \neq 0$ and $ax = b$, then x is uniquely determined. For, if $ax = b$ and $ay = b$, then $a(x - y) = 0$ which implies that $x - y = 0$ since **R** contains no divisors of zero.

It is clear that divisibility is a transitive relation, i.e. $a \mid b$ and $b \mid c$ imply that $a \mid c$.

An element u of an integral domain **R** is called a *unit* if $u \mid 1$; in other words, u is a unit if and only if it has a multiplicative inverse in **R**.

The unit element 1 is always a unit, but, in general, 1 is not the only unit. The units in the integral domain of the integers are, for example, 1 and -1.

If u_1 and u_2 are units, then so is u_1u_2. Indeed, from $u_1x_1 = 1$ and $u_2x_2 = 1$ it follows that $(u_1u_2)(x_1x_2) = 1$. Moreover, if u is a unit, then so is u^{-1}. This shows that the units of an integral domain form an abelian group with respect to multiplication.

Let a, a' be elements of an integral domain **R**. Then, a is said to be *associated* to a', or is called an *associate* of a', if **R** contains a unit u such that $au = a'$. From the fact that the units of **R** form a group it follows immediately that being associated is an equivalence relation.

4.3.4. Lemma. *If a divides b, then every associate of a divides every associate of b. If $a \mid b$ and $b \mid a$, then a and b are associated, and conversely.*

Proof. Let $ax = b$ and $a = a'u$, $b = b'v$ with units u, v. We obtain $a'ux = b'v$ and hence $a'uxv^{-1} = b'$, i.e. $a' \mid b'$. The relations $a \mid b$ and $b \mid a$ mean that $ax = b$ and $by = a$. This gives $axy = a$ which implies that $xy = 1$. Hence, x and y are units so that a and b are associated. The converse is obvious. This completes the proof.

4.3.5. Lemma. *Let $b \neq 0$. Then, $(a) \subseteq (b)$ if and only if $b \mid a$. In particular, $(a) = (b)$ if and only if a and b are associated.*

Proof. Suppose that $(a) \subseteq (b)$. Then $a \in (b)$ so that $a = bx$, i.e. $b \mid a$. Conversely, let $a = bx$. Then, every element of (a) has the form $ay = bxy$ and is therefore contained in (b), i.e. $(a) \subseteq (b)$. It follows, in particular, that $(a) = (b)$ if and only if $a \mid b$ and $b \mid a$. By the previous lemma, $a \mid b$ and $b \mid a$ hold if and only if a and b are associated. So, the lemma is proved.

By the *trivial divisors* of a non-zero element a we mean the units and the associates of a.

A non-zero element of an integral domain is called a *prime element* if it is not a unit and has only trivial divisors.[1]

Clearly, if p is a prime element, then so is every associate of p.

The definition of a prime element differs slightly from that of a prime number. For, according to the definition in § 2.2, a prime number p is positive, while by the definition above also $-p$ would be a prime element. The reason is that in the ring of the integers there is a natural way to single out one of the two elements in each class of associated non-zero integers, namely the positive one. In an arbitrary integral domain there is, in general, not such a natural way to single out a particular element in each class of associates. If we count the negative prime numbers among the prime elements of the ring of the integers, then we have to state the Unique Factorization Theorem 2.4.2 in a slightly different form because we have, for example, $15 = 3 \cdot 5 = (-3)(-5)$, but we shall not regard these two factorizations as essentially distinct. The proper formulation of the Unique Factorization Theorem will be given in the next section.

We shall now give an example of an integral domain in which factorization into prime elements is not unique.

Let $\mathbf{R}$ denote the integral domain of all complex numbers $\alpha = a_1 + a_2\sqrt{-5}$ where a_1 and a_2 are integers. Lower case Greek letters will stand for elements of $\mathbf{R}$. We call $N(\alpha) = \alpha\bar{\alpha} = a_1^2 + 5a_2^2$ the norm of α. We find that $N(\alpha)\,N(\beta) = N(\alpha\beta)$ because $(\alpha\bar{\alpha})\,(\beta\bar{\beta}) = (\alpha\beta)\,(\bar{\alpha}\bar{\beta}) = (\alpha\beta)\,(\overline{\alpha\beta})$. If $\alpha \mid \beta$, then $N(\alpha) \mid N(\beta)$, for $\alpha\xi = \beta$ implies that $N(\alpha)\,N(\xi) = N(\beta)$. Since $N(1) = 1$ we conclude that ε is a unit if and only if $N(\varepsilon) = 1$. Writing $\varepsilon = e_1 + e_2\sqrt{-5}$ we obtain $N(\varepsilon) = e_1^2 + 5e_2^2 = 1$. This gives $e_1 = \pm 1$, $e_2 = 0$ so that 1 and -1 are the only units of $\mathbf{R}$. We have

$$6 = 2 \cdot 3 = (1 + \sqrt{-5})\,(1 - \sqrt{-5}). \tag{1}$$

[1]) Sometimes, what we call prime element is called an *irreducible* element. The term prime element is then restricted to those elements p for which Corollary 4.4.4. holds. It is easy to see that every prime element is irreducible. We can disregard the difference between irreducible and prime elements because we shall soon restrict our attention to integral domains in which the two concepts coincide.

We show that the factors 2, 3, $1 + \sqrt{-5}$, $1 - \sqrt{-5}$ are prime elements in **R**. Suppose that, on the contrary, $2 = \alpha\beta$ where neither α nor β is a unit, i.e. $N(\alpha) > 1$, $N(\beta) > 1$. This gives $4 = N(2) = N(\alpha)\, N(\beta)$ so that $N(\alpha) = N(\beta) = 2$. But 2 is not a norm of any element of **R** because there are obviously no integers a_1 and a_2 such that $a_1^2 + 5a_2^2 = 2$. In a similar way one easily verifies that the other three factors are prime elements. Moreover, 2 is not associated to $1 + \sqrt{-5}$ or $1 - \sqrt{-5}$ because the only units are 1 and -1. This shows that 6 has two essentially distinct factorizations into prime elements.

4.4. Principal ideal rings

An integral domain **R** is called a *principal ideal ring* if every ideal of **R** is a principal ideal.

As an example, we show that the integers form a principal ideal ring. The zero ideal (0) is obviously a principal ideal. Let J be any non-zero ideal in the ring of the integers. Then J contains natural numbers. Let a be the least natural number in J. If x is any integer in J, then we divide x by a to obtain $x = aq + r$ where $0 \leqq r < a$. Since J is an ideal and $a \in J$, it follows that $aq \in J$. Hence, $r = x - aq$ belongs to J. Since a is the least natural number in J, we conclude that $r = 0$. This gives $x = aq$ so that J consists of all multiples of a, i.e. $J = (a)$.

The importance of principal ideal rings is due to two facts: First, the arithmetical properties of principal ideal rings are similar to those of the integers; secondly, some rings which frequently occur in algebra are principal ideal rings.

4.4.1. Theorem on the Greatest Common Divisor. *Let $a_1, a_2, \ldots, a_n$ be non-zero elements of a principal ideal ring* **R**. *Then* **R** *contains at least one common divisor d of $a_1, a_2, \ldots, a_n$ with the following property:*

(i) *d is divisible by every common divisor of $a_1, a_2, \ldots, a_n$.*

Every common divisor of $a_1, a_2, \ldots, a_n$ with the property (i) *is called a* greatest common divisor *of $a_1, a_2, \ldots, a_n$. Any two greatest common* *divisors of* $a_1, a_2, \ldots, a_n$ *are associated. Any greatest common divisor* d *of* $a_1, a_2, \ldots, a_n$ *can be represented as a linear combination*

$$\text{(ii)} \quad d = a_1x_1 + a_2x_2 + \ldots + a_nx_n \quad (x_i \in \mathbf{R}).$$

Conversely, every common divisor of $a_1, a_2, \ldots, a_n$ *that can be expressed in the form* (ii) *is a greatest common divisor.*

Proof. By a common divisor we always refer to a common divisor of $a_1, a_2, \ldots, a_n$. If d and d' are common divisors with the property (i), then $d \mid d'$ and $d' \mid d$ so that, by Lemma 4.3.4, d and d' are associated.

Every common divisor that admits a representation (ii) has the property (i). Indeed, any common divisor divides the right-hand side of (ii).

Thus, it remains to prove that there exists a common divisor which has a representation (ii). Let J denote the set of all linear combinations

$$a_1y_1 + a_2y_2 + \ldots + a_ny_n$$

where $y_1, y_2, \ldots, y_n$ are arbitrary elements of $\mathbf{R}$. One easily verifies that J is an ideal in $\mathbf{R}$. Since $\mathbf{R}$ is a principal ideal ring we have $J = (d)$. From the definition of J it follows that $(a_i) \subseteq (d)$ for $i = 1, 2, \ldots, n$ and hence $d \mid a_i$. Therefore d is a common divisor of $a_1, a_2, \ldots, a_n$ and, being an element of J, d can be represented in the form (ii). This completes the proof.

Due to Theorem 4.4.1, all greatest common divisors of $a_1, a_2, \ldots, a_n$ are associated to one of them, say d. Therefore the principal ideal (d) is uniquely determined by $a_1, a_2, \ldots, a_n$. We write

$$(d) = (a_1, a_2, \ldots, a_n).$$

In case $(a_1, a_2, \ldots, q_n) = (1)$, we say that $a_1, a_2, \ldots, a_n$ *are relatively prime.*

4.4.2. Theorem. *Let* $\mathbf{R}$ *be a principal ideal ring. An ideal* (p) *is a prime ideal distinct from* (0) *and* $\mathbf{R}$ *if and only if* p *is a prime element.*

Proof. If $p = 0$, then $(p) = (0)$ and if p is a unit, then $(p) = \mathbf{R}$. Therefore we may assume that p is neither equal to 0 nor a unit. Suppose that p is not a prime element so that $p = rs$ where neither

r nor s is a unit. Then we have $r \not\equiv 0 \pmod{p}$ and $s \not\equiv 0 \pmod{p}$ but $rs \equiv 0 \pmod{p}$.[1]) Thus, the residue class ring $\mathbf{R}/(p)$ has divisors of zero, i.e. (p) is not a prime ideal.

Conversely, suppose that p is a prime element so that the only divisors of p are the units and the associates of p. Therefore $a \not\equiv 0 \pmod{p}$ implies that $(a, p) = (1)$. Hence, by Theorem 4.4.1, there exist $x, y \in \mathbf{R}$ such that

$$ax + py = 1.$$

This equation gives

$$ax \equiv 1 \pmod{p}.$$

This shows that every non-zero element of $\mathbf{R}/(p)$ has a multiplicative inverse, in other words, $\mathbf{R}/(p)$ is a field. Since a field contains no divisors of zero, it follows that (p) is a prime ideal. This completes the proof.

In view of 4.3.2, we obtain

4.4.3. Corollary. *If $\mathbf{R}$ is a principal ideal ring, then every prime ideal of $\mathbf{R}$ other than (0) and $\mathbf{R}$ is maximal.*

The following corollary is only another way to state the result of Theorem 4.4.2 that (p) is a prime ideal if p is a prime element.

4.4.4. Corollary. *If p is a prime element of a principal ideal ring, then $p \mid ab$ implies that $p \mid a$ or $p \mid b$ or both.*

Our next aim is to prove an analogue to the Unique Factorization Theorem for integers.

In the case of the integers, the existence of a factorization into prime numbers was proved by induction. This proof does not apply to arbitrary principal ideal rings. Therefore we first prove a lemma from which the existence of a factorization into prime elements can easily be deduced.

4.4.5. Lemma. *Let $a_1, a_2, \ldots$ be a sequence of non-zero elements of a principal ideal ring $\mathbf{R}$ such that each a_i divides a_{i-1} but is not associated to a_{i-1}. Then the sequence is finite.*

---

[1]) We write mod p instead of mod (p).

Proof. From Lemma 4.3.5 we conclude that the principal ideals (a_i) form a chain with respect to inclusion, i.e.

$$(a_1) \subset (a_2) \subset \dots \tag{1}$$

As in the proof of Theorem 4.2.4 we conclude that the union

$$U = (a_1) \cup (a_2) \cup \dots$$

is an ideal in $\mathbf{R}$. Since $\mathbf{R}$ is a principal ideal ring, we have $U = (c)$. Then, c is contained in some (a_i). Let a_n be the first ideal in the sequence (1) that contains c. We show that a_n is the last element in that sequence. Suppose that, on the contrary, there exists an a_{n+1} in the sequence. Then, (a_n) is properly contained in (a_{n+1}) so that we arrive at the contradiction

$$(c) \subseteq (a_n) \subset (a_{n+1}) \subseteq U = (c).$$

This proves the lemma.

4.4.6. Unique Factorization Theorem. *Any element a of a principal ideal ring which is neither a unit nor equal to zero can be represented as a product of prime elements. This representation is unique in the following sense: If*

$$a = p_1 p_2 \dots p_r = q_1 q_2 \dots q_s$$

are two factorizations into prime elements p_i and q_j, then $r = s$ and p_i is associated to q_i if the factors are suitably arranged.

Proof. Suppose that a cannot be represented as a product of prime elements. Then a cannot be a prime element. Therefore there exists a factorization $a = a'a''$ where neither a' nor a'' is associated to a. Then at least one of the factors a', a'' cannot be represented as a product of prime elements, for otherwise a itself would have such a representation. Suppose that a' admits no representation as a product of prime elements. Then, a' has again a factorization $a' = a^*a^{**}$ where neither a^* nor a^{**} is associated to a' and at least one of them cannot be represented as a product of prime elements. Continuing in this way and taking, at each step, a factor which cannot be represented as a product of prime elements we obtain an infinite sequence of the form considered in Lemma 4.4.5. But, by this lemma, such a sequence cannot be infinite. Hence, it follows that a can be represented as a product of prime elements.

Now let

$$a = p_1 p_2 \cdots p_r = q_1 q_2 \cdots q_s \tag{2}$$

be two factorizations into prime elements p_i, q_j. We have to show that $r = s$ and that, on proper numbering, p_i and q_i are associated. This is certainly true for $r = 1$. We assume that our proposition is true for every product of less than r prime elements. Since the right-hand side of (2) is divisible by p_1 it follows from Corollary 4.4.4 that at least one factor q_i is divisible by p_1. We may assume that the notation is such that $p_1 \mid q_1$. Since q_1 is a prime element, this gives $p_1 = q_1 u$ where u is a unit. Thus we obtain

$$p_1 p_2 \cdots p_r = (q_1 u)(u^{-1} q_2) \cdots q_s .$$

Since a principal ideal ring contains no divisors of zero the last equation gives

$$p_2 \cdots p_r = (u^{-1} q_2) \cdots q_s . \tag{3}$$

Since the left-hand side is a product of $r - 1$ prime elements it follows from the induction assumption that $r - 1 = s - 1$ and that we can establish a one-to-one correspondence between the prime factors on the left-hand side and the prime factors on the right-hand side of (3) such that corresponding prime factors are associated. This completes the proof.

Occasionally, as for instance in the ring of the integers, it is expedient to single out a particular member in each class of associated prime elements. We will call such a member a distinguished prime element. (In the case of the integers, these are the positive prime numbers.) Combining associated prime elements to powers and collecting the units into a single unit factor we obtain a unique factorization

$$a = v \prod_{p \mid a} p^{\alpha_p}$$

where v is a unit and p ranges over all distinguished prime elements that divide a. To simplify the notation we also write

$$a = v \prod_{p} p^{\alpha_p}$$

where p ranges over all distinguished prime elements and only finitely many α_p are positive.

An integral domain in which the Unique Factorization Theorem holds is called a *unique factorization ring*. It turns out that the class of the unique factorization rings is larger than the class of the principal ideal rings (cf. Theorem 5.3.3 and Exercise 20 in Chapter 5).

Let $a_1, \ldots, a_n$ be non-zero elements of a unique factorization ring. Then, the greatest common divisors can be defined in the sense that a greatest common divisor is divisible by every common divisor of $a_1, \ldots, a_n$. But a representation of a greatest common divisor as a linear combination of $a_1, \ldots, a_n$ need not exist.

Suppose that the a_i have the factorizations

$$a_i = u_i \prod_p p^{\alpha_{ip}} \qquad (i = 1, 2, \ldots, n)$$

where p ranges over all distinguished prime elements and the u_i are units. Then every element associated to

$$\prod_p p^{\delta_p} \quad \text{where} \quad \delta_p = \min(\alpha_{1p}, \alpha_{2p}, \ldots, \alpha_{np})$$

is called a greatest common divisor of $a_1, a_2, \ldots, a_n$. In the case of a principal ideal ring this definition agrees with that in Theorem 4.4.1. The associates of

$$\prod_p p^{\lambda_p} \quad \text{where} \quad \lambda_p = \max(\alpha_{1p}, \alpha_{2p}, \ldots, \alpha_{np})$$

are called the *least common multiples* of $a_1, a_2, \ldots, a_n$.

If p is a prime element of a unique factorization ring $\mathbf{R}$, then (p) is a prime ideal. Indeed, if $ab \neq 0$ and $ab \equiv 0 \pmod{p}$, then p must be a prime factor of a or b (or both). This means that $a \equiv 0 \pmod{p}$ or $b \equiv 0 \pmod{p}$. Hence, $\mathbf{R}/(p)$ contains no divisors of zero so that (p) is a prime ideal. But in this more general case, $\mathbf{R}/(p)$ need not be a field. The first part of the proof of Theorem 4.4.2 remains valid for an arbitrary unique factorization ring $\mathbf{R}$. Hence, $\mathbf{R}/(c)$ contains divisors of zero unless c is a prime element.

4.5. Euclidean rings

In § 4.4, it was proved that the integers form a principal ideal ring. The proof was based on the division algorithm for integers. It was also due to the division algorithm that, in the case of the integers, we could not only prove the existence of the greatest common divisor but also provide a method by which the greatest common divisor can be computed and represented as a linear combination (cf. the proof of Theorem 2.3.1). Therefore it is desirable to find a similar division algorithm for other integral domains. In this section we study integral domains in which such an algorithm can be defined.

An integral domain $\mathbf{R}$ is called a *euclidean ring* if it is possible to assign to each non-zero element a of $\mathbf{R}$ a non-negative integer $H(a)$ such that the following conditions are satisfied:

(i) For any two elements $a, b \in \mathbf{R}$ with $a \neq 0$ there exist elements $q, r \in \mathbf{R}$ such that

$$b = aq + r \quad \text{where} \quad r = 0 \text{ or } H(r) < H(a).$$

(ii) For any two elements $a, b \in \mathbf{R}$ with $ab \neq 0$,

$$H(a) \leqq H(ab).$$

We shall show that this generalized division algorithm serves the same purpose as the division algorithm of the integers.

4.5.1. Theorem. *Every euclidean ring is a principal ideal ring.*

Proof. Let $\mathbf{R}$ be a euclidean ring and J a non-zero ideal in $\mathbf{R}$. We choose an element $a \neq 0$ in J such that $H(a)$ is as small as possible. If b is any element of J, then there exist $q, r \in \mathbf{R}$ such that $b = aq + r$ where $r = 0$ or $H(r) < H(a)$. Since J is an ideal, $r = a - bq$ belongs to J. Therefore $H(r) < H(a)$ contradicts the choice of a. Hence, we have $r = 0$ and $b = aq$. This shows that all elements of J are multiples of a, i.e. $J = (a)$, and the theorem is proved.

The converse of this theorem is not true, i.e. the class of the principal ideal rings is larger than the class of the euclidean rings.

We now show that in every euclidean ring there exists a generalized euclidean algorithm which permits to compute a greatest common divisor and to represent it as a linear combination.

4.5.2. THEOREM. *Let a_1, a_2 be two non-zero elements of a euclidean ring. Then there exists an algorithm by which a greatest common divisor of a_1 and a_2 can be found and represented as a linear combination of a_1 and a_2.*

Proof. By (i), there exist two sequences $q_1, q_2, \ldots$ and $r_1, r_2, \ldots$ such that

$$\begin{array}{lll} a_1 = a_2q_1 + r_1, & r_1 \neq 0, & H(r_1) < H(a_2), \\ a_2 = r_1q_2 + r_2, & r_2 \neq 0, & H(r_2) < H(r_1), \\ r_1 = r_2q_3 + r_3, & r_3 \neq 0, & H(r_3) < H(r_2), \\ r_2 = r_3q_4 + r_4, & r_4 \neq 0, & H(r_4) < H(r_3), \\ \cdots\cdots & & \\ r_{k-2} = r_{k-1}q_k + r_k, & r_k \neq 0, & H(r_k) < H(r_{k-1}). \end{array}$$

Since the $H(r_i)$ form a strictly decreasing sequence of non-negative integers, the algorithm must terminate, after a finite number of steps, with a remainder that is equal to zero, i.e. we finally obtain

$$r_{k-1} = r_kq_{k+1}.$$

We can now proceed in the same way as in the proof of Theorem 2.3.1 to show that r_k is a greatest common divisor of a_1 and a_2 and we can also use the algorithm to obtain a representation $r_k = a_1x_1 + a_2x_2$. This completes the proof.

The function H plays also a role in connection with associated elements and units:

4.5.3. *Let a and b be non-zero elements of a euclidean ring* **R**. *If $a \mid b$ and $H(a) = H(b)$, then a and b are associated.*

Proof. Let $ax = b$. There are elements y and s in **R** such that $a = by + s$ where $s = 0$ or $H(s) < H(b)$. If $s \neq 0$, then we obtain

$$s = a - by = a - axy = a(1 - xy)$$

and therefore, by (ii), $H(a) = H(b) \leqq H(s)$, a contradiction. This gives $s = 0$ so that $b \mid a$. Thus we have $a \mid b$ and $b \mid a$ which proves our proposition.

4.5.4. *An element $u \neq 0$ of a euclidean ring is a unit if and only if $H(u) = H(1)$.*

Proof. Suppose that u is a unit so that $uu^{-1} = 1$. By (ii), this gives $H(u) \leqq H(uu^{-1}) = H(1)$. On the other hand, we have $H(1) \leqq H(1u) = H(u)$ and, hence, $H(u) = H(1)$.

Conversely, let $H(u) = H(1)$. Due to $1 \mid u$ it follows from 4.5.3 that 1 and u are associated, i.e. u is a unit. This completes the proof.

As an example of a euclidean ring we consider the ring **G** of the Gaussian integers, i.e. the complex numbers $\alpha = a_1 + a_2 i$ where a_1 and a_2 are integers. Lower case Greek letters will stand for elements of **G**, lower case Latin letters for integers.

We first define a function H and show that it satisfies the conditions (i) and (ii). Let

$$H(\alpha) = a_1^2 + a_2^2 = |\alpha|^2.$$

Note, that in this case $H(0)$ is also defined and $H(\alpha) = 0$ if and only if $\alpha = 0$. Using the conjugate complex number $\bar{\alpha} = a_1 - a_2 i$ we can also write $H(\alpha) = \alpha\bar{\alpha}$. As is well known, $|\alpha\beta|^2 = |\alpha|^2 \, |\beta|^2$, i.e.

$$H(\alpha\beta) = H(\alpha)\, H(\beta). \tag{1}$$

Let $\alpha = a_1 + a_2 i \neq 0$ and $\beta = b_1 + b_2 i$. To prove that (i) is satisfied, we have to show that there exists an element $\varkappa = k_1 + k_2 i$ in **G** such that $\beta - \alpha\varkappa = 0$ or $H(\beta - \alpha\varkappa) < H(\alpha)$. Since $H(0)$ is defined and $H(\alpha) = 0$ if and only if $\alpha = 0$, the case $\beta - \alpha\varkappa = 0$ need not be considered separately. Thus we have only to satisfy the condition $H(\beta - \alpha\varkappa) < H(\alpha)$. We have

$$\frac{\beta}{\alpha} = \frac{a_1 b_1 + a_2 b_2}{a_1^2 + a_2^2} + \frac{a_1 b_2 - a_2 b_1}{a_1^2 + a_2^2}\, i\,.$$

It is obviously possible to find integers k_1 and k_2 such that

$$\left| \frac{a_1 b_1 + a_2 b_2}{a_1^2 + a_2^2} - k_1 \right| \leqq \frac{1}{2}\,, \qquad \left| \frac{a_1 b_2 - a_2 b_1}{a_1^2 + a_2^2} - k_2 \right| \leqq \frac{1}{2}\,.$$

Putting $\varkappa = k_1 + k_2 i$ we obtain

$$\left| \frac{\beta}{\alpha} - \varkappa \right|^2 = \left(\frac{a_1 b_1 + a_2 b_2}{a_1^2 + a_2^2} - k_1 \right)^2 + \left(\frac{a_1 b_2 - a_2 b_1}{a_1^2 + a_2^2} - k_2 \right)^2 \leqq \frac{1}{4} + \frac{1}{4} = \frac{1}{2}$$

and therefore

$$|\alpha|^2 \left| \frac{\beta}{\alpha} - \varkappa \right|^2 = |\beta - \alpha\varkappa|^2 \leqq \frac{1}{2} |\alpha|^2 < |\alpha|^2$$

which means

$$H(\beta - \alpha\varkappa) < H(\alpha).$$

Thus, the function $H(\alpha) = |\alpha|^2$ satisfies condition (i).

For any $\beta \neq 0$ we have $H(\beta) \geqq 1$. Therefore it follows from (1) that condition (ii) is satisfied.

This shows that **G** is a euclidean ring.

By 4.5.4, $\varepsilon = e_1 + e_2 i$ is a unit in **G** if and only if $H(\varepsilon) = e_1^2 + e_2^2 = 1$. Thus **G** contains four units, namely $1, -1, i, -i$.

Our next aim is to determine the prime elements of **G**.

If $H(\alpha) = p$ is a prime number, then α is a prime element of **G**. Indeed, if $\delta \mid \alpha$, then $H(\delta)$ divides $H(\alpha) = p$ so that $H(\delta) = 1$ or $H(\delta) = p$. In the first case, δ is a unit, in the second case, δ is associated to α by 4.5.3.

From $\pi\bar{\pi} = H(\pi)$ we conclude that $\pi \mid H(\pi)$ where $H(\pi)$ has, of course, to be considered as an element of **G**. Suppose now that π is a prime element. Since the Unique Factorization Theorem holds in **G**, it follows tat π divides a prime divisor p of $H(\pi)$. Moreover, π cannot divide two distinct prime numbers. For suppose that $\pi \mid p$ and $\pi \mid q$ where p and q are distinct prime numbers. Then, there exist integers x and y such that $px + qy = 1$. Hence π would be a divisor of 1 which is impossible since π is not a unit. Therefore we obtain precisely one member of each class of associated prime elements of **G** if we factorize all prime numbers into prime elements of **G** and take only non-associated prime factors of each prime number.

First, let $p = 2$. In **G** we obtain the factorization

$$2 = (1 + i)(1 - i).$$

From $H(1 + i) = 2$ we conclude that $1 + i$ is a prime element. The two factors $1 + i$ and $1 - i$ are associated, namely $1 - i = (-1)(1 + i)$.

Next, let p be a prime number such that $p \equiv 3 \pmod 4$. From $\pi \mid p$ it follows that $H(\pi)$ divides $H(p) = p^2$ so that $H(\pi) = p$ or $H(\pi) = p^2$. The first case is impossible. Indeed, $H(\pi)$ is of the form $x^2 + y^2$, and one easily verifies that $x^2 + y^2 \equiv 0, 1,$ or $2 \pmod 4$ whereas $p \equiv 3 \pmod 4$. It follows that $H(\pi) = p^2$. Writing $p = \pi\alpha$ we obtain

$$p^2 = H(p) = H(\pi)\, H(\alpha) = p^2 H(\alpha).$$

This gives $H(\alpha) = 1$ so that α is a unit. Hence π and p are associates, i.e. p is also a prime element in **G**.

Finally, let p be a prime number with $p \equiv 1 \pmod 4$. In § 6.8 we shall prove that the congruence

$$(p-1)! \equiv -1 \pmod p$$

holds for every prime number p. This gives, for $p > 2$,

$$-1 \equiv (p-1)! = \prod_{k=1}^{(p-1)/2} k \prod_{k=1}^{(p-1)/2} (p-k) = (-1)^{(p-1)/2} \left(\left(\frac{p-1}{2}\right)!\right)^2.$$

We write $z = \left(\frac{p-1}{2}\right)!$, and if $p \equiv 1 \pmod 4$ we obtain

$$z^2 \equiv -1 \pmod p$$

or $p \mid (z^2 + 1)$. Now let π be a prime element that divides p. Then π divides $z^2 + 1 = (z + i)(z - i)$ and hence $\pi \mid (z + i)$ or $\pi \mid (z - i)$. If π and p were associated, then it would follow that $p \mid (z + i)$ or $p \mid (z - i)$ which is obviously not true since neither $z/p + i/p$ nor $z/p - i/p$ belongs to **G**. Therefore π and p are not associated so that p is not a prime element. Writing $p = \pi\alpha$ we thus find that $H(\pi) > 1$ and $H(\alpha) > 1$. Then

$$H(p) = p^2 = H(\pi)\, H(\alpha)$$

gives $H(\pi) = p$ and $H(\alpha) = p$ so that α is a prime element. Putting $\pi = x + yi$ we obtain

$$p = H(\pi) = x^2 + y^2 = (x + yi)\, \alpha$$

hence

$$\alpha = x - yi = \bar{\pi}.$$

Therefore $p = \pi\bar{\pi}$ is the product of two prime elements. Since the units of **G** are known it is easy to verify that π and $\bar{\pi}$ are not associated.

Note that occasionally we obtained the following result: Every prime number p with $p \equiv 1 \pmod 4$ can be represented as the sum of two square numbers.

4.6. Fields of quotients

If a ring is a subring of a field then it is necessarily commutative and contains no divisors of zero. So, if the ring has a unit element, it is an integral domain. In this section we shall prove that, conversely, every integral domain is a subring of some field.

4.6.1. THEOREM. *Every integral domain can be embedded in a field. More precisely: If* **R** *is an integral domain, then there exists a field* **F** *such that* **F** *has a subring isomorphic to* **R**.

Proof. We consider the set of all ordered pairs (a, b) with $a, b \in \mathbf{R}$ and $b \neq 0$. For these pairs we define a relation $\sim$ as follows:

$$(a, b) \sim (a', b') \text{ if and only if } a'b = ab'.$$

It is easy to see that this is an equivalence relation. We show, as an example, that the relation is transitive. If

$$(a, b) \sim (a', b') \quad \text{and} \quad (a', b') \sim (a'', b''),$$

then $a'b = ab'$ and $a''b' = a'b''$. Since **R** is commutative, multiplication of the last equations yields $a'b'a''b = a'b'ab''$ and, since **R** contains no divisors of zero, the last equation implies that $a''b = ab''$. This gives $(a, b) \sim (a'', b'')$.

We now define addition and multiplication of the pairs:

$$(a, b) + (c, d) = (ad + bc, bd),$$

$$(a, b)\,(c, d) = (ac, bd).$$

Note that the sum as well as the product are again pairs whose second element is distinct from zero, for $b \neq 0$ and $d \neq 0$ imply that $bd \neq 0$ since **R** has no divisors of zero. To show the reason for this definition we anticipate that in the field, which we are going to construct, (a, b) will play the role of the quotient a/b. Therefore our definitions of addition and multiplication correspond to the usual addition and multiplication of fractions.

A straightforward calculation shows that the relation $\sim$ is compatible with addition and multiplication, i.e. if

$$(a, b) \sim (a', b') \quad \text{and} \quad (c, d) \sim (c', d'), \tag{1}$$

then

$$(a, b) + (c, d) \sim (a', b') + (c', d') \tag{2}$$

and

$$(a, b)\,(c, d) \sim (a', b')\,(c', d').$$

This enables us to define the sum and the product of equivalence classes. Let K and L be two equivalence classes and

$$(a, b) \in K, \quad (c, d) \in L.$$

Then the class that contains the pair

$$(a, b) + (c, d) = (ad + bc, bd)$$

is called the sum $K + L$. Due to the compatibility of the relation $\sim$ with addition, the sum $K + L$ is uniquely determined by K and L and does not depend on the choice of the elements $(a, b) \in K$ and $(c, d) \in L$. Indeed, if we choose other elements

$$(a', b') \in K, \quad (c', d') \in L,$$

then the relations (1) are satisfied, and therefore (2) shows that $(a', b') + (c', d')$ is contained in the class $K + L$. Similarly, we define the product KL to be the class that contains the pair

$$(a, b)\,(c, d) = (ac, bd).$$

Since the relation $\sim$ is compatible with multiplication it follows that KL is uniquely determined by K and L alone, i.e. KL does not depend on the choice of the elements $(a, b) \in K$, $(c, d) \in L$.

We shall now prove that the equivalence classes form a field. The commutative and associative laws for addition and multiplication as well as the distributive law can be verified by a straightforward calculation which will be left to the reader. It is also easy to see that the equation

$$(a, b) + (x, y) = (c, d)$$

has the solution $(x, y) \sim (bc - ad, bd)$. The zero element is the class of all pairs $(0, b)$, and the unit element is the class of the pairs (b, b). Finally, for $a \neq 0$,

$$(a, b)\,(b, a) = (ab, ab)$$

so that every non-zero element has a multiplicative inverse. This shows that the equivalence classes form a field **F**.

We now consider the equivalence classes that contain a pair of the form $(a, 1)$. It is clear that the class of $(a, 1)$ consists precisely of all pairs (ax, x) with $x \neq 0$. We shall refer to these classes as integral classes. It is evident that $(a, 1) \sim (c, 1)$ if and only if $a = c$. This shows that there is a one-to-one correspondence

$$a \leftrightarrow (a, 1)$$

between the elements of **R** and the integral classes. Moreover,

$$(a, 1) + (c, 1) = (a + c, 1), \ (a, 1)\,(c, 1) = (ac, 1).$$

Therefore the integral classes form a subring **R**′ of **F** which is isomorphic to **R**. Finally,

$$(b, 1)\,(a, b) = (ab, b) \sim (a, 1)$$

so that (a, b) is the quotient $(a, 1)/(b, 1)$ of the integral classes $(a, 1)$ and $(b, 1)$ in **F**. This completes the proof.

It is expedient to disregard the distinction between the subring **R**′ of **F** and the given ring **R** and to consider **R** itself as a subring of **F**. We then say that **R** and **R**′ have been identified. The term "embedding of **R** in **F**" refers to this identification. This process is very familiar when **R** stands for the ring of the integers. It simply means that we do not distinguish between an integer a and the fraction $a/1$.

The field **F** is called the *field of quotients* of **R**. The field of quotients of the integers, for instance, is the field of the rational numbers.

We observe that **F** is the smallest field that contains **R** as a subring in the sense that no proper subfield of **F** contains **R**. Indeed if a subfield **F*** of **F** contains **R** as a subring, then **F*** must also contain all quotients of elements of **R**. But since all elements of **F** are such quotients it follows that **F*** = **F**.

4.7. Prime fields. Characteristic

A field is called a *prime field* if it contains no proper subfield.

The field of the rational numbers and the residue class field $\boldsymbol{Z}/(p)$ for a prime number p are examples of prime fields. Indeed, a field of numbers contains the integer 1 and therefore it must contain all integers because they are obtained from 1 by repeated additions and subtractions; moreover, as a field, it must also contain all quotients of non-zero integers which means that it contains all rational numbers. To see that $\boldsymbol{Z}/(p)$ is a prime field we have only to observe that all elements can already be obtained by repeatedly adding the unit element to itself.

4.7.1. *Any field* F *contains precisely one prime field.*

Proof. It is clear that the intersection of an arbitrary set of subfields of F is again a subfield of F. Let P be the intersection of *all* subfields of F. It is evident that P is a prime field because every subfield of P would also be a subfield of F. Moreover, F cannot contain more than one prime field. For, if P and P' were two prime fields contained in F, then $\mathsf{P} \cap \mathsf{P}'$ would be a proper subfield of both P and P' which is impossible. This completes the proof.

Let F be a given field. To determine the prime field of F we proceed as follows. Since we have to distinguish between the unit element of F and the integer 1 we shall denote the unit element of F by e. We consider the integral multiples of e, namely

$$m \cdot e = \begin{cases} e + e + \dots + e & (m \text{ summands}) \text{ for } m > 0, \\ 0 & \text{for } m = 0, \\ -(e + e + \dots + e) & (|m| \text{ summands}) \text{ for } m < 0. \end{cases}$$

These elements of F obviously form an integral domain R. The mapping

$$m \to m \cdot e \tag{1}$$

is a homomorphism of the ring $\boldsymbol{Z}$ of the integers onto R since

$$m \cdot e + n \cdot e = (m + n) \cdot e, \; (m \cdot e)\,(n \cdot e) = (mn) \cdot e.$$

Let J denote the kernel of the homomorphism (1). Then $\boldsymbol{Z}/J$ is isomorphic to R.

We distinguish two cases.

(i) $J = (0)$. Then F and R are said to be of *characteristic* 0. We have $\mathsf{R} \cong \boldsymbol{Z}$. The prime field of F contains R and therefore it also contains the field of quotients of **R** which is isomorphic to the field $\boldsymbol{Q}$ of the rational numbers. Thus the prime field of any field of characteristic 0 is isomorphic to $\boldsymbol{Q}$.

(ii) $J = (p) \neq (0)$. We first observe that $p = 1$, i.e. $J = \boldsymbol{Z}$, is imposible because **R** contains at least two elements, namely e and 0. Since $\boldsymbol{Z}/(p)$ is isomorphic to the integral domain **R** it follows that $\boldsymbol{Z}/(p)$ has no divisors of zero. Therefore p is a prime number. In this case, **F** and **R** are said to be of *characteristic* p. We have $\mathbf{R} \cong \boldsymbol{Z}/(p)$ so that **R** itself is a field. Thus, the prime field of **F** is isomorphic to $\boldsymbol{Z}/(p)$.

It turned out that the examples of prime fields given above exhaust all possible types. Moreover, for any given prime number p there exist fields of characteristic p.

We summarize our results:

4.7.2. THEOREM. *A field is of characteristic* 0 *if and only if its prime field is isomorphic to the field of the rational numbers. The characteristic of a field is a prime number p if and only if its prime field is isomorphic to the residue class field of the integers mod p.*

We conclude this chapter with two fairly obvious but useful theorems.

4.7.3. *If* **F** *is a field of characteristic* $p > 0$, *then the mapping*

$$x \to x^p \quad (x \in \mathbf{F}) \tag{2}$$

is an automorphism of **F**.

Proof. It is clear that $(xy)^p = x^p y^p$. To prove that the mapping (2) has the property of an automorphism also with respect to addition we observe that, for a prime number p, the binomial coefficients

$$\binom{p}{k} = \frac{p(p-1)\dots(p-k+1)}{1\cdot 2\dots k} \qquad (k = 1, 2, \dots, p-1)$$

are divisible by p since all the factors in the denominator are less than p. This gives

$$(x+y)^p = x^p + \binom{p}{1} x^{p-1}y + \dots + \binom{p}{p-1} x^{p-1}y + y^p = x^p + y^p. \tag{3}$$

Hence, the mapping (2) is an endomorphism. Moreover, (3) yields $(x - y)^p = x^p - y^p$ so that the mapping (2) is injective. The kernel is an ideal of **F**, and since **F** is a field it contains only the two trivial ideals (0) and **F**. The kernel is distinct from **F** since $1 \neq 0$. Therefore the mapping (2) is an automorphism of **F** and the proposition is proved.

4.7.4. *A prime field has no automorphism other than the identity.*

Proof. Any automorphism carries the unit element 1 into itself. Since all the elements of a prime field can be obtained from 1 by addition, subtraction, multiplication, and division it follows that any automorphism leaves every element of a prime field fixed. This completes the proof.

Exercises

1. Prove that in a ring with unit element the commutative law of addition is a consequence of each distributive law. (Hint: Apply Exercise 39 of Chapter 3 to the additive group.)
2. Prove that every complete matrix algebra over any field is simple. (Thus 4.3.2 does not hold for arbitrary non-commutative rings.) Determine the divisors of zero in a complete matrix algebra over a field.
3. Show that a complete matrix algebra over $\boldsymbol{Z}$ is not simple.
4. Let **A** be an algebra of finite dimension n over a field **F** and let $v_1, \ldots, v_n$ denote a basis of **A**. To any element $a = \alpha_1 v_1 + \ldots + \alpha_n v_n$ assign the $n \times n$ matrix $M(a)$ whose element in the i-th row and k-th column is

$$\sum_{r=1}^{n} \alpha_r \gamma_{irk}$$

   where the γ_{irk} are defined by (1) in § 4.1. Show that the mapping $a \to M(a)$ is an isomorphism of **A** into the complete matrix algebra of degree n over **F**.
5. Find the centres of a complete matrix algebra and of the quaternion algebra.

6. Let G be a finite group of order greater than 1 with the identity e. Show that the centre of the group algebra of G over a field $\mathbf{F}$ contains elements other than αe, $\alpha \in \mathbf{F}$.
7. Show that every group algebra of a finite group of order greater than 1 contains divisors of zero.
8. Find the divisors of zero in $\boldsymbol{Z}/(12)$.
9. Let J be an ideal of the ring $\mathbf{R}$ and let S be a subring such that $J \subseteq S \subseteq \mathbf{R}$. Prove that S/J is an ideal in $\mathbf{R}/J$ if and only if S is an ideal in $\mathbf{R}$.
10. Prove that every subring of $\boldsymbol{Z}$ is an ideal.
11. Let $G_{\mathbf{F}}$ denote the group algebra of a finite group G over a field $\mathbf{F}$. Show that the mapping

$$\sum_{x \in G} \alpha_x x \to \sum_{x \in G} \alpha_x \quad (\alpha_x \in \mathbf{F})$$

is a homomorphism of $G_{\mathbf{F}}$ onto $\mathbf{F}$. Find the kernel and a basis of the kernel as an algebra over $\mathbf{F}$.
12. Show that the only equivalence relations on a ring $\mathbf{R}$ that are compatible with addition and multiplication are the congruence relations mod J where J is an ideal in $\mathbf{R}$.
13. Construct tables for addition and multiplication in the residue class field of the Gaussian integers mod (3).
14. Prove that a finite integral domain is a field.
15. Prove that every algebra of finite dimension over a field is a division algebra if it contains no divisors of zero.
16. Give an example to show that the existence of the unit element is not redundant for the validity of 4.3.3. (Hint: Consider a suitable zero ring.)
17. Show that $2 + \sqrt{3}$ is a unit of the ring of all numbers $x + y\sqrt{3}$ where $x, y \in \boldsymbol{Z}$. Hence show that this ring contains infinitely many units.
18. Show that every element of $\boldsymbol{Z}/(m)$ is a divisor of zero or a unit.
19. Find the intersection of the ideals (m) and (n) in $\boldsymbol{Z}$.
20. Show that all ideals of $\boldsymbol{Z}/(m)$ are principal ideals. Determine the prime ideals of $\boldsymbol{Z}/(24)$.
21. Show that the following integral domains are euclidean rings

and find the units:

(a) $x + y\sqrt{-2}$; $x, y \in \boldsymbol{Z}$.

(b) $x + y\varrho$, where $x, y \in \boldsymbol{Z}$ and $\varrho = \frac{1}{2}(-1 + \sqrt{3})$.

Is 43 a prime element in any of these rings?

22. Let **R** denote the integral domain of all rational numbers of the form $a/2^n$ where $a, n \in \boldsymbol{Z}$. Discuss divisibility in **R**. Find the units of **R**. Find prime elements of **R** that are not prime numbers.
23. Let p be a prime number such that $p \equiv 1 \pmod 4$. It follows from § 4.5. that there exist integers x, y such that $p = x^2 + y^2$. Use the Unique Factorization Theorem for the Gaussian integers to prove that this representation of p as a sum of two square numbers is unique except for the order of the summands and the signs of x and y.
24. An element a of a ring is said to be nilpotent if $a^n = 0$ for some natural number n.

    (i) Find the natural numbers m such that $\boldsymbol{Z}/(m)$ has no nilpotent elements other than 0.

    (ii) Find the natural numbers m such that any element of $\boldsymbol{Z}/(m)$ is either nilpotent or has a multiplicative inverse.

    (iii) Show that all nilpotent elements of a commutative ring form an ideal.

    (iv) Show that $1 - a$ has a multiplicative inverse if a is nilpotent.
25. Prove that $a^p \equiv a \pmod p$ for every integer a and every prime number p. (Hint: Use 4.7.3 and 4.7.4.)
26. Show that in every field of prime characteristic p

$$(a - b)^{p-1} = a^{p-1} + a^{p-2}b + \dots + ab^{p-2} + b^{p-1}.$$

27. Let **R** be an arbitrary ring and let R_1 denote the set of all pairs (a, α) where $a \in \boldsymbol{Z}$, $\alpha \in \mathbf{R}$. In R_1 define addition and multiplication as follows:

$$(a, \alpha) + (b, \beta) = (a + b, \alpha + \beta),$$

$$(a, \alpha)(b, \beta) = (ab, \alpha\beta + a\beta + b\alpha)$$

where

$$a\beta = \begin{cases} \beta + \ldots + \beta & (a \text{ summands}) \text{ if } a > 0, \\ 0 & \text{if } a = 0, \\ -(\beta + \ldots + \beta) & (|a| \text{ summands}) \text{ if } a < 0. \end{cases}$$

Show that (1, 0) is the unit element of R_1 and that R_1 contains a subring isomorphic to $\mathbf{R}$. (In short: Every ring can be embedded in a ring with unit element.)

28. Let p be a given prime number and let Q_p denote the ring of all rational numbers a/b where p does not divide b. Find the units of Q_p. Show that Q_p is a principal ideal ring and find all the ideals.

29. Let $\mathbf{R}$ be a ring with unit element. Define new operations as follows:

$$a \oplus b = a + b - 1, \qquad a \circ b = a + b - ab. \qquad (a, b \in \mathbf{R})$$

Show that with respect to these operations the elements of $\mathbf{R}$ form a ring $\mathbf{R}_0$. Find the zero and the unit element of $\mathbf{R}_0$. Show that $\mathbf{R}$ and $\mathbf{R}_0$ are isomorphic.

30. A group G is said to be characteristically simple if it contains no subgroup other than G and the identity subgroup that is mapped onto itself by every automorphism of G. Prove that the additive group of any field is characteristically simple.

5

Polynomials

The reader probably knows polynomials from school and from lectures on calculus. There, polynomials are regarded as functions. Though we shall also deal with the zeros of polynomials, our chief aim is not to study the properties of polynomial functions. We are rather concerned with rings whose elements are polynomials.

5.1. Polynomials in one indeterminate

Let $\mathbf{R}$ be an integral domain and S an infinite cyclic semigroup consisting of the identity e and the powers $x, x^2, x^3, \ldots$ Then the semigroup algebra of S over $\mathbf{R}$ is called the ring of the *polynomials over* $\mathbf{R}$ and is denoted by $\mathbf{R}[x]$. (For the definition of a semigroup algebra see Example 11 in § 4.1.) Thus $\mathbf{R}[x]$ consists of all finite linear combinations

$$a_m x^m + a_{m-1} x^{m-1} + \ldots + a_1 x + a_0 e$$

of elements of S with coefficients $a_m, a_{m-1}, \ldots, a_1, a_0$ of $\mathbf{R}$. The elements of the form ae with $a \in \mathbf{R}$ obviously form a subring of $\mathbf{R}\,[x]$ isomorphic to $\mathbf{R}$. We therefore identify this subring with $\mathbf{R}$ and write

$$a_m x^m + a_{m-1} x^{m-1} + \ldots + a_1 x + a_0 .$$

Such an expression is called a *polynomial over* $\mathbf{R}$, and $a_m, \ldots, a_0$ are said to be its *coefficients*. In this context we refer to x as an *indeterminate* and speak of polynomials in one indeterminate.

The zero element of $\mathbf{R}[x]$ is the so-called *zero polynomial* all of whose coefficients are equal to zero. We shall denote the zero polynomial by 0. Due to the embedding of $\mathbf{R}$ in $\mathbf{R}[x]$, the unit element of $\mathbf{R}[x]$ coincides with the unit element of $\mathbf{R}$.

If

$$f(x) = a_m x^m + a_{m-1} x^{m-1} + \dots + a_1 x + a_0$$

is a polynomial over $\mathbf{R}$ and if $a_m \neq 0$, then m is called the *degree* of $f(x)$. We refer to $a_m x^m$ and a_m as the *leading term* and the *leading coefficient*, respectively. A *monic* polynomial is one whose leading coefficient is equal to 1. Note that we do not assign a degree to the zero polynomial. Polynomials of degree 0 are the non-zero elements of $\mathbf{R}$. By *constant polynomials* we understand the polynomials of degree 0 and the zero polynomial. By a *linear* polynomial we understand one of degree 1. Forming the product

$$\begin{aligned} f(x)\, g(x) &= (a_m x^m + \dots + a_0)\,(b_n x^n + \dots + b_0) \\ &= a_m b_n x^{m+n} + \dots + a_0 b_0 \end{aligned}$$

of two polynomials $f(x)$ and $g(x)$ of degrees m and n, respectively, we see that the product is distinct from the zero polynomial unless at least one of the factors is equal to the zero polynomial. Indeed $a_m \neq 0$ and $b_n \neq 0$ imply that $a_m b_n \neq 0$ since $\mathbf{R}$ is an integral domain. This gives

5.1.1. *If* $\mathbf{R}$ *is an integral domain, then so is* $\mathbf{R}[x]$.

Moreover, we observed that the degree of $f(x)\, g(x)$ is equal to the sum of the degrees of $f(x)$ and $g(x)$.

5.1.2. *The units of* $\mathbf{R}[x]$ *coincide with the units of* $\mathbf{R}$.

Proof. It is clear that every unit of $\mathbf{R}$ is a unit of $\mathbf{R}[x]$. Conversely, if $f(x)g(x) = 1$, then both $f(x)$ and $g(x)$ must be of degree 0, i.e. $f(x) = a_0$, $g(x) = b_0$ and $a_0 b_0 = 1$ so that a_0 and b_0 are units in $\mathbf{R}$.

5.2. Polynomials over fields

5.2.1. THEOREM. *The ring* $\mathbf{K}[x]$ *of all polynomials in one indeterminate over a field* $\mathbf{K}$ *is a euclidean ring.*

Proof. We shall show that the function

$$H(f(x)) = \text{degree of } f(x) \quad (f(x) \in \mathbf{K}[x])$$

has the properties (i) and (ii) in § 4.5.

Let

$$f(x) = a_m x^m + \dots + a_0 \qquad (a_m \neq 0),$$

and

$$g(x) = b_n x^n + \dots + b_0$$

be polynomials of $\mathbf{K}[x]$ and $f(x) \neq 0$. To verify (i) we have to show that there exist polynomials $q(x)$ and $r(x)$ in $\mathbf{K}[x]$ such that

$$g(x) = q(x)\, f(x) + r(x)$$

where $r(x) = 0$ or $H(r(x)) < H(f(x))$. If $g(x) = 0$ or $H(g(x)) < H(f(x))$, then $q(x) = 0$ and $r(x) = g(x)$ have the required property. Now suppose that $g(x) \neq 0$, say $b_n \neq 0$, and $n \geqq m$. Then, the degree n_1 of

$$g_1(x) = g(x) - a_m^{-1} b_n x^{n-m} f(x)$$

does not exceed $n - 1$ since both $g(x)$ and $a_m^{-1} b_n x^{n-m} f(x)$ have the same leading terms. If $n_1 \geqq m$, then we can again substract a suitable multiple of $f(x)$ from $g_1(x)$ to obtain a polynomial whose degree does not exceed $n_1 - 1$. After a finite number of steps we arrive at a polynomial

$$q(x) = a_m^{-1} b_n x^{n-m} + \dots$$

such that

$$r(x) = g(x) - q(x)\, f(x)$$

is the zero polynomial or is of degree less than m. This shows that $H(f(x))$ satisfies condition (i).

That $H(f(x))$ also satisfies condition (ii) follows immediately from the fact that the degree of $f(x)\, g(x)$ is equal to the sum of the degrees of $f(x)$ and $g(x)$. This completes the proof.

By 5.1.2, the units of $\mathbf{K}[x]$ are the non-zero elements of the field $\mathbf{K}$.

The prime elements of $\mathbf{K}[x]$ are those polynomials $p(x)$ of positive degree that are only divisible by units of $\mathbf{K}[x]$ and by associates of $p(x)$, i.e. polynomials of the form $cp(x)$ with $c \in \mathbf{K}$, $c \neq 0$. In other words, $p(x)$ is a prime element of $\mathbf{K}[x]$ if there exists no factorization $p(x) = r(x)\, s(x)$ where $r(x), s(x) \in \mathbf{K}[x]$ and both $r(x)$ and $s(x)$ are of positive degrees. Such polynomials $p(x)$ are called *irreducible polynomials* or *prime polynomials* in $\mathbf{K}[x]$.

Whether a polynomial is irreducible depends on $\mathbf{K}$. For instance, $x^2 + 1$ is irreducible in $\boldsymbol{R}[x]$, whereas, in $\boldsymbol{C}[x]$ it has the factorization $x^2 + 1 = (x + i)(x - i)$. If we wish to emphasize that irreducibility refers to $\mathbf{K}[x]$, then we say that a polynomial is irreducible in $\mathbf{K}[x]$ or over $\mathbf{K}$.

In the case of polynomials over a field there is a natural way to single out a particular member of each class of associated prime polynomials, namely the unique monic polynomial in each class. Therefore any $f(x) \in \mathbf{K}[x]$ other than 0 has a unique representation

$$f(x) = ap_1(x)^{n_1}\, p_2(x)^{n_2} \ldots p_r(x)^{n_r}$$

where a is a non-zero element of $\mathbf{K}$ (the leading coefficient of $f(x)$) and $p_1(x), \ldots, p_r(x)$ are distinct monic prime polynomials in $\mathbf{K}[x]$.

Two polynomials $f(x)$, $g(x) \in \mathbf{K}[x]$ are relatively prime if their greatest common divisors are the units of $\mathbf{K}[x]$, i.e. the non-zero elements of $\mathbf{K}$.

A field $\mathbf{E}$ that contains $\mathbf{K}$ as a subfield is called an *extension field* of $\mathbf{K}$. Clearly, a polynomial $f(x) \in \mathbf{K}[x]$ can also be regarded as a polynomial of $\mathbf{E}[x]$. The following proposition will be used later:

5.2.2. Lemma. *Let $f(x)$, $g(x)$ be polynomials in $\mathbf{K}[x]$ and let $\mathbf{E}$ be an extension field of $\mathbf{K}$. Suppose that $f(x)$ and $g(x)$, regarded as polynomials in $\mathbf{E}[x]$ have a greatest common divisor $d_1(x) \in \mathbf{E}[x]$. Then, there exists a polynomial $d(x) \in \mathbf{K}[x]$ that is associated to $d_1(x)$ in $\mathbf{E}[x]$.*

To put it briefly, polynomials in $\mathbf{K}[x]$ have greatest common divisors with coefficients in $\mathbf{K}$ (and not only with coefficients in some extension field of $\mathbf{K}$).

Proof. The greatest common divisor can be obtained by the Euclidean algorithm. When the Euclidean algorithm is started with two polynomials in $\mathbf{K}[x]$, then, evidently, all polynomials that occur in the subsequent steps have coefficients in **K**. Therefore, one finally arrives at a greatest common divisor which belongs to $\mathbf{K}[x]$. This completes the proof.

The field of quotients of $\mathbf{K}[x]$ is denoted by $\mathbf{K}(x)$ and consists of all quotients $f(x)/g(x)$ where $f(x)$, $g(x) \in \mathbf{K}[x]$ and $g(x) \neq 0$. We call $\mathbf{K}(x)$ the field of the *rational functions* in the indeterminate x.

5.3. Polynomials over integral domains

Let **R** be a unique factorization ring. A polynomial

$$a_m x^m + \dots + a_1 x + a_0$$

of positive degree of $\mathbf{R}[x]$ is said to be *primitive* if $a_m, \dots, a_1, a_0$ are relatively prime.

5.3.1. Gauss' Lemma. *The product of two primitive polynomials is primitive.*

Proof. Let $f(x)$, $g(x)$ be primitive polynomials in $\mathbf{R}[x]$. Let us assume that, on the contrary, $h(x) = f(x)\, g(x)$ is not primitive. Then, there exists a prime element p of **R** that divides all the coefficients of $h(x)$. The residue class ring $\mathbf{R}/(p)$ is an integral domain. We consider the ring $\mathbf{R}/(p)\,[y]$ of the polynomials in an indeterminate y with coefficients of $\mathbf{R}/(p)$. Since $\mathbf{R}/(p)$ is an integral domain, so is $\mathbf{R}/(p)\,[y]$. To any polynomial $f(x) \in \mathbf{R}[x]$ we assign a polynomial $f_p(y) \in \mathbf{R}/(p)\,[y]$ by the following rule: if

$$f(x) = a_m x^m + \dots + a_1 x + a_0$$

then

$$f_p(y) = A_m y^m + \dots + A_1 y + A_0$$

where A_k denotes the residue class mod (p) to which a_k belongs. It is evident that the mapping $f(x) \to f_p(y)$ is a homomorphism of $\mathbf{R}[x]$ onto $\mathbf{R}/(p)\,[y]$.

Since $f(x)$ and $g(x)$ are primitive, both $f_p(y)$ and $g_p(y)$ are distinct from the zero polynomial, whereas all coefficients of $h(x)$ were assumed to be divisible by p so that $h_p(y)$ is the zero polynomial. This gives

$$f_p(y)\, g_p(y) = h_p(y) = 0$$

which contradicts the fact that $\mathbf{R}/(p)\,[y]$, as an integral domain, contains no divisors of zero. This proves the lemma.

The units of $\mathbf{R}[x]$ are the units of $\mathbf{R}$. There exist two kinds of prime elements in $\mathbf{R}[x]$: first, the prime elements of $\mathbf{R}$ and, secondly, the primitive polynomials of $\mathbf{R}[x]$ that are irreducible in $\mathbf{R}[x]$.

Let d denote a greatest common divisor of the coefficients of $f(x) \in \mathbf{R}[x]$. Then we have $f(x) = df^*(x)$ where $f^*(x)$ is a primitive polynomial. If $f(x) = df^*(x) = d_1 f_1^*(x)$ where both $f^*(x)$ and $f_1^*(x)$ are primitive, then it is easy to see that d and d_1 are associated in $\mathbf{R}$ and $f^*(x)$ and $f_1^*(x)$ are associated in $\mathbf{R}[x]$.

Let $\mathbf{F}$ denote the field of quotients of $\mathbf{R}$. Any polynomial $\varphi(x) \in \mathbf{F}[x]$ has a representation

$$\varphi(x) = \frac{1}{b} f(x) \tag{1}$$

where $f(x) \in \mathbf{R}[x]$ and $b \in \mathbf{R}$. Indeed, we need only choose b as the least common multiple of the denominators of the coefficients of $\varphi(x)$.

In case the polynomial $f(x)$ is not primitive, we have $f(x) = df^*(x)$ where $f^*(x)$ is primitive. This gives

$$\varphi(x) = \frac{d}{b} f^*(x).$$

5.3.2. Lemma. *Let $f^*(x)$ be a primitive polynomial of $\mathbf{R}[x]$. If $\frac{d}{b} f^*(x)$ has coefficients in $\mathbf{R}$, then b divides d.*

Proof. Since $\frac{d}{b} f^*(x)$ belongs to $\mathbf{R}[x]$, we have

$$\frac{d}{b} f^*(x) = cg^*(x)$$

where $c \in \mathbf{R}$ and $g^*(x)$ is a primitive polynomial in $\mathbf{R}[x]$. This gives $df^*(x) = bcg^*(x)$ so that d and bc are associated in $\mathbf{R}$ which means that b divides d. This completes the proof.

We now come to an important theorem.

5.3.3. Theorem. *If $\mathbf{R}$ is a unique factorization ring, then so is $\mathbf{R}[x]$.*

Before we prove the theorem we shall restate it more explicitely: If $\mathbf{R}$ is a unique factorization ring, then any polynomial $f(x)$ of

$\mathbf{R}[x]$ of positive degree admits a factorization

$$f(x) = \left(\prod_{i=1}^{m} p_i\right)\left(\prod_{i=1}^{r} f_i^*(x)\right) \tag{2}$$

where the p_i are prime elements of $\mathbf{R}$ and the $f_i^*(x)$ primitive, in $\mathbf{R}[x]$ irreducible polynomials of positive degree. If $f(x)$ is primitive, then the factor $\prod_{i=1}^{m} p_i$ does not occur. If

$$f(x) = \left(\prod_{i=1}^{n} q_i\right)\left(\prod_{i=1}^{s} g_i^*(x)\right) \tag{3}$$

is another factorization of the same kind, then, first $m = n$ and the p_i are associated to the q_i in a suitable order, secondly, $r = s$ and the $f_i^*(x)$ are associated to the $g_i^*(x)$ in a suitable order.

Proof. Let $\mathbf{F}$ be the field of quotients of $\mathbf{R}$. The polynomial $f(x)$, regarded as polynomial in $\mathbf{F}[x]$ has a factorization

$$f(x) = \prod_{i=1}^{r} \varphi_i(x)$$

where the $\varphi_i(x)$ are prime polynomials in $\mathbf{F}[x]$. Each $\varphi_i(x)$ can be written in the form

$$\varphi_i(x) = \frac{d_i}{b_i} f_i^*(x) \qquad (i = 1, \ldots, r)$$

where $d_i, b_i \in \mathbf{R}$ and $f_i^*(x)$ is a primitive polynomial in $\mathbf{R}[x]$. Putting $d = \prod_{i=1}^{r} d_i$, $b = \prod_{i=1}^{r} b_i$, we obtain

$$f(x) = \frac{d}{b} \prod_{i=1}^{r} f_i^*(x).$$

It is clear that the $f_i^*(x)$ are irreducible in $\mathbf{F}[x]$ and hence, all the more, irreducible in $\mathbf{R}[x]$. By 5.3.1, the product on the right-hand side of the last equation is primitive. Since $f(x) \in \mathbf{R}[x]$, it follows from 5.3.2 that b divides d, say $d = bc$. This gives

$$f(x) = c \prod_{i=1}^{r} f_i^*(x).$$

We decompose c into a product of prime elements,

$$c = \prod_{i=1}^{m} p_i$$

to obtain the factorization (2).

Now, let (3) be another factorization. By 5.3.1, the products $\prod_{i=1}^{r} f_i^*(x)$ and $\prod_{i=1}^{s} g_i^*(x)$ are primitive. Therefore $\prod_{i=1}^{m} p_i$ and $\prod_{i=1}^{n} q_i$ are associated, say

$$\prod_{i=1}^{m} p_i = u \prod_{i=1}^{n} q_i$$

where u denotes a unit of $\mathbf{R}$. By the Unique Factorization Theorem in $\mathbf{R}$ it follows that $m = n$ and that the p_i are associated to the q_i taken in a suitable order. Moreover,

$$u \prod_{i=1}^{r} f_i^*(x) = \prod_{i=1}^{s} g_i^*(x).$$

From the Unique Factorization Theorem in $\mathbf{F}[x]$ it follows that $r = s$ and that $f_i^*(x)$ is associated in $\mathbf{F}[x]$ to $g_i^*(x)$ where we assume that the numbering is suitably chosen. This means that

$$f_i^*(x) = \frac{h_i}{k_i} g_i^*(x) \quad (i = 1, \ldots, r)$$

where $h_i, k_i \in \mathbf{R}$. We write $k_i f_i^*(x) = h_i g_i^*(x)$ and make use of the fact that $f_i^*(x)$ and $g_i^*(x)$ are primitive to find that k_i and h_i are associated in $\mathbf{R}$, say $h_i = u_i k_i$ where the u_i are units in $\mathbf{R}$. Thus we obtain $f_i^*(x) = u_i g_i^*(x)$ so that $f_i^*(x)$ and $g_i^*(x)$ are associated in $\mathbf{R}[x]$. This completes the proof.

Incidentally, the last proof yields another important theorem:

5.3.4. Gauss' Theorem. *Let $\mathbf{R}$ be a unique factorization ring and $\mathbf{F}$ its field of quotients. If a polynomial of $\mathbf{R}[x]$ is irreducible in $\mathbf{R}[x]$ then it is also irreducible in $\mathbf{F}[x]$.*

Numerous sufficient conditions for a polynomial to be irreducible are known. We shall deal only with one of them.

5.3.5. Eisenstein's Theorem. *Let $\mathbf{R}$ be a unique factorization* *ring and let*

$$f(x) = a_m x^m + \dots + a_1 x + a_0$$

be a polynomial of $\mathbf{R}[x]$ *that satisfies the following conditions: There is a prime element* p *of* $\mathbf{R}$ *such that*

$$a_m \not\equiv 0 \pmod{p},$$

$$a_i \equiv 0 \pmod{p} \text{ for } i = 0, 1, \dots, m-1,$$

$$a_0 \not\equiv 0 \pmod{p^2}.$$

Then $f(x)$ *is irreducible in* $\mathbf{R}[x]$, *and, by* 5.3.4, *also irreducible over the field of quotients of* $\mathbf{R}$.

Proof. Suppose that, on the contrary, there is a factorization $f(x) = g(x)\,h(x)$ where

$$g(x) = b_r x^r + \dots + b_1 x + b_0,$$

$$h(x) = c_s x^s + \dots + c_1 x + c_0$$

and $b_r \neq 0$, $c_s \neq 0$, $r > 0$, $s > 0$, $r + s = m$. From

$$a_0 = b_0 c_0, \quad a_0 \equiv 0 \pmod{p}, \quad a_0 \not\equiv 0 \pmod{p^2}$$

we conclude that precisely one of the elements b_0, c_0 is divisible by p. We may assume that the notation is such that

$$b_0 \equiv 0 \pmod{p}, \qquad c_0 \not\equiv 0 \pmod{p}.$$

Not all the b_i are divisible by p since, otherwise, all the a_i would be divisible by p whereas $a_m \not\equiv 0 \pmod{p}$. Let k denote the least subscript such that b_k is not divisible by p so that

$$b_0 \equiv b_1 \equiv \dots \equiv b_{k-1} \equiv 0 \pmod{p}, \quad b_k \not\equiv 0 \pmod{p}.$$

By the well known rule for multiplication of polynomials we have

$$a_k = b_k c_0 + b_{k-1} c_1 + b_{k-2} c_2 + \dots$$

This gives

$$a_k \equiv b_k\, c_0 \not\equiv 0 \pmod{p}$$

which contradicts the hypotheses that $a_k \equiv 0 \pmod{p}$ for $k < m$. So the theorem is proved.

As an application of Eisenstein's Theorem we prove that, for a prime number p,

$$f(x) = \frac{x^p - 1}{x - 1} = x^{p-1} + x^{p-2} + \dots + x + 1$$

is irreducible in $\mathbf{Z}[x]$ where $\mathbf{Z}$ denotes the ring of the integers. We put $y = x - 1$ and $f(x) = f(y + 1) = g(y)$. It is clear that $f(x)$ is irreducible if and only if $g(y)$ is irreducible. We obtain

$$g(y) = \frac{(y + 1)^p - 1}{y + 1 - 1}$$

$$= y^{p-1} + \binom{p}{1} y^{p-2} + \dots + \binom{p}{p-2} y + \binom{p}{p-1}.$$

Since the binomial coefficients are divisible by p but not by p^2 the irreducibility of $g(y)$ follows immediately from Theorem 5.3.5.

5.4. Roots. The derivative

Let $\mathbf{R}$ be an integral domain and

$$f(x) = a_m x^m + \dots + a_1 x + a_0$$

a polynomial of $\mathbf{R}[x]$. By substituting an element c of $\mathbf{R}$ for the indeterminate x we obtain the element

$$f(c) = a_m c^m + \dots + a_1 c + a_0$$

of $\mathbf{R}$. We call $f(c)$ the *value* of $f(x)$ at $x = c$. If $g(x) \in \mathbf{R}[x]$ and

$$f(x) + g(x) = h(x), \ f(x)\, g(x) = k(x)$$

then it is obvious that

$$f(c) + g(c) = h(c), \ f(c)\, g(c) = k(c).$$

In other words, the mapping

$$f(x) \to f(c) \quad \big(f(x) \in \mathbf{R}[x]\big) \tag{1}$$

is a homomorphism of $\mathbf{R}[x]$ onto $\mathbf{R}$.

If $\mathbf{S}$ is a ring that contains $\mathbf{R}$ and if γ is an element of $\mathbf{S}$ that commutes with every $a \in \mathbf{R}$, then we can also substitute γ for x

to obtain

$$f(\gamma) = a_m\gamma^m + \ldots + a_1\gamma + a_0, \tag{2}$$

the value of $f(x)$ at $x = \gamma$. As above, it is evident that the mapping

$$f(x) \to f(\gamma) \quad (f(x) \in \mathbf{R}[x])$$

is a homomorphism of $\mathbf{R}[x]$ into $\mathbf{S}$. An expression of the form (2) is called a *polynomial* in γ though γ need not be an indeterminate.

An element $\alpha \in \mathbf{S}$ is called a *zero* or a *root* of the polynomial $f(x)$ if $f(\alpha) = 0$.

We now assume that $\mathbf{S}$ itself is an integral domain and that $\alpha \in \mathbf{S}$ is a root of $f(x)$. We divide $f(x)$ by $x - \alpha$. Since the leading coefficient of $x - \alpha$ is 1, the quotient $f(x)/(x - \alpha)$ has coefficients in $\mathbf{S}$ and the remainder is an element of $\mathbf{S}$, i.e. the division does not involve elements of the quotient field of $\mathbf{S}$. We find

$$f(x) = (x - \alpha)\, f_1(x) + \varrho$$

where $f_1(x) \in \mathbf{S}[x]$ and $\varrho \in \mathbf{S}$. Substituting α for x we obtain $0 = f(\alpha) = \varrho$. This gives:

5.4.1. *If α is a root of $f(x)$, then $f(x)$ is divisible by $x - \alpha$.*

The converse of 5.4.1 is trivial, i.e. if $f(x) = (x - \alpha)\, f_1(x)$, then α is a root of $f(x)$.

If the factor $f_1(x)$ in $f(x) = (x - \alpha)\, f_1(x)$ has again the root α, then we obtain $f_1(x) = (x - \alpha)\, f_2(x)$ and hence $f(x) = (x - \alpha)^2\, f_2(x)$. In case $f_2(\alpha) = 0$ we obtain a third factor $x - \alpha$. Continuing in this way we finally arrive at a factorization

$$f(x) = (x - \alpha)^k\, f_k(x), \quad f_k(\alpha) \neq 0.$$

We say that α is a root of *multiplicity* k. A root of multiplicity 1 is also called a *simple* root. By a *multiple* root we mean a root whose multiplicity is greater than 1.

Let us assume that $\mathbf{S}$ contains another root β of $f(x)$ and let l denote the multiplicity of β. Then $f(x) = (x - \alpha)^k\, f_k(x)$ is divisible by $(x - \beta)^l$. The factor $(x - \alpha)^k$ is not divisible by $x - \beta$ since $(\beta - \alpha)^k \neq 0$. Therefore $(x - \beta)^l$ divides $f_k(x)$, i.e.

$$f(x) = (x - \alpha)^k\, (x - \beta)^l\, f_r(x).$$

If **S** contains roots of $f(x)$ other than α and β, then we obtain more powers of linear factors that divide $f(x)$. Since $f(x)$ is of degree m, the number of linear factors, each counted according to its multiplicity, cannot exceed m. This gives:

5.4.2. *A polynomial of degree m over an integral domain* **R** *cannot have more than m roots in any integral domain containing* **R**. *Here each root may be counted according to its multiplicity.*

By the *derivative* of the polynomial

$$f(x) = a_m x^m + \dots + a_2 x^2 + a_1 x + a_0$$

we mean the polynomial

$$f'(x) = m a_m x^{m-1} + \dots + 2a_2 x + a_1.$$

In case the coefficients of $f(x)$ are numbers, the derivative coincides with the derivative in the sense of differential calculus. But we have to use this formal definition and must not refer to limits because it is impossible to define a limit within the framework of our concepts and axioms. Our formal derivative satisfies the familiar rules, namely

$$\big(f(x) + g(x)\big)' = f'(x) + g'(x), \quad \big(f(x)\, g(x)\big)' = f'(x)\, g(x) + f(x)\, g'(x).$$

For the proof of these rules it is expedient to write

$$f(x) = \sum_{i=0}^{\infty} a_i x^i, \quad g(x) = \sum_{i=0}^{\infty} b_i x^i$$

where x^0 stands for 1 and at most finitely many coefficients a_i, b_i are distinct from zero. We obtain

$$\big(f(x) + g(x)\big)' = \left(\sum_{i=0}^{\infty} (a_i + b_i)\, x^i\right)' = \sum_{i=1}^{\infty} i(a_i + b_i)\, x^{i-1}$$

$$= \sum_{i=1}^{\infty} i a_i x^{i-1} + \sum_{i=1}^{\infty} i b_i x^{i-1} = f'(x) + g'(x).$$

Suppose that the second rule is satisfied for two polynomials $g_1(x)$ and $g_2(x)$. Then it also holds for $g(x) = g_1(x) + g_2(x)$. For, by the first rule, we have

$$[f(g_1 + g_2)]' = (fg_1 + fg_2)' = (fg_1)' + (fg_2)'$$

$$= f'g_1 + fg_1' + f'g_2 + fg_2' = f'(g_1 + g_2) + f(g_1' + g_2')$$

$$= f'(g_1 + g_2) + f(g_1 + g_2)'.$$

Therefore it is sufficient to prove the second rule for $g(x) = bx^k$. For $k = 0$ the rule holds trivially. For $k > 0$ we obtain

$$(f(x)\, g(x))' = \left(\sum_{i=0}^{\infty} a_i b x^{i+k}\right)' = \sum_{i=0}^{\infty} (i + k)\, a_i b x^{i+k-1}$$

$$= \sum_{i=0}^{\infty} i a_i x^{i-1} b x^k + \sum_{i=0}^{\infty} a_i x^i k b x^{k-1} = f'(x)\, g(x) + f(x)\, g'(x).$$

From these rules we infer, in particular, that $[(x - \alpha)^k]' = k(x - \alpha)^{k-1}$.

We now determine the polynomials whose derivative is the zero polynomial. The polynomial

$$f(x) = a_m x^m + \ldots + a_1 x + a_0$$

of $\mathbf{R}[x]$ satisfies the condition $f'(x) = 0$ if and only if

$$n a_n = 0 \text{ for } n = 1, 2, \ldots, m. \tag{3}$$

We have to distinguish two cases:

(i) $\mathbf{R}$ is of *characteristic* 0. Then it follows from (3) that

$$a_n = 0 \text{ for } n = 1, 2, \ldots, m.$$

Hence, $f(x)$ is a constant polynomial. Conversely, by the definition, the derivative of a constant polynomial is the zero polynomial.

(ii) $\mathbf{R}$ *is of characteristic* $p > 0$. In this case (3) yields only

$$a_n = 0 \text{ if } n \not\equiv 0 \pmod{p}.$$

So, we obtain

$$f(x) = a_{lp} x^{lp} + a_{(l-1)p} x^{(l-1)p} + \ldots + a_{2p} x^{2p} + a_p x^p + a_0.$$

It is evident that, conversely, every polynomial of this form has the zero polynomial as its derivative.

The derivative is closely connected with the multiplicity of roots. Let α be a root of $f(x)$ of multiplicity k, i.e.

$$f(x) = (x - \alpha)^k f_k(x), \quad f_k(\alpha) \neq 0.$$

Forming the derivative we obtain

$$\begin{aligned} f'(x) &= k(x - \alpha)^{k-1} f_k(x) + (x - \alpha)^k f_k'(x) \\ &= (x - \alpha)^{k-1} [k f_k(x) + (x - \alpha) f_k'(x)] \end{aligned} \tag{4}$$

so that $f'(x)$ is divisible by $(x - \alpha)^{k-1}$. This gives:

5.4.3. *If α is a root of multiplicity $k > 1$ of $f(x)$, then α is a root of multiplicity at least $k - 1$ of $f'(x)$.*

In general, the adjunct "at least" must not be omitted in the last statement as the following example shows: If **R** is of characteristic $p > 0$, then the derivative of $(x - \alpha)^p$ is the zero polynomial. But if **R** is of characteristic 0, then 5.4.3 can be sharpened as follows:

5.4.4. *If the coefficients of a polynomial $f(x)$ belong to an integral domain of characteristic 0, then a root of multiplicity k of $f(x)$ is a root of multiplicity $k - 1$ of $f'(x)$. (For $k = 1$ this means that a simple root of $f(x)$ is not a root of $f'(x)$.)*

Proof. If **R** is of characteristic 0, then the factor in brackets on the right-hand side of (4) has the value $k f_k(\alpha) \neq 0$ at $x = \alpha$. Therefore $(x - \alpha)^{k-1}$ is the highest power of $x - \alpha$ that divides $f'(x)$. This completes the proof.

The higher derivatives are defined in the usual way, namely

$$\begin{aligned} f^{(0)}(x) &= f(x), \\ f^{(n)}(x) &= \left(f^{(n-1)}(x)\right)', \quad n = 1, 2, \ldots \end{aligned}$$

From 5.4.4 we obtain the following:

5.4.5. *If the coefficients of a polynomial $f(x)$ belong to an integral domain of characteristic 0, then a root α of multiplicity k of $f(x)$ is a root of multiplicity $k - n$ of $f^{(n)}(x)$ for $n = 0, 1, \ldots, k$. In particular,*

$$f(\alpha) = f'(\alpha) = \ldots = f^{(k-1)}(\alpha) = 0, \quad f^{(k)}(\alpha) \neq 0.$$

Conversely, the last equations imply that α is a root of multiplicity k of $f(x)$.

We now consider polynomials whose coefficients belong to a field $\mathbf{K}$. Any field that contains $\mathbf{K}$ as a subfield is called an *extension field* of $\mathbf{K}$.

5.4.6. *Let $q(x)$, $f(x)$ be polynomials of $\mathbf{K}[x]$ and let $q(x)$ be irreducible in $\mathbf{K}[x]$. If $q(x)$ and $f(x)$ have a common root in some extension field $\mathbf{E}$ of $\mathbf{K}$, then $f(x)$ is divisible by $q(x)$.*

Proof. Let $\alpha \in \mathbf{E}$ be a common root of $q(x)$ and $f(x)$. Then, $q(x)$ and $f(x)$, regarded as polynomials in $\mathbf{E}[x]$ have at least the common linear factor $x - \alpha$ so that they have a non-constant greatest common divisor in $\mathbf{E}[x]$. It follows from Lemma 5.2.2 that they have a greatest common divisor $d(x) \in \mathbf{K}[x]$ of positive degree. Since $q(x)$ is irreducible in $\mathbf{K}[x]$ we conclude that $d(x)$ is associated to $q(x)$ and, hence, $q(x) \mid f(x)$, which proves our proposition.

We now come to a remarkable property of irreducible polynomials over fields of characteristic 0.

5.4.7. *Let $\mathbf{K}$ be a field of characteristic 0. Then an irreducible polynomial of $\mathbf{K}[x]$ has no multiple root in any extension field of $\mathbf{K}$.*

Proof. Let $q(x)$ be an irreducible polynomial of $\mathbf{K}[x]$. Due to the hypothesis on the characteristic of $\mathbf{K}$, the derivative $p'(x)$ is not the zero polynomial. Suppose that, on the contrary, $q(x)$ has a multiple root α in an extension field $\mathbf{E}$ of $\mathbf{K}$. Then, α is also a root of $q'(x)$. By 5.4.6, we conclude that $q(x)$ divides $q'(x)$. But this is impossible because the degree of $q'(x)$ is by 1 less than the degree of $q(x)$. This proves our proposition.

If $\mathbf{K}$ is of characteristic $p > 0$, then the argument in the last proof fails if and only if $q'(x)$ is the zero polynomial. We shall show that, over fields of characteristic $p > 0$, there may exist irreducible polynomials with multiple roots. Let a be an element of $\mathbf{K}$ that is not the p-th power of any element of $\mathbf{K}$. We shall show later (§ 6.3) that there exists an extension field $\mathbf{E}$ of $\mathbf{K}$ which contains an element $\varkappa$ such that $\varkappa^p = a$. By 4.7.3, we obtain

$$x^p - a = x^p - \varkappa^p = (x - \varkappa)^p$$

so that $x^p - a$ has $\varkappa$ as a root of multiplicity p. It remains to show that $x^p - a$ is irreducible in $\mathbf{K}[x]$. Suppose that, on the contrary, $x^p - a$ is reducible in $\mathbf{K}[x]$. Then some power $(x - \varkappa)^r$ where $1 \leqq r \leqq p - 1$ is a polynomial of $\mathbf{K}[x]$. We conclude that, in particular, $\varkappa^r$ belongs to $\mathbf{K}$. Due to $(r, p) = 1$ there exist integers u, v such that $ur + vp = 1$. This gives

$$\varkappa = \varkappa^{ur+vp} = (\varkappa^r)^u (\varkappa^p)^v = (\varkappa^r)^u a^v$$

so that $\varkappa \in \mathbf{K}$ which contradicts our assumption.

This process of constructing an irreducible polynomial with multiple roots is essentially based on the existence of an element $a \in \mathbf{K}$ for which $\sqrt[p]{a}$ is not contained in $\mathbf{K}$. However, if $\mathbf{K}$ contains the p-th root of each element, then our construction does not only fail but there do not exist irreducible polynomials with multiple roots. Indeed, if $f(x)$ is a polynomial of $\mathbf{K}[x]$ such that $f'(x)$ is the zero polynomial, then $f(x)$ is of the form

$$f(x) = \sum_{i=0}^{l} a_{ip} x^{ip}.$$

Since $\mathbf{K}$ contains the $\sqrt[p]{a_{ip}}$ we obtain

$$f(x) = \sum_{i=0}^{l} \left(\sqrt[p]{a_{ip}}\, x^i\right)^p = \left(\sum_{i=0}^{l} \sqrt[p]{a_{ip}}\, x^i\right)^p = f_0(x)^p$$

where $f_0(x) \in \mathbf{K}[x]$. This shows that, in our particular case, a polynomial $f(x)$ with $f'(x) = 0$ is reducible.

We observe that for an element a of a field $\mathbf{F}$ of characteristic $p > 0$ the p-th root $\varkappa = \sqrt[p]{a}$ is unique. For, suppose that $\varkappa_1^p = a$. Then we have $0 = \varkappa^p - \varkappa_1^p = (\varkappa - \varkappa_1)^p$ so that $\varkappa = \varkappa_1$.

The existence of irreducible polynomials with multiple roots gives rise to complications in the study of extension fields. Therefore one singles out those fields for which these complications cannot occur. A field $\mathbf{K}$ is said to be *perfect* if every irreducible polynomial of $\mathbf{K}[x]$ has simple roots in any extension field of $\mathbf{K}$.

Summarizing our last results we obtain:

5.4.8. Theorem. *Every field of characteristic* 0 *is perfect. A field* $\mathbf{K}$ *of characteristic* $p > 0$ *is perfect if and only if* $\mathbf{K}$ *contains the* p*-th root of each of its elements.*

We mention the following corollary:

5.4.9. *Every finite field is perfect.*

Proof. If **K** is a finite field, then the characteristic of **K** is necessarily a prime number p. If $a, b \in \mathbf{K}$ and $a \neq b$, then $a^p \neq b^p$. For, otherwise, we would obtain $0 = a^p - b^p = (a - b)^p$, i.e. $a = b$. Now, let $a_1, \ldots, a_n$ be the elements of **K**. Then $a_1^p, \ldots, a_n^p$ are all distinct so that they coincide with $a_1, \ldots, a_n$ but for the order. Therefore, every element of **K** is a p-th power or, in other words, **K** contains the p-th root of each element. Thus, our proposition follows from 5.4.8.

Let $f(x)$ be a polynomial of degree $m > 0$ with coefficients in a field **K** of characteristic 0. To simplify the notation we shall assume that $f(x)$ is monic. Moreover, in every class of associated divisors of $f(x)$ we shall always choose the monic one in our next considerations.

Later (§ 6.3) we shall prove that there exists an extension field **E** of **K** such that $f(x)$ has m roots in **E** where each root is counted according to its multiplicity. Let $\alpha_1, \ldots, \alpha_r$ denote the distinct roots of $f(x)$ and $k_1, \ldots, k_r$ their respective multiplicities. Hence, in $\mathbf{E}[x]$ we have the following factorization

$$f(x) = (x - \alpha_1)^{k_1} (x - \alpha_2)^{k_2} \ldots (x - \alpha_r)^{k_r}. \tag{5}$$

Let $g_k(x)$ stand for the product of all factors $x - \alpha_i$ such that $k_i = k$ and put $g_k(x) = 1$ if $f(x)$ has no root of multiplicity k. This gives

$$f(x) = g_1(x)\, g_2(x)^2\, g_3(x)^3 \ldots g_t(x)^t.$$

Since a root of multiplicity k of $f(x)$ is a root of multiplicity $k - 1$ of $f'(x)$, we find that

$$\big(f(x), f'(x)\big) = g_2(x)\, g_3(x)^2 \ldots g_t(x)^{t-1}.$$

Thus we obtain

$$\frac{f(x)}{(f(x), f'(x))} = g_1(x)\, g_2(x) \ldots g_t(x) = g(x).$$

The polynomial $g(x)$ has the same roots as $f(x)$ but each of them as a simple root. Since booth $f(x)$ and $f'(x)$ have coefficients in **K** and since the greatest common divisor $(f(x), f'(x))$ can be computed by the Euclidean algorithm we conclude that $(f(x), f'(x)) \in \mathbf{K}[x]$ so that also the coefficients of $g(x)$ belong to $\mathbf{K}$. Moreover, the computation of $g(x)$ does not require the knowledge of the roots of $f(x)$.

5.5. Polynomials in several indeterminates

Let S be the direct product of n cyclic semigroups $S_1, \ldots, S_n$ where S_k consists of the identity and the powers $x_k, x_k^2, \ldots$ The group algebra of S over an integral domain $\mathbf{R}$ is called the ring of the *polynomials in the indeterminates* $x_1, x_2, \ldots, x_n$ over $\mathbf{R}$ and is denoted by $\mathbf{R}[x_1, x_2, \ldots, x_n]$. The elements of $\mathbf{R}[x_1, x_2, \ldots, x_n]$ are the sums

$$f(x_1, x_2, \ldots, x_n) = \sum_{i_1, i_2, \ldots, i_n} a_{i_1, i_2, \ldots, i_n} x_1^{i_1} x_2^{i_2} \ldots x_n^{i_n} \tag{1}$$

where $i_1, i_2, \ldots, i_n$ ranges over a finite system of n-tuples of non-negative integers and x_k^0 stands for the identity of S. The *coefficients* $a_{i_1, i_2, \ldots, i_n}$ are elements of $\mathbf{R}$. As in the case of one indeterminate we identify each element $a x_1^0 x_2^0 \ldots x_n^0$ with its coefficient a so that $\mathbf{R}$ becomes a subring of $\mathbf{R}[x_1, x_2, \ldots, x_n]$.

If $a_{i_1, i_2, \ldots, i_n} \neq 0$, then the sum $i_1 + i_2 + \ldots + i_n$ is called the *degree* of the term $a_{i_1, i_2, \ldots, i_n} x_1^{i_1} x_2^{i_2} \ldots x_n^{i_n}$. By the degree of $f(x_1, x_2, \ldots, x_n)$ we understand the highest degree that occurs among the terms on the right-hand side of (1). As in the case of one indeterminate we do not define the degree of the zero polynomial. If all the terms of $f(x_1, x_2, \ldots, x_n)$ have one and the same degree, say m, then $f(x_1, x_2, \ldots, x_n)$ is called a *homogeneous* polynomial or a *form* of degree m. It is evident that every non-zero polynomial has a unique representation as a sum of homogeneous polynomials.

As an expedient tool for proofs we define an order for the terms as follows: If the coefficients a and b are distinct from zero, then the term $a x_1^{i_1} x_2^{i_2} \ldots x_n^{i_n}$ is said to be higher than $b x_1^{j_1} x_2^{j_2} \ldots x_n^{j_n}$ if the first non-zero element in the sequence

$$i_1 - j_1, i_2 - j_2, \ldots, i_n - j_n$$

is positive. We refer to this order as the *lexicographic order*. The highest term of a polynomial $f(x_1, x_2, \ldots, x_n)$ is called the *leading term* and is denoted by $LT(f)$. For $n = 1$ this definition agrees with that in § 5.1.

5.5.1. Lemma. *If* $\mathbf{R}$ *is an integral domain and* $f(x_1, x_2, \ldots, x_n)$ *and* $g(x_1, x_2, \ldots, x_n)$ *are polynomials of* $\mathbf{R}[x_1, x_2, \ldots, x_n]$ *then*

$$LT(fg) = LT(f)\,LT(g).$$

Proof. We saw in § 5.1 that the proposition is true for $n = 1$. Let us assume that the lemma holds for polynomials in fewer than n indeterminates. Collecting the terms in which x_1 occurs with the same exponent we obtain

$$f(x_1, x_2, \ldots, x_n)$$
$$= x_1^r\varphi_r(x_2, \ldots, x_n) + x_1^{r-1}\varphi_{r-1}(x_2, \ldots, x_n) + \ldots + \varphi_0(x_2, \ldots, x_n),$$
$$g(x_1, x_2, \ldots, x_n)$$
$$= x_1^s\psi_s(x_2, \ldots, x_n) + x_1^{s-1}\psi_{s-1}(x_2, \ldots, x_n) + \ldots + \psi_0(x_2, \ldots, x_n)$$

where $\varphi_i(x_2, \ldots, x_n)$ and $\psi_i(x_2, \ldots, x_n)$ are polynomials in $\mathbf{R}[x_2, \ldots, x_n]$. In $LT(fg)$ the indeterminate x_1 must occur with the highest possible exponent. This gives

$$LT(fg) = x_1^{r+s}LT(\varphi_r\psi_s).$$

From the induction assumption we conclude that

$$LT(\varphi_r\psi_s) = LT(\varphi_r)\,LT(\psi_s).$$

Thus, we obtain

$$LT(fg) = x_1^{r+s}LT(\varphi_r)\,LT(\psi_s) = \left(x_1^rLT(\varphi_r)\right)\left(x_1^sLT(\psi_s)\right) = LT(f)\,LT(g)$$

which proves the lemma.

As a corollary we obtain:

5.5.2. *If* $\mathbf{R}$ *is an integral domain, then so is* $\mathbf{R}[x_1, x_2, \ldots, x_n]$.

If $\mathbf{R}^*$ stands for $\mathbf{R}[x_1, \ldots, x_{n-1}]$, then it is evident that $\mathbf{R}[x_1, \ldots, x_{n-1}, x_n] = \mathbf{R}^*[x_n]$. In view of Theorem 5.3.3 we thus obtain by induction on the number of indeterminates the following

5.5.3. Theorem. *If* $\mathbf{R}$ *is a unique factorization ring, then so is* $\mathbf{R}[x_1, x_2, \ldots, x_n]$.

If $\mathbf{K}$ is a field, then it follows from Theorem 5.2.1 that $\mathbf{K}[x_1]$ is a unique factorization ring. Therefore we obtain as a corollary of the previous theorem:

5.5.4. THEOREM. *If $\mathbf{K}$ is a field, then $\mathbf{K}[x_1, x_2, \ldots, x_n]$ is a unique factorization ring.*

Let $\mathbf{T}$ be a commutative ring which contains the integral domain $\mathbf{R}$ as a subring. By substituting elements $c_1, c_2, \ldots, c_n$ of $\mathbf{T}$ for the indeterminates $x_1, x_2, \ldots, x_n$ in the polynomial $f(x_1, x_2, \ldots, x_n) \in \mathbf{R}[x_1, x_2, \ldots, x_n]$ we obtain its *value* $f(c_1, c_2, \ldots, c_n)$ at $c_1, c_2, \ldots, c_n$.

5.5.5. *Let $\mathbf{R}$ be an integral domain that contains infinitely many elements and let $R_1, R_2, \ldots, R_n$ denote n infinite subsets of $\mathbf{R}$. If $f(x_1, x_2, \ldots, x_n)$ is a polynomial in the indeterminates $x_1, x_2, \ldots, x_n$ with coefficients in $\mathbf{R}$ such that $f(a_1, a_2, \ldots, a_n) = 0$ for every choice of the elements $a_i \in R_i$, then f is the zero polynomial.*

Proof. The proposition is true for polynomials in one indeterminate since a polynomial of degree r cannot have more than r roots in $\mathbf{R}$. We use induction on the number of indeterminates and assume that our proposition holds for $n - 1$ indeterminates. We write

$$f(x_1, x_2^r, \ldots, x_n)$$
$$= x_1\varphi_r(x_2, \ldots, x_n) + x_1^{r-1}\varphi_{r-1}(x_2, \ldots, x_n) + \ldots + \varphi_0(x_2, \ldots, x_n)$$

where $\varphi_i(x_2, \ldots, x_n) \in \mathbf{R}[x_2, \ldots, x_n]$. For every choice of elements a_i of R_i, $i = 2, \ldots, n$, we have $f(x_1, a_2, \ldots, a_n) = 0$ if we substitute any element of R_1 for x_1. Since R_1 is infinite it follows that $\varphi_j(a_2, \ldots, a_n) = 0$ for $j = 0, 1, \ldots, r$. By the induction assumption we conclude that all the φ_j are zero polynomials so that f itself is the zero polynomial as was to be proved.

Even in the case of a single indeterminate the condition that $\mathbf{R}$ contains infinitely many elements is not redundant in 5.5.5. For example it follows from 4.7.3 and 4.7.4 that all elements of the residue class field $\mathbf{Z}/(p)$ are zeros of the polynomial $x^p - x$.

5.6. Symmetric polynomials

A polynomial in the indeterminates $x_1, x_2, \ldots, x_n$ is said to be *symmetric* if it remains unchanged under every permutation of the indeterminates. Let us consider some examples:

The *power sums*

$$S_k = x_1^k + x_2^k + \ldots + x_n^k \quad (k = 0, 1, 2, \ldots),$$

Wronski's polynomials

$$P = \sum_{i_1+i_2+\ldots+i_n=k} x_1^{i_1} x_2^{i_2} \ldots x_n^{i_n} \quad (k = 0, 1, 2, \ldots),$$

the discriminant

$$D = \prod_{i<k} (x_i - x_k)^2 \quad (i, k = 1, \ldots, n).$$

The most important symmetric polynomials are the so-called *elementary symmetric polynomials* which are defined as follows:

$$C_1 = x_1 + x_2 + \ldots + x_n,$$

$$C_2 = x_1 x_2 + x_1 x_3 + \ldots + x_{n-1} x_n,$$

$$\ldots\ldots\ldots\ldots\ldots\ldots\ldots\ldots\ldots\ldots$$

$$C_n = x_1 x_2 \ldots x_n,$$

i.e., in a more explicit notation,

$$C_k = \sum_{i_1<i_2<\ldots<i_k} x_{i_1} x_{i_2} \ldots x_{i_k}.$$

where the sum is to be taken over all the $\binom{n}{k}$ systems of subscripts $i_1, i_2, \ldots, i_k$ such that $1 \leqq i_1 < i_2 < \ldots < i_k \leqq n$. Let z be another indeterminate. One readily verifies that

$$(z - x_1)(z - x_2) \ldots (z - x_n) =$$

$$z^n - C_1 z^{n-1} + C_2 z^{n-2} - \ldots + (-1)^n C_n.$$

The following theorem states that, roughly speaking, every symmetric polynomial can be expressed in terms of the elementary symmetric polynomials.

5.6.1. THEOREM. *Let* **R** *be an integral domain and* $s(x_1, x_2, \ldots, x_n)$ *a symmetric polynomial of* $\mathbf{R}[x_1, x_2, \ldots, x_n]$. *Then there exists one and only one polynomial* f *with coefficients in* **R** *such that* $s(x_1, x_2, \ldots, x_n) = f(C_1, C_2, \ldots, C_n)$.

Proof. If $ax_1^{k_1}x_2^{k_2} \ldots x_n^{k_n}$ is the leading term of $s(x_1, x_2, \ldots, x_n)$, then the exponents satisfy the condition $k_1 \geqq k_2 \geqq \ldots \geqq k_n$. Indeed, suppose that, on the contrary, $k_i < k_j$ for some $i < j$. Since $s(x_1, x_2, \ldots, x_n)$ is symmetric, it also contains the term

$$ax_1^{k_1} \ldots x_i^{k_j} \ldots x_j^{k_i} \ldots x_n^{k_n}.$$

But this term is higher (in the lexicographic order) than

$$ax_1^{k_1} \ldots x_i^{k_i} \ldots x_j^{k_j} \ldots x_n^{k_n}$$

so that we arrive at a contradiction.

Let $k_1 \geqq k_2 \geqq \ldots \geqq k_n$. By Lemma 5.5.1, the product

$$C_1^{k_1-k_2}C_2^{k_2-k_3} \ldots C_{n-1}^{k_{n-1}-k_n}C_n^{k_n}$$

has the leading term

$$x_1^{k_1}x_2^{k_2} \ldots x_n^{k_n}.$$

In particular, distinct products of powers of the $C_1, C_2, \ldots, C_n$ have distinct leading terms.

After these remarks we prove our theorem by induction on the lexicographic order of the leading terms. If all the exponents in the leading term of a symmetric polynomial are equal to zero, then the polynomial is an element of **R** and there is nothing to prove. Suppose that every symmetric polynomial of $\mathbf{R}[x_1, x_2, \ldots, x_n]$ whose leading term is lower than that of $s(x_1, x_2, \ldots, x_n)$ can be expressed as a polynomial in $C_1, C_2, \ldots, C_n$. We consider the polynomial

$$t(x_1, x_2, \ldots, x_n) = s(x_1, x_2, \ldots, x_n) - aC_1^{k_1-k_2}C_2^{k_2-k_3} \ldots C_{n-1}^{k_{n-1}-k_n}C_n^{k_n}. \quad (1)$$

It is clear that $t(x_1, x_2, \ldots, x_n)$ is a symmetric polynomial. As we observed above, the leading terms of $s(x_1, x_2, \ldots, x_n)$ and $aC_1^{k_1-k_2} \ldots C_n^{k_n}$ coincide so that the leading term of $t(x_1, x_2, \ldots, x_n)$ is lower than that of $s(x_1, x_2, \ldots, x_n)$. By the induction assumption, $t(x_1, x_2, \ldots, x_n)$ can therefore be expressed as a polynomial in $C_1, C_2, \ldots, C_n$. Consequently, it follows from (1) that $s(x_1, x_2, \ldots, x_n)$ can also be represented as a polynomial in $C_1, C_2, \ldots, C_n$.

It remains to show that this representation is unique. Suppose that

$$s(x_1, x_2, \ldots, x_n) = f(C_1, C_2, \ldots, C_n) = g(C_1, C_2, \ldots, C_n)$$

where f and g are polynomials. Then the difference

$$\begin{aligned} f(C_1, C_2, \ldots, C_n) - g(C_1, C_2, \ldots, C_n) &= d(C_1, C_2, \ldots, C_n) \\ &= \varphi(x_1, x_2, \ldots, x_n), \end{aligned}$$

considered as a polynomial $\varphi(x_1, x_2, \ldots, x_n)$ is the zero polynomial. But as we observed above, two distinct products of powers of $C_1, C_2, \ldots, C_n$ have distinct leading terms. Therefore, the highest leading term in all the products of powers of $C_1, C_2, \ldots, C_n$ in d occurs only once so that it is not cancelled. Consequently, φ cannot be the zero polynomial unless d is the zero polynomial. This completes the proof.

The proof of the last theorem yields an algorithm for computing the representation in question though the actual computation will soon become cumbersome when the number of the indeterminates and the degree of the polynomials increase. The computation can be simplified by the following remarks. Since every symmetric polynomial is the sum of unique homogeneous symmetric polynomials it is sufficient to find a method for representing a homogeneous symmetric polynomial as a polynomial in $C_1, C_2, \ldots, C_n$. It is evident that

$$C_1^{l_1} C_2^{l_2} \ldots C_n^{l_n}$$

is homogeneous of degree $l_1 + 2l_2 + \ldots + nl_n$. Let $s(x_1, x_2, \ldots, x_n)$ be a homogeneous symmetric polynomial of degree g. If

$$s(x_1, x_2, \ldots, x_n) = \sum_{l_1, l_2, \ldots, l_n} a_{l_1, l_2, \ldots, l_n} C_1^{l_1} C_2^{l_2} \ldots C_n^{l_n}, \tag{2}$$

then the exponents $l_1, l_2, \ldots, l_n$ satisfy the equation

$$l_1 + 2l_2 + \ldots + nl_n = g.$$

To determine the coefficients on the right-hand side of (2), one can substitute suitable elements of $\mathbf{R}$ for the indeterminates.

As an example we express the discriminants for $n = 2$ and 3 as polynomials in the elementary symmetric polynomials, where $\mathbf{R}$ is the ring of the integers.

For $n = 2$, one easily verifies that

$$D = (x_1 - x_2)^2 = C_1^2 - 4C_2. \tag{3}$$

For $n = 3$, we have

$$D = (x_1 - x_2)^2 (x_1 - x_3)^2 (x_2 - x_3)^2$$

so that D is homogeneous of degree 6. Therefore the exponents l_1, l_2, l_3 in (2) are subject to the condition $l_1 + 2l_2 + 3l_3 = 6$. The leading term of D is $x_1^4 x_2^2$. This gives $l_1 + l_2 + l_3 \leqq 4$. Therefore the following values of l_1, l_2, l_3 can occur

l_1	l_2	l_3
3	0	1
2	2	0
1	1	1
0	3	0
0	0	2

This gives

$$D = aC_1^3C_3 + bC_1^2C_2^2 + cC_1C_2C_3 + dC_2^3 + eC_3^2.$$

Putting $x_3 = 0$ we obtain

$$D = (x_1 - x_2)^2 x_1^2x_2^2 = bC_1^2C_2^2 + dC_2^3.$$

Comparing the leading terms on both sides we find that $b = 1$, and putting $x_1 = -x_2$ we get $d = -4$. Further

$$x_1 = x_2 = 1, \quad x_3 = -2 \text{ yields } e = -27,$$

$$x_1 = x_2 = 2, \quad x_3 = -1 \text{ yields } a = -4,$$

$$x_1 = x_2 = x_3 = 1 \qquad \text{yields } c = 18.$$

Thus, we obtain

$$D = -4C_1^3C_3 + C_1^2C_2^2 + 18C_1C_2C_3 - 4C_2^3 - 27C_3^2. \tag{4}$$

For particular symmetric polynomials there are more expedient methods to derive the representation in terms of the elementary symmetric polynomials, e.g. for the power sums.

The elementary symmetric polynomials C_i and the power sums S_i satisfy the following system of equations:

5.6.2. Newton's Relations.

$$S_k - C_1 S_{k-1} + C_2 S_{k-2} - \dots + (-1)^{k-1} C_{k-1} S_1 + (-1)^k k C_k = 0$$

$$\text{for } k \leqq n, \qquad (5)$$

$$S_l - C_1 S_{l-1} + C_2 S_{l-2} - \dots + (-1)^n C_n S_{l-n} = 0 \text{ for } l > n. \qquad (6)$$

Proof. In

$$z^n - C_1 z^{n-1} + C_2 z^{n-2} + \dots + (-1)^n C_n = (z - x_1)(z - x_2) \dots (z - x_n) \qquad (7)$$

we substitute $x_1, x_2, \dots, x_n$ for z and add all these equations. This gives (5) for $k = n$. To obtain (6), we multiply (7) by z^{l-n} before substituting $x_1, x_2, \dots, x_n$ for z and adding. It remains to prove (5) for $k < n$. Let k be a fixed natural number. We proceed by induction on the number of indeterminates. We already know that (5) holds for k indeterminates. Assuming that (5) is satisfied for the C_i and S_i in n indeterminates, we shall prove that (5) holds for the elementary symmetric polynomials and sums of powers of $n + 1$ indeterminates. Writing $C_i(x_1, \dots, x_n)$ and $C_i(x_1, \dots, x_n, x_{n+1})$ for the elementary symmetric polynomials in n and $n + 1$ indeterminates, respectively, we find that

$$C_i(x_1, \dots, x_n) = C_i(x_1, \dots, x_n, 0) \quad (i = 1, \dots, n).$$

Similarly we obtain

$$S_i(x_1, \dots, x_n) = S_i(x_1, \dots, x_n, 0) \quad (i = 1, \dots, n).$$

We now form the left-hand side of (5) for $n + 1$ indeterminates. So we obtain a symmetric polynomial $f(x_1, \dots, x_n, x_{n+1})$ which is obviously homogeneous of degree k. From the induction assumption it follows that $f(x_1, \dots, x_n, 0) = 0$. Hence, by 5.4.1, $f(x_1, \dots, x_n, x_{n+1})$ is divisible by x_{n+1}. Since $f(x_1, \dots, x_n, x_{n+1})$ is symmetric we conclude that it is also divisible by $x_1, x_2, \dots,$ and x_n. But as the degree k of f is less than $n + 1$ it follows that $f(x_1, \dots, x_n, x_{n+1})$ is the zero polynomial. This completes the proof.

Newton's Relations can be used to represent the S_i as polynomials in the C_j, namely

$$S_1 = C_1,$$
$$S_2 = C_1S_1 - 2C_2 = C_1^2 - 2C_2,$$
$$S_3 = C_1S_2 - C_2S_1 + 3C_3 = C_1^3 - 3C_1C_2 + 3C_3,$$
$$\dots\dots\dots\dots\dots\dots\dots\dots$$

If the coefficients belong to a field of characteristic 0, we can also express the C_i in terms of S_j:

$$C_1 = S_1,$$
$$C_2 = \frac{1}{2}S_1^2 - \frac{1}{2}S_2,$$
$$C_3 = \frac{1}{6}S_1^3 - \frac{1}{2}S_1S_2 + \frac{1}{3}S_3,$$
$$\dots\dots\dots\dots\dots\dots$$

This gives the following corollary to Theorem 5.6.1:

5.6.3. *Let* $\mathbf{R}$ *be an integral domain of characteristic* 0 *and* $\mathbf{F}$ *the field of quotients of* $\mathbf{R}$. *For every symmetric polynomial* $s(x_1, \dots, x_n)$ *of* $\mathbf{R}[x_1, \dots, x_n]$ *there exists a polynomial* $g \in \mathbf{F}[x_1, \dots, x_n]$ *such that* $s(x_1, \dots, x_n) = g(S_1, \dots, S_n)$.

5.7. The resultant and the discriminant

Let

$$f(z) = a_0z^m + a_1z^{m-1} + \dots + a_m \quad (a_0 \neq 0),$$
$$g(z) = b_0z^n + b_1z^{n-1} + \dots + b_n \quad (b_0 \neq 0)$$

be polynomials in an indeterminate z whose coefficients belong to an arbitrary field $\mathbf{K}$.

In § 6.3 we shall prove that there exist extension fields of $\mathbf{K}$ in which $f(z)$ and $g(z)$ have m and n roots, respectively, where every root is counted according to its multiplicity. Let $\mathbf{E}$ denote such an extension field.

Our aim is to derive a criterion for $f(z)$ and $g(z)$ to have at least one common root in $\mathbf{E}$.

If this is the case, then $f(z)$ and $g(z)$ have a greatest common divisor of positive degree in $\mathbf{E}[z]$. By Lemma 5.2.2, $f(z)$ and $g(z)$ then also have a greatest common divisor of positive degree in $\mathbf{K}[z]$.

5.7.1. Lemma. *The polynomials $f(z), g(z) \in \mathbf{K}[z]$ of degrees m and n, respectively, have a greatest common divisor of positive degree if and only if there exist two non-zero polynomials*

$$h(z) = c_0 z^{m-1} + c_1 z^{m-2} + \dots + c_{m-1},$$
$$k(z) = d_0 z^{n-1} + d_1 z^{n-2} + \dots + d_{n-1}$$

with coefficients in $\mathbf{K}$ whose degrees do not exceed $m - 1$ and $n - 1$, respectively, such that

$$k(z)\, f(z) = h(z)\, g(z). \tag{1}$$

Proof. Suppose that (1) holds with polynomials $h(z)$, $k(z)$ that satisfy the given conditions. We factorize both sides of (1) into prime polynomials of $\mathbf{K}[z]$. It is evident that not all the prime factors of $f(z)$ can divide $h(z)$ since the degree of $h(z)$ is less than that of $f(z)$. Therefore at least one prime factor of $f(z)$ divides $g(z)$.

Conversely, suppose that $f(z)$ and $g(z)$ have a common divisor $t(z) \in \mathbf{K}[z]$ of positive degree so that

$$f(z) = t(z)\, h(z), \quad g(z) = t(z)\, k(z).$$

Then, we obtain

$$k(z)\, f(z) = h(z)\, g(z)$$

where $h(z)$ and $k(z)$ satisfy the conditions of the lemma. This completes the proof.

Computing the products on both sides of (1) and equating the coefficients of corresponding powers of z we find

$$\begin{aligned}
d_0 a_0 &= c_0 b_0,\\
d_0 a_1 + d_1 a_0 &= c_0 b_1 + c_1 b_0,\\
d_0 a_2 + d_1 a_1 + d_2 a_0 &= c_0 b_2 + c_1 b_1 + c_2 b_0,\\
&\dots\dots\dots\dots\dots\dots\\
d_{n-2} a_m + d_{n-1} a_{m-1} &= c_{m-2} b_n + c_{m-1} b_{n-1},\\
d_{n-1} a_m &= c_{m-1} b_n.
\end{aligned}$$

We regard these equations as a system of $m+n$ linear homogeneous equations for the unknowns $-c_i$ and d_i. As is well known, there exists a non-zero solution if and only if the determinant is equal to zero. One usually consideres the transposed determinant, namely

$$R(f, g) = \begin{vmatrix} a_0 & a_1 & \dots & a_m & & & \\ & a_0 & a_1 & \dots & a_m & & \\ \dots & \dots & \dots & \dots & \dots & & \\ & & & a_0 & a_1 & \dots & a_m \\ b_0 & b_1 & \dots & b_n & & & \\ & b_0 & b_1 & \dots & b_n & & \\ & \dots & \dots & \dots & \dots & \dots & \\ & & & b_0 & b_1 & \dots & b_n \end{vmatrix} \begin{matrix} \left.\vphantom{\begin{matrix}a\\a\\a\\a\end{matrix}}\right\} n \text{ rows} \\ \left.\vphantom{\begin{matrix}a\\a\\a\\a\end{matrix}}\right\} m \text{ rows} \end{matrix}$$

where the blank spaces stand for zeros. This determinant $R(f, g)$ is called the *resultant* of $f(z)$ and $g(z)$. The determinant is also referred to as *Sylvester's determinant.*

As an immediate consequence of Lemma 5.7.1 we obtain:

5.7.2. THEOREM. *The polynomials $f(z)$ and $g(z)$ with coefficients in a field $\mathbf{K}$ have at least one common root in a suitable extension field of $\mathbf{K}$ if and only if the resultant $R(f, g) = 0$.*

Let us suppose that the a_i and b_j in the determinant $R(f, g)$ are indeterminates. From the elementary properties of determinants it follows that $R(f, g)$ is homogeneous of degree n in the a_i and homogeneous of degree m in the b_j. Therefore we can write

$$R(f, g) = a_0^n b_0^m P\left(\frac{a_1}{a_0}, \dots, \frac{a_m}{a_0}, \frac{b_1}{b_0}, \dots, \frac{b_n}{b_0}\right)$$

where P is a polynomial.

Now, let $x_1, \dots, x_m, y_1, \dots, y_n$ be indeterminates. We consider the polynomials

$$f(z) = a_0 \prod_{i=1}^{m} (z - x_i) = a_0 z^m + a_1 z^{m-1} + \dots + a_m,$$

$$g(z) = b_0 \prod_{j=1}^{n} (z - y_j) = b_0 z^n + b_1 z^{n-1} + \dots + b_n.$$

The quotients $\frac{a_1}{a_0}, \dots, \frac{a_m}{a_0}$ and $\frac{b_1}{b_0}, \dots, \frac{b_n}{b_0}$ are, except for the signs, the elementary symmetric polynomials of $x_1, \dots, x_m$ and $y_1, \dots, y_n$, respectively. Therefore $P\left(\frac{a_1}{a_0}, \dots, \frac{a_m}{a_0}, \frac{b_1}{b_0}, \dots, \frac{b_n}{b_0}\right)$ can be regarded as a polynomial in $x_1, \ldots, x_m, y_1, \ldots, y_n$. This polynomial is equal to zero if $f(z)$ and $g(z)$ have a common root, i.e. if $x_i = y_j$ for some pair i, j. This shows that P is divisible by all the differences $x_i - y_j$ so that P is a multiple of

$$\prod_{i=1}^{m} \prod_{j=1}^{n} (x_i - y_j).$$

It is readily seen that

$$S = a_0^n b_0^m \prod_{i=1}^{m} \prod_{j=1}^{n} (x_i - y_j) = a_0^n \prod_{i=1}^{m} g(x_i) = (-1)^{mn} b_0^m \prod_{j=1}^{n} f(y_j).$$

The representation of S as the product of the $g(x_i)$ shows that S is homogeneous of degree m in the b_j, and from the last product we infer that S is homogeneous of degree n in the a_i. Hence we obtain $P = cS$ where c is independent of the a_i and b_j. Comparing the terms $a_0^n b_n^m$ in P and S we find that $c = 1$. This gives the following expression for the resultant:

$$R(f, g) = a_0^n \prod_{i=1}^{m} g(x_i) = (-1)^{mn} b_0^m \prod_{j=1}^{n} f(y_j). \tag{2}$$

From this representation of $R(f, g)$, Theorem 5.7.2. becomes once more evident.

We now form the resultant of

$$f(z) = a_0 z^m + a_1 z^{m-1} + \ldots + a_m = a_0 \prod_{i=1}^{m} (z - x_i)$$

and its derivative. By (2) we obtain

$$R(f, f') = a_0^{m-1} \prod_{i=1}^{m} f'(x_i). \tag{3}$$

It follows from Theorem 5.7.2 that $R(f, f') = 0$ if and only if $f(z)$ has multiple roots.

The *discriminant* $D(f)$ of the polynomial $f(z)$ is defined as follows

$$D(f) = a_0^{2m-2} \prod_{i<j} (x_i - x_j)^2. \tag{4}$$

In view of

$$f'(x_i) = a_0 \prod_{\substack{j=1 \\ j \neq i}}^{m} (x_i - x_j)$$

we find

$$D(f) = (-1)^{m(m-1)/2} a_0^{m-2} \prod_{i=1}^{m} f'(x_i)$$

so that, by (3),

$$R(f, f') = (-1)^{m(m-1)/2} a_0 D(f). \tag{5}$$

For $a_0 = 1$, $D(f)$ is the discriminant

$$D = \prod_{i<j} (x_i - x_j)^2$$

as defined in § 5.6.

Exercises

1. Find the greatest common divisor of

$$x^8 + 3x^7 + 3x^6 + 3x^5 + x^4 + 2x^3 + 2x^2 + 2x + 1$$

and

$$x^4 + x^3 + 2x^2 + x + 1$$

and represent it as a linear combination.

2. Show that $(x^m - 1,\ x^n - 1) = x^d - 1$ where $(m, n) = d$. (Give a proof that does not refer to the roots.)
3. Let $f(x), g(x) \in \mathbf{K}[x]$ where $\mathbf{K}$ is any field. Prove the "chain rule"

$$[f(g(x))]' = f'(g(x))\, g'(x).$$

4. Express the polynomials $g_1(x), \ldots, g_t(x)$ introduced at the end of § 5.4 as quotients of greatest common divisors of $f(x)$ and its derivatives.
5. Find all polynomials with coefficients in a field that are divisible by their derivative.
6. Let $a_1, \ldots, a_n$ be distinct elements of a field $\mathbf{K}$.
   (i) Find polynomials $g_1(x), \ldots, g_n(x) \in \mathbf{K}[x]$ whose degrees do not exceed $n - 1$ such that

$$g_i(a_i) = 1, g_i(a_j) = 0 \text{ if } j \neq i \quad (i = 1, \ldots, n).$$

(ii) Let $b_1, \ldots, b_n$ be arbitrary elements of $\mathbf{K}$. Show that there is one and only one polynomial $g(x) \in \mathbf{K}[x]$ whose degree does not exceed $n - 1$ such that $g(a_i) = b_i$ for $i = 1, \ldots, n$.

7. (i) Let $\mathbf{K}$ be a finite field and let

$$\varphi: a \rightarrow a\varphi \quad (a \in \mathbf{K})$$

be a mapping of $\mathbf{K}$ into $\mathbf{K}$. Show that there is a polynomial $f(x) \in \mathbf{K}[x]$ such that $f(a) = a\varphi$ for every $a \in \mathbf{K}$.

(ii) Give an example to show that the finiteness condition in (i) cannot be dropped.

8. Let k be a natural number and put

$$\binom{x}{k} = \frac{x(x-1)\ldots(x-k+1)}{k!}, \quad \binom{x}{0} = 1.$$

(i) Show that any polynomial $f(x) \in \mathbf{Q}[x]$ of degree m has a unique representation

$$f(x) = c_0\binom{x}{0} + c_1\binom{x}{1} + \ldots + c_m\binom{x}{m}$$

with coefficients $c_i \in \mathbf{Q}$.

(ii) Show that $\binom{a}{k}$ is an integer for every integer a.

(iii) Prove that the coefficients c_i in (i) are integers if and only if $f(a)$ is an integer for every integer a.

9. Let $f(x) \in \mathbf{C}[x]$ be a polynomial of degree m and let $a_1, \ldots, a_{m+1}$ be distinct rational numbers such that $f(a_i)$ is a rational number. Show that the coefficients of $f(x)$ are rational numbers.

10. If $a_nx^n + a_{n-1}x^{n-1} + \ldots + a_0$ is an irreducible polynomial with coefficients in any field show that $a_0x^n + a_1x^{n-1} + \ldots + a_n$ is also irreducible.

11. Find all irreducible polynomials of degree 2 and 3 with coefficients in $\mathbf{Z}/(3)$.

12. Let $f(x) \in \mathbf{Z}[x]$. Show that no integer is a root of $f(x)$ if both $f(0)$ and $f(1)$ are odd. Prove that no rational number is a root of $f(x)$ if, moreover, the leading coefficient is odd.

13. Let $a_0, \ldots, a_m, b, c$ be elements of a unique factorization ring and $(b, c) = 1$. Show that $a_mx^m + \ldots + a_0$ is not divisible by $bx + c$ unless $c \mid a_0$ and $b \mid a_m$.

14. Let $f(x)$ be a polynomial of degree n with coefficients in $\boldsymbol{Z}$. Suppose that for $2n+1$ integers a_i the value $f(a_i)$ is equal to a positive or negative prime number or to ± 1. Prove that $f(x)$ is irreducible in $\boldsymbol{Q}[x]$.
15. Let $a_1, \ldots, a_n$ be distinct integers. Prove that

$$(x-a_1)\ldots(x-a_n)-1$$

is irreducible in $\boldsymbol{Q}[x]$.
16. Let $\mathbf{K}$ be any field and let $f(x) \in \mathbf{K}[x]$ be irreducible over $\mathbf{K}$. If z is an indeterminate prove that $f(x)$ is also irreducible over $\mathbf{K}(z)$.
17. If p is a prime number and n is any natural number $\geqq 2$ prove that $\sqrt[n]{p}$ is not rational.
18. Let $f(x)$ be an irreducible polynomial in $\boldsymbol{Q}[x]$. Suppose that in some extension field of $\boldsymbol{Q}$ there is an element α such that $f(\alpha) = f(-\alpha) = 0$. Prove that there exists a polynomial $g(x)$ in $\boldsymbol{Q}[x]$ such that $f(x) = g(x^2)$.
19. (i) Evaluate Vandermonde's determinant, namely

$$\begin{vmatrix} 1 & 1 & \ldots 1 \\ x_1 & x_2 & \ldots x_n \\ x_1^2 & x_2^2 & \ldots x_n^2 \\ \ldots & \ldots & \ldots \\ x_1^{n-1} & x_2^{n-1} & \ldots x_n^{n-1} \end{vmatrix} = \prod_{i>k}(x_i - x_k).$$

(ii) Show that

$$\prod_{i>k}(x_i - x_k)^2 = \begin{vmatrix} n & S_1 & S_2 & \ldots S_{n-1} \\ S_1 & S_2 & S_3 & \ldots S_n \\ S_2 & S_3 & S_4 & \ldots S_{n+1} \\ \ldots & \ldots & \ldots & \ldots \\ S_{n-1} & S_n & S_{n+1} & \ldots S_{2n-2} \end{vmatrix}$$

where the S_i denote the power sums of $x_1, x_2, \ldots, x_n$.
20. Let $\mathbf{P}$ be the integral domain of the polynomials in the indeterminates x, y with coefficients in any field. Let (x, y) be the ideal of $\mathbf{P}$ generated by x and y, i.e. the intersection of all ideals of $\mathbf{P}$ containing x and y. Show that both (x) and (x, y) are prime ideals and determine $\mathbf{P}/(x)$ and $\mathbf{P}/(x, y)$. (This yields an example of a prime ideal which is not maximal for obviously (x) is properly contained in (x, y).)

6

Fields

In this chapter we shall study and construct extension fields. We shall be concerned primarily with finite extensions, i.e. those which are algebras of finite rank. The automorphisms of such extensions will play an important role.

6.1. Adjunction

If F is a subfield of a field E, then E is said to be an *extension field* of F. Let S be a set of elements of E. It is clear that there are subfields of E that contain S, e.g. E itself. By $\mathsf{F}(S)$ we denote the intersection of all subfields of E that contain S. It is obvious that $\mathsf{F}(S)$ is a subfield of E. We say that $\mathsf{F}(S)$ is obtained from F by the *adjunction* of S to F.

If a subfield of E contains S, then it contains all polynomials in finitely many elements of S with coefficients in F and the quotients of such polynomials. We shall refer to these elements as the rational expressions of elements of S with coefficients in F. It is evident that these rational expressions form a field. Therefore $\mathsf{F}(S)$ coincides with the set of all rational expressions of elements of S with coefficients in F.

If S consists of finitely many elements, $\alpha_1, \ldots, \alpha_n$ say, then we write $\mathsf{F}(\alpha_1, \ldots, \alpha_n)$.

For two sets S_1, S_2 of elements of E we have

$$\mathsf{F}(S_1 \cup S_2) = (\mathsf{F}(S_1))\,(S_2).$$

Indeed, from

$$\mathbf{F}(S_1) \subseteq \mathbf{F}(S_1 \cup S_2), \quad S_2 \subseteq \mathbf{F}(S_1 \cup S_2)$$

we conclude that

$$(\mathbf{F}(S_1))\,(S_2) \subseteq \mathbf{F}(S_1 \cup S_2);$$

on the other hand

$$\mathbf{F} \subseteq (\mathbf{F}(S_1))\,(S_2), \quad S_1 \cup S_2 \subseteq (\mathbf{F}\,(S_1))\,(S_2)$$

imply that

$$\mathbf{F}(S_1 \cup S_2) \subseteq (\mathbf{F}(S_1))\,(S_2).$$

Instead of $\mathbf{F}(S_1 \cup S_2)$ we also write $\mathbf{F}(S_1, S_2)$.

An extension is called *simple* if it is obtained by adjoining a single element.

Let $\vartheta \in \mathbf{E}$. Then ϑ is called a *generating* or *primitive* element of the simple extension $\mathbf{F}(\vartheta)$. The set of all polynomials

$$g(\vartheta) = c_0 + c_1\vartheta + \ldots + c_k\vartheta^k$$

with coefficients $c_i \in \mathbf{F}$ form an integral domain $\mathbf{F}[\vartheta]$ contained in $\mathbf{F}(\vartheta)$. It is evident that $\mathbf{F}(\vartheta)$ is the field of quotients of $\mathbf{F}[\vartheta]$.

We now consider the integral domain $\mathbf{F}[x]$ of all polynomials in an indeterminate x over $\mathbf{F}$. It is clear that the mapping $g(x) \to g(\vartheta)$, i.e.

$$c_0 + c_1x + \ldots + c_kx^k \to c_0 + c_1\vartheta + \ldots + c_k\vartheta^k$$

is a homomorphism of $\mathbf{F}[x]$ onto $\mathbf{F}[\vartheta]$. The kernel of this homomorphism is an ideal P of $\mathbf{F}[x]$, and

$$\mathbf{F}[x]/P \cong \mathbf{F}[\vartheta].$$

The ideal P consists of all polynomials $h(x) \in \mathbf{F}[x]$ for which $h(\vartheta) = 0$. Since $\mathbf{F}[\vartheta]$ is an integral domain, P is a prime ideal. Moreover, $P \neq \mathbf{F}[x]$ since the polynomials of degree 0 in $\mathbf{F}[x]$, i.e. the non-zero elements of $\mathbf{F}$, do certainly not belong to P. Since $\mathbf{F}[x]$ is a principal ideal ring we conclude that $P = (0)$ or $P = (p(x))$ where $p(x)$ is an irreducible polynomial in $\mathbf{F}[x]$. We consider these two cases separately.

(i) $P = (0)$. Then we have $\mathbf{F}[x] \cong \mathbf{F}[\vartheta]$. Forming the fields of quotients of $\mathbf{F}[x]$ and $\mathbf{F}[\vartheta]$ we infer that

$$\mathbf{F}(x) \cong \mathbf{F}(\vartheta).$$

In this case, ϑ is called *transcendental* over $\mathbf{F}$, and $\mathbf{F}(\vartheta)$ is said to be a *simple transcendental extension* of $\mathbf{F}$. Thus, an element ϑ of $\mathbf{E}$ that is transcendental over $\mathbf{F}$ is characterized by the property that the powers of ϑ are linearly independent over $\mathbf{F}$ or, in other words, that ϑ is not a root of any polynomial of $\mathbf{F}[x]$.

(ii) $P = (p(x))$ where $p(x) \in \mathbf{F}[x]$ is irreducible in $\mathbf{F}[x]$ and $p(\vartheta) = 0$. By 4.4.3, $\mathbf{F}[x]/(p(x))$ is a field so that $\mathbf{F}[\vartheta]$ coincides with its field of quotients, i.e.

$$\mathbf{F}[\vartheta] = \mathbf{F}(\vartheta) \cong \mathbf{F}[x]/(p(x)).$$

In this case, ϑ is called *algebraic* over $\mathbf{F}$, and we say that $\mathbf{F}(\vartheta)$ is a *simple algebraic extension of* $\mathbf{F}$. From $\mathbf{F}(\vartheta) = \mathbf{F}[\vartheta]$ it follows, in particular, that every rational expression in ϑ with coefficients in $\mathbf{F}$ is equal to a polynomial in ϑ with coefficients in $\mathbf{F}$.

The polynomial $p(x)$ is not uniquely determined by the ideal P but only up to non-zero factors of $\mathbf{F}$, i.e. $p(x)$ may be replaced by any associate in $\mathbf{F}[x]$. Therefore we may assume that $p(x)$ is a monic polynomial. This monic polynomial is called the *minimal polynomial* of ϑ over $\mathbf{F}$.

The minimal polynomial $p(x)$ of ϑ over $\mathbf{F}$ can also be characterized as follows: $p(x)$ is *the* monic polynomial of least degree in $\mathbf{F}[x]$ with the root ϑ. Indeed, since $p(x)$ is irreducible it follows from 5.4.6 that every polynomial in $\mathbf{F}[x]$ with the root ϑ must be divisible by $p(x)$ so that its degree is at least equal to that of $p(x)$. Moreover, there cannot exist two monic polynomials $p(x)$ and $p_1(x)$ of the same degree in $\mathbf{F}[x]$ such that $p(\vartheta) = p_1(\vartheta) = 0$; for otherwise, the difference $p(x) - p_1(x)$ would be a non-zero polynomial with the root ϑ whose degree is less than the degree of $p(x)$.

If

$$p(x) = x^n + a_1 x^{n-1} + \dots + a_n,$$

then every residue class mod $p(x)$ can be represented by one and only one polynomial

$$b_0 + b_1 x + \dots + b_{n-1} x^{n-1} \quad (b_i \in \mathbf{F})$$

whose degree does not exceed $n - 1$. Due to $\mathsf{F}(\vartheta) \cong \mathsf{F}[x]/(p(x))$, every element of $\mathsf{F}(\vartheta)$ has a unique representation

$$b_0 + b_1\vartheta + \dots + b_{n-1}\vartheta^{n-1}$$

where $b_i \in \mathsf{F}$. If we wish to emphasize that the minimal polynomial of ϑ over F is of degree n, then we say that ϑ is *algebraic of degree n over* F.

The terms *transcendental* and *algebraic number* without reference to a particular field always mean a real or complex number that is transcendental or algebraic, respectively, over the field of the rational numbers. One can prove that, for instance, the numbers π and e are transcendental numbers, whereas i, $\sqrt{2}$, and $\sqrt[5]{3 + i}$ are obviously algebraic numbers.

To give an example of case (i) above, we make use of the fact that π is a transcendental number. The field $\boldsymbol{Q}(x)$ of all rational functions in an indeterminate x with coefficients in the field $\boldsymbol{Q}$ of the rational numbers is isomorphic to the field $\boldsymbol{Q}(\pi)$ of all rational expressions of π with rational coefficients. An isomorphism of $\boldsymbol{Q}(x)$ onto $\boldsymbol{Q}(\pi)$ is obtained by substituting π for x in every rational function in $\boldsymbol{Q}(x)$.

We now consider an example of a finite algebraic extension of the field $\boldsymbol{R}$ of the real numbers. The minimal polynomial of i over $\boldsymbol{R}$ is $x^2 + 1$. Hence, by the arguments in case (ii), the extension field $\boldsymbol{C} = \boldsymbol{R}(i)$ is isomorphic to the residue class field $\boldsymbol{R}[x]/(x^2 + 1)$. We shall exhibit this isomorphism once more without referring to the general case. Every residue class of $\boldsymbol{R}[x]$ mod $(x^2 + 1)$ can be represented by one and only one linear polynomial $a + bx$. To the residue class of $a + bx$ mod $(x^2 + 1)$ we assign the complex number $a + bi$:

$$a + bx \bmod (x^2 + 1) \to a + bi. \tag{1}$$

It is clear that this is a one-to-one mapping of $\boldsymbol{R}[x]/(x^2 + 1)$ onto $\boldsymbol{R}(i) = \boldsymbol{C}$, and it is also evident that to the sum of two residue classes mod $(x^2 + 1)$ there corresponds the sum of the respective complex numbers. As to multiplication, we have

$$\begin{aligned}(a + bx)(c + dx) &= ac + (ad + bc)\,x + bdx^2 \\ &\equiv (ac - bd) + (ad + bc)\,x \,\big(\mathrm{mod}\ (x^2 + 1)\big)\end{aligned}$$

and, on the other hand,

$$(a + bi)(c + di) = (ac - bd) + (ad + bc)\,i\,.$$

This shows that (1) is an isomorphism of $\boldsymbol{R}[x]/(x^2 + 1)$ onto $\boldsymbol{R}(i)$.

Two extension fields E and E' of F are called *isomorphic over* F if there is an isomorphism σ of E onto E' which leaves every element of F fixed, i.e. which carries every element of F into itself. We refer to σ as an *isomorphism of* E *onto* E' *over* F or, briefly, as an isomorphism of E/F onto E'/F.

6.1.1. *Any two simple transcendental extensions of* F *are isomorphic over* F.

Proof. Every simple transcendental extension of F is isomorphic to the field $\mathsf{F}(x)$ of the rational functions of an indeterminate x.

6.1.2. *If* ϑ *and* ϑ' *are roots of one and the same irreducible polynomial of* $\mathsf{F}[x]$, *then there exists an isomorphism of* $\mathsf{F}(\vartheta)/\mathsf{F}$ *onto* $\mathsf{F}(\vartheta')/\mathsf{F}$ *which carries* ϑ *into* ϑ'. *Conversely, if* $\mathsf{F}(\vartheta)$ *and* $\mathsf{F}(\vartheta')$ *are simple algebraic extensions such that there is an isomorphism of* $\mathsf{F}(\vartheta)/\mathsf{F}$ *onto* $\mathsf{F}(\vartheta')/\mathsf{F}$ *which carries* ϑ *into* ϑ', *then* ϑ *and* ϑ' *are roots of the same irreducible polynomial in* $\mathsf{F}[x]$.

Proof. Suppose that ϑ and ϑ' are roots of the irreducible polynomial $p(x) \in \mathsf{F}[x]$. Then there exist isomorphisms σ and σ' of $\mathsf{F}[x]/(p(x))$ onto $\mathsf{F}(\vartheta)$ and $\mathsf{F}(\vartheta')$, respectively. Both σ and σ' leave F fixed, and the residue class of x mod $p(x)$ is mapped onto ϑ and ϑ', respectively. Therefore $\sigma^{-1}\sigma'$ is an isomorphism with the required property. Conversely, suppose that there exists an isomorphism τ of $\mathsf{F}(\vartheta)/\mathsf{F}$ onto $\mathsf{F}(\vartheta')/\mathsf{F}$ such that $\vartheta^\tau = \vartheta'$. By applying τ to an equation $p(\vartheta) = 0$, where $p(x) \in \mathsf{F}[x]$, we obtain $(p(\vartheta))^\tau = p(\vartheta^\tau) = 0$. This completes the proof.

6.2. Algebraic extension fields

Let E be an extension field of F. The elements $\omega_1, \ldots, \omega_r$ of E are called *linearly dependent over* F if there are elements $a_1, \ldots, a_r$ of F

of which at least one is distinct from zero such that

$$a_1\omega_1 + \dots + a_r\omega_r = 0.$$

Otherwise, $\omega_1, \dots, \omega_r$ are said to be *linearly independent* over $\mathbf{F}$.

An infinite set of elements of $\mathbf{E}$ is called linearly independent over $\mathbf{F}$ if every finite subset is linearly independent over $\mathbf{F}$. If $\mathbf{E}$ contains an infinite set of elements which are linearly independent over $\mathbf{F}$, then $\mathbf{E}$ is said to be an extension of $\mathbf{F}$ of *infinite degree*. For example, every simple transcendental extension $\mathbf{F}(x)$ is of infinite degree over $\mathbf{F}$ because $1, x, x^2, \dots$ are linearly independent over $\mathbf{F}$.

If there exists a natural number n such that $\mathbf{E}$ contains n elements that are linearly independent over $\mathbf{F}$ whereas every set of more than n elements is linearly dependent over $\mathbf{F}$, then $\mathbf{E}$ is called an extension field of *degree* n of $\mathbf{F}$. We write $[\mathbf{E}:\mathbf{F}] = n$.

An extension field $\mathbf{E}$ of $\mathbf{F}$ with $[\mathbf{E}:\mathbf{F}] = n$ can be regarded as an n-dimensional vector space over $\mathbf{F}$ and is therefore an algebra of rank n over $\mathbf{F}$. Indeed, one readily verifies that the postulates for a vector space and for an algebra are satisfied because $\mathbf{E}$ is a commutative ring and $\mathbf{F}$ a subfield of $\mathbf{E}$. Every basis of $\mathbf{E}$ as an algebra over $\mathbf{F}$ is called a *basis of* $\mathbf{E}$ *over* $\mathbf{F}$. It is evident that $[\mathbf{E}:\mathbf{F}] = 1$ if and only if $\mathbf{E} = \mathbf{F}$.

6.2.1. *If* $\mathbf{F}(\vartheta)$ *is a simple algebraic extension and if* ϑ *is of degree* n *over* $\mathbf{F}$, *then* $[\mathbf{F}(\vartheta):\mathbf{F}] = n$.

Proof. Every element of $\mathbf{F}(\vartheta)$ has a unique representation

$$b_0 + b_1\vartheta + \dots + b_{n-1}\vartheta^{n-1}$$

where the b_i are elements of $\mathbf{F}$. Therefore $1, \vartheta, \dots, \vartheta^{n-1}$ form a basis of $\mathbf{F}(\vartheta)$ over $\mathbf{F}$, and our proposition is proved.

An extension field $\mathbf{E}$ of $\mathbf{F}$ is said to be *algebraic* over $\mathbf{F}$ if every element of $\mathbf{E}$ is algebraic over $\mathbf{F}$.

6.2.2. *Every extension field* $\mathbf{E}$ *of finite degree over* $\mathbf{F}$ *is algebraic over* $\mathbf{F}$.

Proof. Let $[\mathbf{E}:\mathbf{F}] = n$. Then, for any element α of $\mathbf{E}$, the powers $1 = \alpha^0, \alpha, \alpha^2, \dots, \alpha^n$ are linearly dependent over $\mathbf{F}$. Therefore there exist elements $c_0, c_1, c_2, \ldots, c_n$ of F, not all equal to zero, such that

$$c_0 + c_1\alpha + c_2\alpha^2 + \ldots + c_n\alpha^n = 0.$$

In other words, α is a root of a polynomial with coefficients in F. This completes the proof.

Moreover, we have obtained the additional result that the degree of any element of E over F does not exceed $n = [\mathsf{E}: \mathsf{F}]$.

6.2.3. Theorem. *Let K be an extension field of E and E an extension field of F such that $[\mathsf{K}: \mathsf{E}]$ and $[\mathsf{E}: \mathsf{F}]$ are finite. Then*

$$[\mathsf{K}: \mathsf{F}] = [\mathsf{K}: \mathsf{E}]\,[\mathsf{E}: \mathsf{F}].$$

Proof. Let $[\mathsf{K}: \mathsf{E}] = m$, $[\mathsf{E}: \mathsf{F}] = n$. We choose bases

$$\alpha_1, \ldots, \alpha_m \text{ and } \beta_1, \ldots, \beta_n$$

of K over E and E over F, respectively. Any element $\varkappa \in \mathsf{K}$ has a unique representation

$$\varkappa = \varepsilon_1\alpha_1 + \ldots + \varepsilon_m\alpha_m$$

where $\varepsilon_1, \ldots, \varepsilon_m$ belong to E. Further, each ε_i can be uniquely expressed in the form

$$\varepsilon_i = c_{i1}\beta_1 + \ldots + c_{in}\beta_n$$

with elements c_{ij} of F. Therefore we obtain

$$\varkappa = \sum_{i=1}^{m} \sum_{j=1}^{n} c_{ij}\alpha_i\beta_j.$$

Moreover, the mn products $\alpha_i\beta_j$ are linearly independent over F. For suppose that

$$\sum_{i=1}^{m} \sum_{j=1}^{n} d_{ij}\alpha_i\beta_j = 0$$

where $d_{ij} \in \mathsf{F}$. This gives

$$\sum_{i=1}^{m} \left(\sum_{j=1}^{n} d_{ij}\beta_j \right) \alpha_i = \sum_{i=1}^{m} \delta_i\alpha_i = 0$$

where the elements

$$\delta_i = \sum_{j=1}^{n} d_{ij}\beta_j$$

belong to $\mathbf{E}$. Since $\alpha_1, \ldots, \alpha_m$ are linearly independent over $\mathbf{E}$ it follows that $\delta_i = 0$ for $i = 1, \ldots, m$. But since $\beta_1, \ldots, \beta_n$ are linearly independent over $\mathbf{F}$ we conclude that $d_{ij} = 0$ for $i = 1, \ldots, m$ and $j = 1, \ldots, n$. This shows that the mn elements $\alpha_i\beta_j$ form a basis of $\mathbf{K}$ over $\mathbf{F}$ so that

$$[\mathbf{K}:\mathbf{F}] = mn = [\mathbf{K}:\mathbf{E}]\,[\mathbf{E}:\mathbf{F}]$$

as was to be proved.

6.2.4. THEOREM. *Every extension of finite degree over* $\mathbf{F}$ *can be obtained by the adjunction of finitely many algebraic elements. Conversely, every extension field of* $\mathbf{F}$ *that is obtained by the adjunction of finitely many algebraic elements to* $\mathbf{F}$ *is of finite degree and, hence, algebraic over* $\mathbf{F}$.

Proof. Let $\mathbf{E}$ be of finite degree over $\mathbf{F}$. Then $\mathbf{E}$ is algebraic over $\mathbf{F}$ and can be obtained by adjoining a basis of $\mathbf{E}$ over $\mathbf{F}$ to $\mathbf{F}$. This proves the first part of the theorem.

Conversely, let $\alpha_1, \ldots, \alpha_r$ be elements of some extension field of $\mathbf{F}$ that are algebraic over $\mathbf{F}$. We have to prove that $\mathbf{F}(\alpha_1, \ldots, \alpha_r)$ is of finite degree over $\mathbf{F}$. The field $\mathbf{F}(\alpha_1, \ldots, \alpha_r)$ can be obtained by successive adjunction of $\alpha_1, \ldots, \alpha_r$ to $\mathbf{F}$. So we consider the sequence

$$\mathbf{F} \subseteq \mathbf{F}(\alpha_1) \subseteq \mathbf{F}(\alpha_1, \alpha_2) \subseteq \ldots \subseteq \mathbf{F}(\alpha_1, \ldots, \alpha_{r-1}) \subseteq \mathbf{F}(\alpha_1, \ldots, \alpha_r) \qquad (1)$$

of extension fields. It is clear that

$$[\mathbf{F}(\alpha_1, \ldots, \alpha_{i-1}, \alpha_i) : \mathbf{F}(\alpha_1, \ldots, \alpha_{i-1})] \leqq [\mathbf{F}(\alpha_i) : \mathbf{F}]$$

because the minimal polynomial of α_i over $\mathbf{F}$ might be reducible over $\mathbf{F}(\alpha_1, \ldots, \alpha_{i-1})$. This shows that in the sequence (1) each field is an extension of finite degree of the preceding one. Thus, it follows from Theorem 6.2.3 that the degree of $\mathbf{F}(\alpha_1, \ldots, \alpha_r)$ over $\mathbf{F}$ is finite. So the second part of the theorem is proved.

6.2.5. COROLLARY. *Let* α *and* β *be elements of an extension field* $\mathbf{E}$ *of* $\mathbf{F}$. *If* α *and* β *are algebraic over* $\mathbf{F}$, *then so are* $\alpha \pm \beta$, $\alpha\beta$ *and* α/β *if* $\beta \neq 0$.

Proof. By Theorem 6.2.4, $\mathbf{F}(\alpha, \beta)$ is algebraic over $\mathbf{F}$ if both α and β are algebraic over $\mathbf{F}$. Now, $\alpha \pm \beta$, $\alpha\beta$, and α/β if $\beta \neq 0$ are contained in $\mathsf{F}(\alpha, \beta)$ and hence algebraic over F. This proves the corollary.

It follows that the elements of any extension field E of F which are algebraic over F form a subfield of E. Of course, this subfield coincides with E if E itself is algebraic over F. In particular, the algebraic numbers form a subfield of the field C of all complex numbers.

Another consequence of Theorem 6.2.4 is the following:

6.2.6. THEOREM. *If α is algebraic over the field E and if E is algebraic over F, then α is algebraic over F.*

Proof. Since α is algebraic over E, there are polynomials in $\mathsf{E}[x]$ which have α as a root, e.g. the minimal polynomial

$$x^m + \gamma_1 x^{m-1} + \ldots + \gamma_m$$

of α over E. Therefore α is algebraic over the field $\mathsf{F}(\gamma_1, \ldots, \gamma_m)$ so that the degree

$$[\mathsf{F}(\gamma_1, \ldots, \gamma_m, \alpha) : \mathsf{F}(\gamma_1, \ldots, \gamma_m)]$$

is finite. The elements $\gamma_1, \ldots, \gamma_m$ are algebraic over F because E is algebraic over F. It follows from Theorem 6.2.4 that $[\mathsf{F}(\gamma_1, \ldots, \gamma_m) : \mathsf{F}]$ is finite. Hence, by Theorem 6.2.3, $\mathsf{F}(\gamma_1, \ldots, \gamma_m, \alpha)$ is of finite degree over F. Due to 6.2.2, this implies that α is algebraic over F as was to be proved.

6.3. Construction of extension fields

So far we assumed that we are given an extension field E of F, and adjunction of elements to F took place within E. Therefore all adjunctions led to well-defined extension fields of F. On the other hand, we could not expect that within E there exist extension fields of F with given properties. For instance, if E happens to be algebraic over F, then there is no simple transcendental extension of F within E. Or if $g(x)$ is a given polynomial in $\mathsf{F}[x]$, then there need not exist a simple algebraic extension $\mathsf{F}(\vartheta)$ within E such that $g(\vartheta) = 0$, for E need not contain a root of $g(x)$.

We now take up the question whether extension fields of F with given properties exist.

6.3.1. THEOREM. *If F is an arbitrary field, then there exists a simple transcendental extension of F. Any two simple transcendental extensions of F are isomorphic over F.*

Proof. Let $\mathsf{F}(x)$ be the field of all rational functions of an indeterminate x with coefficients in F, i.e. the field of quotients of the ring $\mathsf{F}[x]$ of polynomials in x with coefficients of F. Since x is an indeterminate it is evident that $\mathsf{F}(x)$ is a simple transcendental extension of F. For the uniqueness of $\mathsf{F}(x)$ up to isomorphism over F we refer to 6.1.1. This proves our theorem.

We now consider algebraic extensions. As we saw in § 6.2, to a simple algebraic extension $\mathsf{F}(\vartheta)$ there belongs an irreducible polynomial $p(x) \in \mathsf{F}[x]$ such that $p(\vartheta) = 0$. To prove the existence of such an extension we start from a given irreducible polynomial $p(x) \in \mathsf{F}[x]$ and set out to construct an extension field of F in which $p(x)$ has a root.

6.3.2. THEOREM. *Let $p(x)$ be an irreducible polynomial of degree $n > 1$ in $\mathsf{F}[x]$. There exists a simple algebraic extension $\mathsf{F}(\vartheta)$ such that $p(\vartheta) = 0$. If $\mathsf{F}(\vartheta)$ and $\mathsf{F}(\vartheta')$ are two extension fields such that ϑ and ϑ' are roots of the same irreducible polynomial in $\mathsf{F}[x]$, then $\mathsf{F}(\vartheta)$ and $\mathsf{F}(\vartheta')$ are isomorphic over F.*

Proof. Since $p(x)$ is irreducible the residue class ring $\mathsf{F}[x]/(p(x))$ is a field. We shall denote the residue class mod $p(x)$ of a polynomial $f(x) \in \mathsf{F}[x]$ by $\overline{f(x)}$. If a and b are any two distinct elements of F, then their residue classes $\bar{a}$ and $\bar{b}$ are distinct because $a - b$ is not divisible by $p(x)$. Therefore the residue classes mod $p(x)$ that are represented by elements of F form a subfield $\bar{\mathsf{F}}$ of $\mathsf{F}[x]/(p(x))$ which is isomorphic to F. We now form the extension field $\bar{\mathsf{F}}(\bar{x})$ by adjoining the residue class $\bar{x}$ to $\bar{\mathsf{F}}$ within $\mathsf{F}[x]/(p(x))$. If

$$p(x) = a_0x^n + a_1x^{n-1} + \ldots + a_n,$$

then we obtain

$$\begin{aligned}\bar{p}(\bar{x}) &= \bar{a}_0\bar{x}^n + \bar{a}_1\bar{x}^{n-1} + \ldots + \bar{a}_n \\ &= \overline{a_0x^n + a_1x^{n-1} + \ldots + a_n} = \overline{p(x)} = \bar{0}.\end{aligned}$$

This shows that the element $\bar{x}$ of $\mathsf{F}[x]/(p(x))$ is a root of the polynomial

$$\bar{p}(x) = \bar{a}_0 x^n + \bar{a}_1 x^{n-1} + \dots + \bar{a}_n .$$

Consequently $\bar{\mathsf{F}}(\bar{x})$ is a simple algebraic extension of $\bar{\mathsf{F}}$ such that $\bar{x}$ is a root of the polynomial $\bar{p}(x)$ which corresponds to $p(x)$ under the isomorphism of F onto $\bar{\mathsf{F}}$.

Due to the isomorphism between F and $\bar{\mathsf{F}}$ we may identify these fields. Denoting the residue class of x mod $p(x)$ by ϑ we then obtain an extension $\mathsf{F}(\vartheta)$ of F such that

$$p(\vartheta) = a_0 \vartheta^n + a_1 \vartheta^{n-1} + \dots + a_n = 0 .$$

The last proposition of the theorem follows from 6.1.2. This completes the proof.

We shall refer to this construction of $\mathsf{F}(\vartheta)$ as the adjunction of a root ϑ of $p(x)$ to F.

Next we shall deal with the construction of two other extension fields of F with given properties. To prove the uniqueness up to isomorphism over F of these fields we need a lemma about the possibility of extending an isomorphism between two fields to an isomorphism between simple algebraic extensions.

Let K and $\bar{\mathsf{K}}$ be two fields and let τ_0 be an isomorphism of K onto $\bar{\mathsf{K}}$. Then τ_0 induces an isomorphism of $\mathsf{K}[x]$ onto $\bar{\mathsf{K}}[x]$ if we apply τ_0 to the coefficients of the polynomials in $\mathsf{K}[x]$. If $p(x)$ is an irreducible polynomial in $\mathsf{K}[x]$, then its image $\bar{p}(x)$ under τ_0 is obviously an irreducible polynomial in $\bar{\mathsf{K}}[x]$.

Let E and $\bar{\mathsf{E}}$ be two isomorphic extension fields of K and $\bar{\mathsf{K}}$, respectively. An isomorphism τ of E onto $\bar{\mathsf{E}}$ is called an *extension* of τ_0 if τ coincides with τ_0 for the elements of K.

6.3.3. THEOREM. *Let τ_0 be an isomorphism of the field K onto the field $\bar{\mathsf{K}}$. Let $\mathsf{K}(\vartheta)$ and $\bar{\mathsf{K}}(\vartheta)$ be two simple algebraic extensions such that ϑ is a root of an irreducible polynomial $p(x) \in \mathsf{K}[x]$ and $\bar{\vartheta}$ a root of the image $\bar{p}(x)$ of $p(x)$ under τ_0. Then it is possible to extend τ_0 to an isomorphism τ_1 of $\mathsf{K}(\vartheta)$ onto $\bar{\mathsf{K}}(\bar{\vartheta})$ such that τ_1 carries ϑ into $\bar{\vartheta}$.*

Proof. Evidently τ_0 induces an isomorphism τ_0^* of $\mathbf{K}[x]/(p(x))$ onto $\overline{\mathbf{K}}[x]/(\overline{p}(x))$. Note that τ_0^* carries the residue class mod $p(x)$ of any element $a \in \mathbf{K}$ into the residue class mod $\overline{p}(x)$ of the image $\overline{a}$ of a under τ_0. By 6.1.2, there exist isomorphisms

$$\sigma\colon \mathbf{K}[x]/(p(x)) \to \mathbf{K}(\vartheta),$$
$$\overline{\sigma}\colon \overline{\mathbf{K}}[x]/(\overline{p}(x)) \to \overline{\mathbf{K}}(\overline{\vartheta}).$$

It is now easy to see that the mapping $\tau_1 = \sigma^{-1}\tau_0^*\overline{\sigma}$ is an isomorphism of $\mathbf{K}(\vartheta)$ onto $\overline{\mathbf{K}}(\overline{\vartheta})$ with the required property. First, it is evident that τ_1 is an isomorphism of $\mathbf{K}(\vartheta)$ onto $\overline{\mathbf{K}}(\overline{\vartheta})$. Let $a \in \mathbf{K}$. Then σ^{-1} carries a into the residue class of $a \bmod p(x)$; next, τ_0^* carries the residue class of $a \bmod p(x)$ into the residue class of $\overline{a} \bmod \overline{p}(x)$ where $\overline{a}$ denotes the image of a under τ_0; finally, $\overline{\sigma}$ carries the residue class of $\overline{a} \bmod \overline{p}(x)$ into the element $\overline{a}$ of $\overline{\mathbf{K}}$. Hence, τ_1 is an extension of τ_0. The image of ϑ under σ^{-1} is the residue class of $x \bmod p\,(x)$; this residue class is mapped under τ_0^* onto the residue class of $x \bmod \overline{p}(x)$, and $\overline{\sigma}$ carries the residue class of $x \bmod \overline{p}(x)$ into $\overline{\vartheta}$. This completes the proof.

Despite this formal proof the reader should realize that the last theorem is actually evident. Indeed, if the field $\mathbf{K}$ is given then all information about $\mathbf{K}(\vartheta)$ can be derived from the single equation

$$p(\vartheta) = a_0\vartheta^n + a_1\vartheta^{n-1} + \ldots + a_n = 0. \tag{1}$$

The field $\overline{\mathbf{K}}$ is isomorphic to $\mathbf{K}$, and all calculations in $\overline{\mathbf{K}}(\overline{\vartheta})$ are completely determined by

$$\overline{p}(\overline{\vartheta}) = \overline{a}_0\overline{\vartheta}^n + \overline{a}_1\overline{\vartheta}^{n-1} + \ldots + \overline{a}_n = \overline{0}$$

which differs from (1) only by the notation.

Now let $\mathbf{F}$ be a field and $f(x)$ a polynomial of degree $n > 0$ in $\mathbf{F}[x]$. Our aim is to construct an extension field of $\mathbf{F}$ which contains all the roots of $f(x)$. We decompose $f(x)$ in $\mathbf{F}[x]$ into irreducible factors:

$$f(x) = f_1(x)\, f_2(x) \ldots f_r(x).$$

If all these irreducible factors are of degree 1, then $\mathbf{F}$ already contains all the roots of $f(x)$. Otherwise, suppose that the degree of $f_1(x)$ is greater than 1. We then adjoin a root α_1 of $f_1(x)$ to $\mathbf{F}$. In

$\mathbf{F}(\alpha_1)\,[x]$ we obtain the following factorization:

$$f(x) = (x - \alpha_1)\, g_1(x)\, g_2(x) \ldots g_s(x)$$

where $g_1(x), g_2(x), \ldots, g_s(x)$ are irreducible over $\mathbf{F}(\alpha_1)$. If all the $g_i(x)$ are of degree 1, then $\mathbf{F}(\alpha_1)$ contains all the roots of $f(x)$. Otherwise, let $g_1(x)$ be of a degree greater than 1. By adjoining a root α_2 of $g_1(x)$ to $\mathbf{F}(\alpha_1)$ we obtain the field $\mathbf{F}(\alpha_1, \alpha_2)$, and the factorization of $f(x)$ into irreducible polynomials in $\mathbf{F}(\alpha_1, \alpha_2)\,[x]$ has the following form:

$$f(x) = (x - \alpha_1)\,(x - \alpha_2)\, h_1(x) \ldots h_t(x).$$

If not all the irreducible factors $h_i(x)$ are of degree 1 we continue in the same way. Each adjunction of a root of an irreducible factor of $f(x)$ gives rise to at least one new linear factor. Hence, after a finite number of adjunctions we arrive at a field $\mathbf{F}(\alpha_1, \ldots, \alpha_m)$ such that in $\mathbf{F}(\alpha_1, \ldots, \alpha_m)\,[x]$ the polynomial $f(x)$ splits into linear factors, say

$$f(x) = c_0(x - \alpha_1)\,(x - \alpha_2) \ldots (x - \alpha_n).$$

In other words, the field $\mathbf{S} = \mathbf{F}(\alpha_1, \ldots, \alpha_m)$ contains all the roots of $f(x)$, i.e.

$$\mathbf{S} = \mathbf{F}(\alpha_1, \ldots, \alpha_m) = \mathbf{F}(\alpha_1, \ldots, \alpha_m, \alpha_{m+1}, \ldots, \alpha_n).$$

Clearly, the number of adjunctions which is necessary to arrive at $\mathbf{S} = \mathbf{F}(\alpha_1, \ldots, \alpha_n)$ does not exceed $n - 1$. The field $\mathbf{S}$ is called the *splitting field* of $f(x)$ over $\mathbf{F}$. It is clear that $\mathbf{S}$ is in the following sense the smallest field which contains all the roots of $f(x)$: No proper subfield of $\mathbf{S}$ that is an extension of $\mathbf{F}$ contains all the roots of $f(x)$.

6.3.4. THEOREM. *If* $\mathbf{F}$ *is a field and* $f(x)$ *any polynomial of positive degree in* $\mathbf{F}[x]$, *then there exists a splitting field of* $f(x)$ *over* $\mathbf{F}$. *Any two splitting fields of* $f(x)$ *over* $\mathbf{F}$ *are isomorphic over* $\mathbf{F}$.

Proof. The existence of a splitting field has just been proved. It remains to prove its uniqueness up to isomorphism over $\mathbf{F}$.

In order to apply induction we shall prove the following slightly more general proposition:

Let τ_0 be an isomorphism of the field $\mathbf{F}$ onto the field $\bar{\mathbf{F}}$ and let $f(x)$ be a polynomial of $\mathbf{F}[x]$. By applying τ_0 to the coefficients of

$f(x)$ we obtain a polynomial $\bar{f}(x)$ of $\bar{\mathbf{F}}[x]$. Let $\mathbf{S}$ and $\bar{\mathbf{S}}$ be splitting fields of $f(x)$ over $\mathbf{F}$ and of $\bar{f}(x)$ over $\bar{\mathbf{F}}$, respectively. Then there exists an extension of τ_0 to an isomorphism τ of $\mathbf{S}$ onto $\bar{\mathbf{S}}$.

For $\mathbf{F} = \bar{\mathbf{F}}$ and τ_0 the identity automorphism of $\mathbf{F}$, this proposition obviously yields the uniqueness of the splitting field of $f(x)$ up to isomorphism over $\mathbf{F}$.

We prove our proposition by induction on the number k of roots of $f(x)$ outside $\mathbf{F}$.

In case $k = 0$, $\mathbf{F}$ contains all the roots of $f(x)$, in other words, $f(x)$ splits into linear factors in $\mathbf{F}[x]$. Then $\bar{f}(x)$ splits into linear factors in $\bar{\mathbf{F}}[x]$. Hence, $\mathbf{S} = \mathbf{F}$ and $\bar{\mathbf{S}} = \bar{\mathbf{F}}$ so that τ_0 itself is the desired isomorphism.

We now make the following induction assumption: Let $\mathbf{K}$ and $\bar{\mathbf{K}}$ be two fields such that $\mathbf{F} \subset \mathbf{K} \subseteq \mathbf{S}$ and $\bar{\mathbf{F}} \subset \bar{\mathbf{K}} \subseteq \bar{\mathbf{S}}$, respectively, and let τ_1 be an extension of τ_0 to an isomorphism of $\mathbf{K}$ onto $\bar{\mathbf{K}}$. Suppose that fewer than k roots of $f(x)$ are outside $\mathbf{K}$. Then there exists an extension of τ_1 to an isomorphism τ of $\mathbf{S}$ onto $\bar{\mathbf{S}}$.

Let

$$f(x) = f_1(x)\, f_2(x) \ldots f_r(x)$$

be the factorization of $f(x)$ into irreducible factors in $\mathbf{F}[x]$. The corresponding factorization of $\bar{f}(x)$ in $\bar{\mathbf{F}}[x]$ has the form

$$\bar{f}(x) = \bar{f}_1(x)\, \bar{f}_2(x) \ldots \bar{f}_r(x)$$

where $\bar{f}_i(x)$ is obtained by applying τ_0 to the coefficients of $f_i(x)$. It is evident that the $\bar{f}_i(x)$ are irreducible in $\bar{\mathbf{F}}[x]$.

Since we have only to consider the case $k > 0$, we may assume that $f_1(x)$ and, hence, $\bar{f}_1(x)$ are of degree greater than 1. Let α_1 and $\bar{\alpha}_1$ denote roots of the irreducible factors $f_1(x)$ and $\bar{f}_1(x)$, respectively. We form the extensions $\mathbf{K} = \mathbf{F}(\alpha_1)$ and $\bar{\mathbf{K}} = \bar{\mathbf{F}}(\bar{\alpha}_1)$. By Theorem 6.3.3, τ_0 can be extended to an isomorphism τ_1 of $\mathbf{K}$ onto $\bar{\mathbf{K}}$. We now regard $f(x)$ and $\bar{f}(x)$ as polynomials in $\mathbf{K}[x]$ and $\bar{\mathbf{K}}[x]$, respectively. Then the number of roots of $f(x)$ outside $\mathbf{K}$ is less than k. Therefore our induction assumption yields the possibility of extending τ_1 to an isomorphism τ of $\mathbf{S}$ onto $\bar{\mathbf{S}}$. This completes the proof.

As to the number of adjunctions which are necessary to arrive at the splitting field, let us exhibit two extreme cases. Suppose that $f(x)$ is irreducible in $\mathbf{F}[x]$.

It may happen that the adjunction of a single root α_1 of $f(x)$ already yields the entire splitting field, i.e. $\mathbf{F}(\alpha_1) = \mathbf{F}(\alpha_1, \ldots, \alpha_n)$. Then the splitting field is of degree n over $\mathbf{F}$. Let p be a prime number, $p > 2$. As we observed in § 5.3, the polynomial

$$f(x) = x^{p-1} + x^{p-2} + \ldots + x + 1 = \frac{x^p - 1}{x - 1}$$

is irreducible in $\mathbf{Q}[x]$. The roots of $f(x)$ are

$$\varepsilon = \cos\frac{2\pi}{p} + i \sin\frac{2\pi}{p}, \quad \varepsilon^2, \ldots, \quad \varepsilon^{p-1}.$$

Therefore all the roots are contained in $\mathbf{Q}(\varepsilon)$ so that $\mathbf{Q}(\varepsilon)$ is the splitting field of $f(x)$. We observe that $\mathbf{Q}(\varepsilon) = \mathbf{Q}(\varepsilon^k)$ for any exponent k with $1 \leqq k \leqq p - 1$; indeed, $1, \varepsilon, \varepsilon^2, \ldots, \varepsilon^{p-1}$ form a cyclic group of order p, and any element other than 1 generates the whole group.

The other extreme case occurs if each adjunction of a root yields precisely one new linear factor. After the adjunction of $\alpha_1, \ldots, \alpha_i$ we have the factorization

$$f(x) = (x - \alpha_1) \ldots (x - \alpha_i)\, f_i(x)$$

where $f_i(x)$ is irreducible over $\mathbf{F}(\alpha_1, \ldots, \alpha_i)$. If $f(x)$ is of degree n we have therefore to adjoin $n - 1$ roots to arrive at the splitting field. It follows from Theorem 6.2.3 that the degree of the splitting field over $\mathbf{F}$ is equal to $n!$.

The polynomial $x^3 - 2$ is irreducible in $\mathbf{Q}[x]$ since 2 is not a cube of a rational number. After adjoining the real number $\sqrt[3]{2}$ we obtain the factorization

$$x^3 - 2 = (x - \sqrt[3]{2})\,(x^2 + \sqrt[3]{2}x + \sqrt[3]{2^2}).$$

The quadratic factor is irreducible over $\mathbf{Q}(\sqrt[3]{2})$ since its roots are complex numbers whereas $\mathbf{Q}(\sqrt[3]{2})$ consists only of real numbers. The other roots of $x^3 - 2$ are $\varrho\sqrt[3]{2}$ and $\bar{\varrho}\sqrt[3]{2}$ where

$$\varrho = \frac{1}{2}(-1 + \sqrt{-3}), \; \varrho^3 = 1, \; \bar{\varrho} = \varrho^2.$$

So we have to adjoin one of these roots to obtain the factorization

$$x^3 - 2 = (x - \sqrt[3]{2})\,(x - \varrho\sqrt[3]{2})\,(x - \bar{\varrho}\sqrt[3]{2}).$$

The splitting field $\boldsymbol{Q}(\sqrt[3]{2}, \varrho\sqrt[3]{2})$ is of degree 6 over $\boldsymbol{Q}$.

A field $\mathbf{K}$ is called *algebraically closed* if every polynomial of positive degree of $\mathbf{K}[x]$ has a root in $\mathbf{K}$. In other words, $\mathbf{K}$ is algebraically closed if and only if there exists no proper algebraic extension of $\mathbf{K}$.

6.3.5. THEOREM. *For any field* $\mathbf{F}$, *there exists an algebraically closed extension field of* $\mathbf{F}$.

Proof.[1]) We first construct an extension field $\mathbf{K}_1$ of $\mathbf{F}$ such that every polynomial of positive degree of $\mathbf{F}[x]$ has a root in $\mathbf{K}_1$. To each polynomial $f(x) \in \mathbf{F}[x]$ we assign an indeterminate x_f. Let X denote the set of all these indeterminates. We now form the ring $\mathbf{F}[X]$ of all polynomials in the indeterminates x_f; i.e. we first form the direct product D of the cyclic multiplicative semigroups $\{1, x_f, x_f^2, x_f^3, \ldots\}$, then the semigroup algebra of D over $\mathbf{F}$. In $\mathbf{F}[X]$ we consider the ideal J generated by all polynomials $f(x_f)$, i.e. the ideal that consists of all finite linear combinations of the $f(x_f)$ with coefficients of $\mathbf{F}[X]$. We first show that J does not coincide with the entire ring $\mathbf{F}[X]$. Suppose that, on the contrary, $J = \mathbf{F}[X]$. Then there exists a finite linear combination

$$g_1 f_1(x_{f_1}) + \ldots + g_r f_r(x_{f_r}) = 1$$

where $g_i \in \mathbf{F}[X]$. The indeterminates that actually occur in the polynomials $g_1, \ldots, g_r$ form a finite subset of X, say $\{x_1, \ldots, x_n\}$. This gives

$$\sum_{i=1}^{r} g_i(x_1, \ldots, x_n)\, f_i(x_{f_i}) = 1. \tag{2}$$

Let $\mathbf{E}$ be an extension field of $\mathbf{F}$ in which each of the polynomials $f_1(x), f_2(x), \ldots, f_r(x)$ has a root, e.g. the splitting field of the product

$$f_1(x)\, f_2(x) \ldots f_r(x).$$

For $i = 1, 2, \ldots, r$, let $\alpha_i \in \mathbf{E}$ be a root of $f_i(x)$. In (2) we substitute α_i for x_{f_i} and 0 for all other indeterminates. So the left-hand side

[1]) This proof is due to E. Artin.

of (2) becomes equal to 0, a contradiction. Hence, $J \neq \mathbf{F}[X]$. By Theorem 4.2.4, there exists a maximal ideal M of $\mathbf{F}[X]$ that contains J. Then $\mathbf{F}[X]/M$ is a field. We consider the natural homomorphism

$$\sigma\colon \mathbf{F}[X] \to \mathbf{F}[X]/M.$$

It is clear that the image $\mathbf{F}^\sigma$ of $\mathbf{F}$ under σ is a subfield of $\mathbf{F}[X]/M$ isomorphic to $\mathbf{F}$. From the definition of J and since $J \subseteq M$ it follows that every polynomial of $\mathbf{F}^\sigma[x]$ has a root in $\mathbf{F}[X]/M$. Due to the isomorphism of $\mathbf{F}$ and $\mathbf{F}^\sigma$ this proves the existence of an extension field $\mathbf{K}_1$ of $\mathbf{F}$ such that every polynomial of positive degree of $\mathbf{F}[x]$ has a root in $\mathbf{K}_1$.

We now repeat the construction to obtain a sequence

$$\mathbf{K}_0 = \mathbf{F} \subset \mathbf{K}_1 \subset \mathbf{K}_2 \subset \mathbf{K}_3 \subset \cdots$$

of extension fields such that every polynomial of positive degree of $\mathbf{K}_i[x]$ has a root in $\mathbf{K}_{i+1}$. Let $\mathbf{K}$ be the union of all the fields $\mathbf{K}_i$, $i = 1, 2, \ldots$ It is clear that $\mathbf{K}$ is a field. For if $a, b \in \mathbf{K}$, then both a and b are contained in some field $\mathbf{K}_m$ and therefore $a + b$ and ab are defined within $\mathbf{K}_m$, and $a + b$ and ab are obviously independent of the choice of m such that $a, b \in \mathbf{K}_m$. All the field axioms can be verified within some $\mathbf{K}_m$. Any polynomial of $\mathbf{K}[x]$ has its coefficients in some $\mathbf{K}_n$. Therefore it has a root in $\mathbf{K}_{n+1}$ and hence in $\mathbf{K}$. This proves our theorem.

6.3.6. Corollary. *For any field* $\mathbf{F}$, *there exists an algebraically closed extension field* $\mathbf{A}$ *that is algebraic over* $\mathbf{F}$.

Proof. Let $\mathbf{K}$ be any algebraically closed extension field of $\mathbf{F}$ and let $\mathbf{A}$ denote the set of all elements of $\mathbf{K}$ that are algebraic over $\mathbf{F}$. From Corollary 6.2.5 it follows that $\mathbf{A}$ is a subfield of $\mathbf{K}$. Let $f(x)$ be any polynomial of $\mathbf{A}[x]$. Then a root α of $f(x)$ in $\mathbf{K}$ is algebraic over $\mathbf{A}$. Hence, by Theorem 6.2.6, α is algebraic over $\mathbf{F}$, i.e. α belongs to $\mathbf{A}$. Therefore $\mathbf{A}$ is algebraically closed, and the corollary is proved.

We now prove that the field $\mathbf{A}$ in the last corollary is uniquely determined up to an isomorphism.

6.3.7. THEOREM. *Any two algebraically closed extension fields of* $\mathbf{F}$ *that are algebraic over* $\mathbf{F}$ *are isomorphic.*

Proof. Let $\mathbf{A}$ and $\mathbf{A}'$ be two algebraically closed extension fields of $\mathbf{F}$ that are algebraic over $\mathbf{F}$. We consider the set $\mathfrak{S}$ of all pairs $(\mathbf{E}, \tau)$ where $\mathbf{E}$ is a subfield of $\mathbf{A}$ and τ an isomorphism of $\mathbf{E}$ onto a subfield $\mathbf{E}'$ of $\mathbf{A}'$. It is clear that $\mathfrak{S}$ is not empty, for $(\mathbf{F}, \varepsilon) \in \mathfrak{S}$ where ε is the identity automorphism of $\mathbf{F}$. In $\mathfrak{S}$ we define a partial order as follows: $(\mathbf{E}_1, \tau_1) \leqq (\mathbf{E}_2, \tau_2)$ if $\mathbf{E}_1 \subseteq \mathbf{E}_2$ and τ_2 coincides with τ_1 for the elements of $\mathbf{E}_1$. For a chain

$$(\mathbf{K}_1, \tau_1) \leqq (\mathbf{K}_2, \tau_2) \leqq (\mathbf{K}_3, \tau_3) \leqq \dots$$

in $\mathfrak{S}$ we put $\mathbf{K} = \bigcup \mathbf{K}_i$ and define τ to coincide with τ_i for each $\mathbf{K}_i$. Then $(\mathbf{K}, \tau)$ is an upper bound of the chain. By Zorn's Lemma there exists a maximal element $(\mathbf{M}, \sigma)$ in $\mathfrak{S}$. We shall show that $\mathbf{M} = \mathbf{A}$. Suppose that, on the contrary, $\mathbf{M}$ is a proper subfield of $\mathbf{A}$. Then there exists an element $\alpha \in \mathbf{A}$ that is not contained in $\mathbf{M}$. Since $\mathbf{A}$ is algebraic over $\mathbf{F}$, α is algebraic over $\mathbf{M}$. Let $p(x)$ denote the minimal polynomial of α in $\mathbf{M}[x]$. Then the image $p(x)^\sigma$ of $p(x)$ is an irreducible polynomial with coefficients in the image $\mathbf{M}^\sigma \subseteq \mathbf{A}'$. By Theorem 6.3.3, it is possible to extend σ to an isomorphism of $\mathbf{M}(\alpha)$ onto a subfield $\mathbf{M}^\sigma(\alpha^\sigma)$ of $\mathbf{A}'$ where α^σ is a root of $p(x)^\sigma$ in $\mathbf{A}'$. This contradicts the hypothesis that $(\mathbf{M}, \sigma)$ is maximal. Therefore $\mathbf{M}$ coincides with $\mathbf{A}$, and σ is an isomorphism of $\mathbf{A}$ onto a subfield $\mathbf{A}^\sigma$ of $\mathbf{A}'$. Since $\mathbf{A}$ is algebraically closed and algebraic over $\mathbf{F}$, the same holds for $\mathbf{A}^\sigma$. Hence, there exists no proper algebraic extension of $\mathbf{A}^\sigma$ so that necessarily $\mathbf{A}^\sigma = \mathbf{A}'$. Therfeore σ is an isomorphism of $\mathbf{A}/\mathbf{F}$ onto $\mathbf{A}'/\mathbf{F}$ which proves our theorem.

We refer to an algebraically closed extension field of $\mathbf{F}$ which is algebraic over $\mathbf{F}$ as the *algebraic closure* of $\mathbf{F}$, where the definite article is justified by the last theorem.

6.4. Normal extensions

Two elements of an extension field $\mathbf{E}$ of $\mathbf{F}$ are said to be *conjugate over* $\mathbf{F}$ if they are roots of one and the same irreducible polynomial of $\mathbf{F}[x]$.

It turns out that those algebraic extension fields of $\mathbf{F}$ are of particular importance which contain with any element all its conjugates. This property is obviously equivalent to the following one which we shall use as a definition of such extension fields.

An extension field $\mathbf{E}$ of $\mathbf{F}$ is called *normal over* $\mathbf{F}$ if it is algebraic over $\mathbf{F}$ and if every irreducible polynomial of $\mathbf{F}[x]$ that has a root in $\mathbf{E}$ has all its roots in $\mathbf{E}$.

6.4.1. LEMMA. *If* $\mathbf{E}$ *is normal over* $\mathbf{F}$ *and if* $\mathbf{K}$ *is a field such that* $\mathbf{F} \subseteq \mathbf{K} \subseteq \mathbf{E}$, *then* $\mathbf{E}$ *is also normal over* $\mathbf{K}$.

Proof. Let $\varphi(x)$ be an irreducible polynomial of $\mathbf{K}[x]$ that has a root α in $\mathbf{E}$. Then α is algebraic over $\mathbf{F}$. Let $p(x)$ denote the minimal polynomial of α over $\mathbf{F}$. Since $\mathbf{E}$ is normal over $\mathbf{F}$, all the roots of $p(x)$ are contained in $\mathbf{E}$. Because $p(x)$ and $\varphi(x)$ have a common root and $\varphi(x)$ is irreducible in $\mathbf{K}[x]$ it follows that $\varphi(x)$ divides $p(x)$. Hence all the roots of $\varphi(x)$ are also roots of $p(x)$. Therefore $\mathbf{E}$ contains all the roots of $\varphi(x)$ as was to be proved.

In the next theorem we give two other characterizations of normal extension fields of finite degree.

6.4.2. THEOREM. *An extension field* $\mathbf{E}$ *of finite degree is normal over* $\mathbf{F}$ *if and only if it satisfies one of the following conditions.*

(i) $\mathbf{E}$ *is the splitting field of a polynomial of* $\mathbf{F}[x]$.

(ii) *If* $\mathbf{A}$ *is any extension field of* $\mathbf{F}$ *that contains* $\mathbf{E}$, *then any isomorphism of* $\mathbf{E}/\mathbf{F}$ *into* $\mathbf{A}/\mathbf{F}$ *is an automorphism of* $\mathbf{E}/\mathbf{F}$.

Proof. Suppose that $\mathbf{E}$ is normal over $\mathbf{F}$. We show that $\mathbf{E}$ is a splitting field of a polynomial of $\mathbf{F}[x]$. Since $[\mathbf{E}:\mathbf{F}]$ is finite, it follows from Theorem 6.2.4 that $\mathbf{E} = \mathbf{F}(\alpha_1, \ldots, \alpha_r)$ where each α_i is algebraic over $\mathbf{F}$. Let $p_i(x)$ denote the minimal polynomial of α_i over $\mathbf{F}$. Since $\mathbf{E}$ is normal over $\mathbf{F}$ we conclude that all the roots of each $p_i(x)$ are contained in $\mathbf{E}$. Therefore $\mathbf{E}$ is the splitting field of the product

$$f(x) = \prod_{i=1}^{r} p_i(x).$$

Next we assume that $\mathbf{E}$ is the splitting field of a polynomial $f(x) \in \mathbf{F}[x]$ and prove that $\mathbf{E}$ satisfies condition (ii). If $\gamma_1, \ldots, \gamma_n$ are the roots of $f(x)$, then we have $\mathbf{E} = \mathbf{F}(\gamma_1, \ldots, \gamma_n)$. Now let $\mathbf{A}$ be a field such that $\mathbf{F} \subseteq \mathbf{E} \subseteq \mathbf{A}$ and let σ denote an isomorphism of $\mathbf{E}$

into A that leaves F elementwise fixed. Since $f(x)$ has coefficients in F, $f(\gamma_i) = 0$ implies that $f(\gamma_i^\sigma) = 0$ where γ_i^σ denotes the image of γ_i under σ. Therefore γ_i^σ is again a root of $f(x)$. Since σ is a one-to-one mapping we conclude that $\gamma_1^\sigma, \ldots, \gamma_n^\sigma$ is a permutation of $\gamma_1, \ldots, \gamma_n$. Denoting the image of E under σ by E^σ we thus obtain that

$$\mathsf{E}^\sigma = \mathsf{F}(\gamma_1^\sigma, \ldots, \gamma_n^\sigma) = \mathsf{F}(\gamma_1, \ldots, \gamma_n) = \mathsf{E}.$$

Hence σ is an automorphism of E.

Finally, we assume that E satisfies condition (ii) and prove that E is normal over F. By Theorem 6.2.4, we have $\mathsf{E} = \mathsf{F}(\alpha_1, \ldots, \alpha_r)$. Let $p_i(x)$ denote the minimal polynomial of α_i over F. Now let $g(x)$ be any polynomial of $\mathsf{F}[x]$ that is irreducible in $\mathsf{F}[x]$ and has a root γ in E. We form the product

$$h(x) = g(x) \prod_{i=1}^{r} p_i(x)$$

and denote by A the splitting field of $h(x)$ over F. We have to show that any other root γ' of $g(x)$ is contained in E. By Theorem 6.3.3, there exists an isomorphism τ_0 of $\mathsf{F}(\gamma)/\mathsf{F}$ onto $\mathsf{F}(\gamma')/\mathsf{F}$ such that $\gamma' = \gamma^{\tau_0}$. Since $h(x)$ has coefficients in F it can be regarded as a polynomial both in $\mathsf{F}(\gamma)[x]$ and $\mathsf{F}(\gamma')[x]$. Moreover, A is the splitting field of $f(x)$ over $\mathsf{F}(\gamma)$ and over $\mathsf{F}(\gamma')$. In the same way as in the proof of Theorem 6.3.4 we conclude that it is possible to extend τ_0 to an automorphism τ of A/F. By restricting τ to E we obtain an isomorphism σ of E/F into A/F. Since E satisfies condition (ii) it follows that σ is an automorphism of E/F. Thus, from $\gamma \in \mathsf{E}$ it follows that $\gamma' = \gamma^\sigma$ belongs to E.

This completes the proof.

6.5. Separable and inseparable extensions

Let F be any field. An irreducible polynomial $f(x)$ is called *separable* if it has no multiple roots in any extension field of F. An arbitrary polynomial of $\mathsf{F}[x]$ is said to be separable if all its irreducible factors are separable.

Let E be any extension field of F and let α be an element of E which is algebraic over F. Then α is called separable over F if the minimal polynomial of α over $\mathbf{F}$ is separable. An algebraic extension field $\mathbf{A}$ of $\mathbf{F}$ is said to be separable over $\mathbf{F}$ if every element of $\mathbf{A}$ is separable over $\mathbf{F}$.

Polynomials, elements and algebraic extension fields that are not separable are called *inseparable*.

As we observed in § 5.4, there are no inseparable polynomials in $\mathbf{F}[x]$ if and only if $\mathbf{F}$ is perfect. Moreover, every field of characteristic 0 is perfect.

6.5.1. LEMMA. *If $\mathbf{E}$ is separable over $\mathbf{F}$ and if $\mathbf{K}$ is a field such that $\mathbf{F} \subseteq \mathbf{K} \subseteq \mathbf{E}$, then $\mathbf{E}$ is also separable over $\mathbf{K}$.*

Proof. Let $\alpha \in \mathbf{E}$. Then the minimal polynomial $f(x)$ of α over $\mathbf{F}$ has only simple roots. The minimal polynomial $\varphi(x)$ of α over $\mathbf{K}$ is a divisor of $f(x)$ so that all the roots of $\varphi(x)$ are simple. Hence, α is separable over $\mathbf{K}$, as was to be proved.

Separable extension fields of finite degree have the following remarkable property:

6.5.2. THEOREM. *Every separable extension field of finite degree is a simple extension, i.e. it is obtained by adjoining a single element.*

Proof. Let $\mathbf{E} = \mathbf{F}(\gamma_1, \gamma_2, \ldots, \gamma_m)$ be a separable extension field of finite degree over $\mathbf{F}$. We have to show that $\mathbf{E}$ contains an element ϑ such that $\mathbf{E} = \mathbf{F}(\vartheta)$.

We shall prove the theorem under the additional assumption that $\mathbf{F}$ contains infinitely many elements. The theorem is also valid if $\mathbf{F}$ consists of finitely many elements. For the proof in this case we refer to § 6.8 which is devoted to the study of finite fields.

Suppose first that $m = 2$, i.e. let $\mathbf{F}(\alpha, \beta)$ be a separable extension of $\mathbf{F}$. Let $f(x)$ and $g(x)$ denote the minimal polynomials of α and β, respectively, over $\mathbf{F}$. In a suitable extension field of $\mathbf{F}$ we obtain the factorizations

$$f(x) = \prod_{i=1}^{r} (x - \alpha_i), \quad g(x) = \prod_{k=1}^{s} (x - \beta_k)$$

where we assume that $\alpha_1 = \alpha$, $\beta_1 = \beta$. Since β is separable over $\mathbf{F}$ we have $\beta_k \neq \beta_l$ for $k \neq l$. We consider the $r(s-1)$ linear equations

$$\alpha_i + y\beta_k = \alpha_1 + y\beta_1 \quad (i = 1, \ldots, r;\, k = 2, \ldots, s).$$

Each equation has at most a single solution y in F. Since we assumed that F contains infinitely many elements we can choose an element c of F that is distinct from all the solutions of our equations. This gives

$$\alpha_i + c\beta_k \neq \alpha_1 + c\beta_1 \quad (i = 1, \ldots, r; k = 2, \ldots, s).$$

We shall prove that

$$\vartheta = \alpha_1 + c\beta_1 = \alpha + c\beta$$

is a primitive element of $\mathsf{F}(\alpha, \beta)$, i.e. $\mathsf{F}(\alpha, \beta) = \mathsf{F}(\vartheta)$. Clearly, ϑ is contained in $\mathsf{F}(\alpha, \beta)$ because c is an element of F. We have $g(\beta) = 0$ and

$$f(\vartheta - c\beta) = f(\alpha) = 0.$$

Therefore β is a common root of the polynomials $g(x)$ and $f(\vartheta - cx)$. Moreover, β is the only common root of these two polynomials. For substituting one of the other roots $\beta_2, \ldots, \beta_s$ of $g(x)$ for x we obtain $f(\vartheta - c\beta_k)$ which is distinct from zero since $\vartheta - c\beta_k \neq \alpha_i$ for $k = 2, \ldots, s$. Because β is a simple root of $g(x)$, the greatest common divisor of $g(x)$ and $f(\vartheta - cx)$ is $x - \beta$. Now $g(x)$ and $f(\vartheta - cx)$ can both be regarded as polynomials with coefficients in $\mathsf{F}(\vartheta)$. Applying the Euclidean algorithm we find that their greatest common divisor $x - \beta$ has also coefficients in $\mathsf{F}(\vartheta)$. Therefore β is contained in $\mathsf{F}(\vartheta)$ and, hence, $\alpha = \vartheta - c\beta$ also belongs to $\mathsf{F}(\vartheta)$. This implies that $\mathsf{F}(\alpha, \beta) = \mathsf{F}(\vartheta)$.

To prove that $\mathsf{F}(\gamma_1, \gamma_2, \ldots, \gamma_m)$ is simple we proceed by induction on m. Let us assume that $\mathsf{F}(\gamma_1, \ldots, \gamma_{m-1})$ is a simple extension, i.e. $\mathsf{F}(\gamma_1, \ldots, \gamma_{m-1}) = \mathsf{F}(\gamma)$. From our previous result it follows that

$$\mathsf{F}(\gamma_1, \ldots, \gamma_{m-1}, \gamma_m) = \mathsf{F}(\gamma, \gamma_m) = \mathsf{F}(\vartheta).$$

This completes the proof.

Remark. In the proof for $m = 2$ we had only to use the separability of β. So we have actually obtained the sharper result that $\mathsf{F}(\alpha, \beta)$ is a simple extension if at least one of the elements α or β is separable over F. Similarly, $\mathsf{F}(\gamma_1, \ldots, \gamma_m)$ is a simple extension of F if at least $m - 1$ of the elements $\gamma_1, \ldots, \gamma_m$ are separable over F.

As an example we consider $\boldsymbol{Q}(\sqrt{2}, \sqrt{3})$. The sum $\vartheta = \sqrt{2} + \sqrt{3}$ is a primitive element since $\sqrt{2} + \sqrt{3}$, $-\sqrt{2} + \sqrt{3}$, $\sqrt{2} - \sqrt{3}$,

$-\sqrt{2}-\sqrt{3}$ are distinct. We find that

$$\sqrt{2}+\sqrt{3}=\vartheta, \quad 11\sqrt{2}+9\sqrt{3}=\vartheta^3.$$

Therefore $\sqrt{2}$ and $\sqrt{3}$ can be represented as linear combinations of ϑ and ϑ^3 with rational coefficients which shows once more directly that ϑ is a primitive element.

We now use the fact that a separable extension of finite degree is simple to determine the number of distinct isomorphisms of such an extension into an algebraically closed field.

6.5.3. THEOREM. *Let* $\mathbf{E}$ *be a separable extension of* $\mathbf{F}$ *of finite degree* $[\mathbf{E}:\mathbf{F}]=n$ *and let* $\mathbf{C}$ *be the algebraic closure of* $\mathbf{F}$. *Then there are n distinct isomorphisms of* $\mathbf{E}/\mathbf{F}$ *into* $\mathbf{C}/\mathbf{F}$.

Proof. By Theorem 6.5.2, we have $\mathbf{E}=\mathbf{F}(\vartheta)$ where the minimal polynomial of ϑ over $\mathbf{F}$ is of degree n and has n distinct roots in $\mathbf{C}$, say $\vartheta_1=\vartheta, \vartheta_2, \ldots, \vartheta_n$. Clearly, every isomorphism of $\mathbf{E}/\mathbf{F}$ into $\mathbf{C}/\mathbf{F}$ carries ϑ_1 into some ϑ_i. Conversely, Theorem 6.3.2 shows that, for $i=1, 2, \ldots, n$, the mapping

$$c_0+c_1\vartheta_1+\ldots+c_{n-1}\vartheta_1^{n-1}\to c_0+c_1\vartheta_i+\ldots+c_{n-1}\vartheta_i^{n-1}$$

is an isomorphism of $\mathbf{F}(\vartheta_1)/\mathbf{F}$ into $\mathbf{C}/\mathbf{F}$. This completes the proof.

We now turn to algebraic extensions which are not necessarily separable. Since there are no inseparable polynomials in $\mathbf{F}[x]$ unless the characteristic of $\mathbf{F}$ is a prime number we shall assume until the end of this section that $\mathbf{F}$ is of characteristic $p>0$.

From the results of § 5.4 it follows that an irreducible polynomial $f(x)\in\mathbf{F}[x]$ is separable unless the derivative $f'(x)$ is the zero polynomial or, what amounts to the same, if $f(x)$ can be written as a polynomial in x^p, i.e.

$$f(x)=f_1(x^p).$$

It may happen that $f_1(y)$ is again a polynomial in y^p, i.e. $f_1(y)=f_2(y^p)$. This gives

$$f(x)=f_1(x^p)=f_2(x^{p^2}).$$

Continuing in this way we finally arrive at an exponent e such that

$$f(x)=g\left(x^{p^e}\right) \tag{1}$$

whereas $f(x)$ cannot be written as a polynomial in $x^{p^{e+1}}$. Clearly, $g(x)$ is irreducible in $\mathbf{F}[x]$. Moreover, $g(x)$ is separable since $g(x) = g_1(x^p)$ would imply that $f(x) = g_1\left(x^{p^{e+1}}\right)$. If n and n_0 denote the degrees of $f(x)$ and $g(x)$, respectively, then (1) yields

$$n = n_0 p^e. \tag{2}$$

In a suitable extension field $\mathbf{K}$ of $\mathbf{F}$ we obtain the factorization

$$g(x) = c(x - \gamma_1)(x - \gamma_2) \dots (x - \gamma_{n_0}).$$

Since $g(x)$ is separable, all its roots $\gamma_1, \gamma_2, \dots, \gamma_{n_0}$ are distinct. By (1) we have the factorization

$$f(x) = c(x^{p^e} - \gamma_1)(x^{p^e} - \gamma_2) \dots (x^{p^e} - \gamma_{n_0}).$$

For $i = 1, 2, \dots, n_0$ we now adjoin the p^e-th root ϑ_i of γ_i to $\mathbf{K}$ to obtain a field $\mathbf{E}$ such that in $\mathbf{E}[x]$ the polynomial $f(x)$ splits as follows:

$$f(x) = c(x^{p^e} - \vartheta_1^{p^e})(x^{p^e} - \vartheta_2^{p^e}) \dots (x^{p^e} - \vartheta_{n_0}^{p^e}).$$

We recall the fact that the p^e-th roots ϑ_i are unique as we observed in § 5.4. Due to 4.7.3 the last equation yields

$$f(x) = c(x - \vartheta_1)^{p^e}(x - \vartheta_2)^{p^e} \dots (x - \vartheta_{n_0})^{p^e}.$$

This shows that $f(x)$ has n_0 distinct roots all of which have the same multiplicity p^e.

The separable case can be included by taking $e = 0$.

The number n_0 is called the *reduced degree* or the *separable degree* of $\mathbf{F}(\vartheta_i)$ over $\mathbf{F}$. We write

$$n_0 = [\mathbf{F}(\vartheta_i) : \mathbf{F}]_s.$$

Clearly, ϑ is separable over $\mathbf{F}$ if and only if

$$[\mathbf{F}(\vartheta) : \mathbf{F}] = [\mathbf{F}(\vartheta) : \mathbf{F}]_s.$$

We now consider the simple extension $\mathbf{F}(\vartheta_1)$ and determine the number of distinct isomorphisms of $\mathbf{F}(\vartheta_1)/\mathbf{F}$ into an algebraically closed extension field $\mathbf{C}$ of $\mathbf{F}$. It is evident that such an isomorphism carries ϑ_1 into some ϑ_i. Conversely, any mapping $\vartheta_1 \to \vartheta_i$ gives rise to an isomorphism of $\mathbf{F}(\vartheta_1)/\mathbf{F}$ into $\mathbf{C}/\mathbf{F}$. So we obtain the following result:

6.5.4. *If* $\mathbf{C}$ *is the algebraic closure of* $\mathbf{F}$, *then the number of distinct isomorphisms of* $\mathbf{F}(\vartheta)/\mathbf{F}$ *into* $\mathbf{C}/\mathbf{F}$ *is equal to the separable degree of* $\mathbf{F}(\vartheta)$ *over* $\mathbf{F}$.

The numbers e and p^e are called the *exponent of inseparability* and the *inseparable degree* of $\mathbf{F}(\vartheta_i)$ over $\mathbf{F}$, respectively. We write

$$p^e = [\mathbf{F}(\vartheta_i) : \mathbf{F}]_i.$$

In the case of inseparable extensions we cannot confine ourselves to simple extensions. We define the separable degree of an arbitrary extension of finite degree in a way which is suggested by 6.5.4.

Let $\mathbf{E}$ be an arbitrary extension field of finite degree over $\mathbf{F}$ and let $\mathbf{C}$ denote the algebraic closure of $\mathbf{F}$. Then the *separable degree* $[\mathbf{E} : \mathbf{F}]_s$ is defined as the number of distinct isomorphisms of $\mathbf{E}/\mathbf{F}$ into $\mathbf{C}/\mathbf{F}$.

6.5.5. Theorem. *Let* $\mathbf{E}$ *and* $\mathbf{K}$ *be extension fields of* $\mathbf{F}$ *such that* $\mathbf{F} \subseteq \mathbf{E} \subseteq \mathbf{K}$ *and* $[\mathbf{K} : \mathbf{F}]$ *is finite. Then*

$$[\mathbf{K} : \mathbf{F}]_s = [\mathbf{K} : \mathbf{E}]_s \, [\mathbf{E} : \mathbf{F}]_s.$$

Proof. We first consider the following special case: If $\mathbf{F}(\alpha, \beta)$ is an extension of finite degree, then

$$[\mathbf{F}(\alpha, \beta) : \mathbf{F}]_s = [\mathbf{F}(\alpha, \beta) : \mathbf{F}(\alpha)]_s \, [\mathbf{F}(\alpha) : \mathbf{F}]_s. \tag{3}$$

We put $k = [\mathbf{F}(\alpha) : \mathbf{F}]_s$, $l = [\mathbf{F}(\alpha, \beta) : \mathbf{F}(\alpha)]_s$ and contend to show that there are kl distinct isomorphisms of $\mathbf{F}(\alpha, \beta)/\mathbf{F}$ into $\mathbf{C}/\mathbf{F}$. Due to 6.5.4, there are k distinct isomorphisms of $\mathbf{F}(\alpha)/\mathbf{F}$ into $\mathbf{C}/\mathbf{F}$. Let σ_0 be one of these isomorphisms. Then σ_0 maps $\mathbf{F}(\alpha)$ onto some subfield $\mathbf{F}(\bar{\alpha})$ of $\mathbf{C}$ where $\bar{\alpha}$ is a conjugate of α over $\mathbf{F}$. If $\varphi(x)$ is the minimal polynomial of β over $\mathbf{F}(\alpha)$, then the image of $\varphi(x)$ under σ_0 is an irreducible polynomial $\bar{\varphi}(x) \in \mathbf{F}(\bar{\alpha})\,[x]$. Clearly, both $\varphi(x)$ and $\bar{\varphi}(x)$ have the same number of distinct roots in $\mathbf{C}$, namely l. By Theorem 6.3.3, σ_0 can be extended to an isomorphism of $\mathbf{F}(\alpha, \beta)/\mathbf{F}$ into $\mathbf{C}/\mathbf{F}$. Since $\bar{\varphi}(x)$ has l distinct roots in $\mathbf{C}$ there are precisely l distinct extensions of σ_0. This applies to each of the k distinct isomorphisms of $\mathbf{F}(\alpha)/\mathbf{F}$ into $\mathbf{C}/\mathbf{F}$. Therefore the number of distinct isomorphisms of $\mathbf{F}(\alpha, \beta)/\mathbf{F}$ into $\mathbf{C}/\mathbf{F}$ is equal to kl, which proves (3).

By induction it follows from (3) that

$$[\mathbf{F}(\alpha_1, \alpha_2, \ldots, \alpha_r) : \mathbf{F}]_s = \prod_{i=1}^{r} [\mathbf{F}(\alpha_1, \ldots, \alpha_i) : \mathbf{F}(\alpha_1, \ldots, \alpha_{i-1})]_s \tag{4}$$

if $\alpha_1, \ldots, \alpha_r$ are algebraic over $\mathbf{F}$. Here and in the sequel the factor for $i = 1$ on the right-hand side is to be understood as $[\mathbf{F}(\alpha_1) : \mathbf{F}]_s$.

Finally, since $[\mathbf{K} : \mathbf{F}]$ is finite, we can choose algebraic elements $\alpha_1, \ldots, \alpha_r, \alpha_{r+1}, \ldots, \alpha_t$ such that

$$\mathbf{E} = \mathbf{F}(\alpha_1, \ldots, \alpha_r),\ \mathbf{K} = \mathbf{E}(\alpha_{r+1}, \ldots, \alpha_t) = \mathbf{F}(\alpha_1, \ldots, \alpha_r, \alpha_{r+1}, \ldots, \alpha_t).$$

Applying (4) we now obtain the desired result.

Let $\mathbf{E} = \mathbf{F}(\alpha_1, \ldots, \alpha_r)$ be an extension of finite degree. By Theorem 6.2.3, we have

$$[\mathbf{E} : \mathbf{F}] = \prod_{i=1}^{r} [\mathbf{F}(\alpha_1, \ldots, \alpha_i) : \mathbf{F}(\alpha_1, \ldots, \alpha_{i-1})]. \tag{5}$$

In the case of a simple extension, (2) shows that the separable degree is a divisor of the degree. Applying this to he right-hand sides of (4) and (5) we find that $[\mathbf{E} : \mathbf{F}]_s$ divides $[\mathbf{E} : \mathbf{F}]$. The quotient

$$[\mathbf{E} : \mathbf{F}]_i = [\mathbf{E} : \mathbf{F}]\,[\mathbf{E} : \mathbf{F}]_s^{-1}$$

is called the *inseparable degree* of $\mathbf{E}$ over $\mathbf{F}$. For a simple extension this agrees with our previous definition.

From Theorem 6.2.3 and the last theorem we conclude that

$$[\mathbf{K} : \mathbf{F}]_i = [\mathbf{K} : \mathbf{E}]_i\,[\mathbf{E} : \mathbf{F}]_i.$$

Since the inseparable degree of a simple extension is a power of p, the last equation shows that, for every finite extension, $[\mathbf{E} : \mathbf{F}]_i$ is a power of p.

As we defined above, an algebraic extension $\mathbf{E}$ of $\mathbf{F}$ is called separable over $\mathbf{F}$ if every element of $\mathbf{E}$ is separable over $\mathbf{F}$. We shall now derive a criterion for the separability of a finite extension.

6.5.6. Theorem. *An extension* $\mathbf{E}$ *of* $\mathbf{F}$ *of finite degree is separable over* $\mathbf{F}$ *if and only if*

$$[\mathbf{E} : \mathbf{F}] = [\mathbf{E} : \mathbf{F}]_s. \tag{6}$$

Proof. First, suppose that (6) is satisfied. We have to show that any element α_1 of $\mathbf{E}$ is separable over $\mathbf{F}$. Since $[\mathbf{E} : \mathbf{F}]$ is finite, we can choose elements $\alpha_2, \ldots, \alpha_r$ such that $\mathbf{E} = \mathbf{F}(\alpha_1, \alpha_2, \ldots, \alpha_r)$. Due to Theorem 6.2.3, Theorem 6.5.5, and (6) we have

$$\prod_{i=1}^{r} [\mathbf{F}(\alpha_1, \ldots, \alpha_i) : \mathbf{F}(\alpha_1, \ldots, \alpha_{i-1})] = [\mathbf{E} : \mathbf{F}]$$

$$= [\mathbf{E} : \mathbf{F}]_s = \prod_{i=1}^{r} [\mathbf{F}(\alpha_1, \ldots, \alpha_i) : \mathbf{F}(\alpha_1, \ldots, \alpha_{i-1})]_s.$$

Since the separable degree of a simple extension divides its degree we conclude that

$$[\mathbf{F}(\alpha_1, \ldots, \alpha_i) : \mathbf{F}(\alpha_1, \ldots, \alpha_{i-1})] = [\mathbf{F}(\alpha_1, \ldots, \alpha_i) : \mathbf{F}(\alpha_1, \ldots, \alpha_{i-1})]_s$$

for $i = 1, \ldots, r$. Hence, in particular,

$$[\mathbf{F}(\alpha_1) : \mathbf{F}] = [\mathbf{F}(\alpha_1) : \mathbf{F}]_s$$

which means that α_1 is separable over $\mathbf{F}$.

Conversely, suppose that $\mathbf{E} = \mathbf{F}(\alpha_1, \ldots, \alpha_r)$ is separable over $\mathbf{F}$. Then each α_i is separable over $\mathbf{F}$ and, hence, also separable over $\mathbf{F}(\alpha_1, \ldots, \alpha_{i-1})$. This gives

$$[\mathbf{F}(\alpha_1, \ldots, \alpha_{i-1}, \alpha_i) : \mathbf{F}(\alpha_1, \ldots, \alpha_{i-1})]$$

$$= [\mathbf{F}(\alpha_1, \ldots, \alpha_{i-1}; \alpha_i) : \mathbf{F}(\alpha_1, \ldots, \alpha_{i-1})]_s.$$

From Theorem 6.2.3 and Theorem 6.5.5 it follows that

$$[\mathbf{E} : \mathbf{F}] = [\mathbf{E} : \mathbf{F}]_s.$$

This completes the proof. An immediate consequence is the following:

6.5.7. Corollary. *An algebraic extension* $\mathbf{F}(\alpha_1, \ldots, \alpha_r)$ *is separable over* $\mathbf{F}$ *if each* α_i *is separable over* $\mathbf{F}$.

Another consequence of Theorem 6.5.6 is the following theorem:

6.5.8. Theorem. *If* $\mathbf{E}$ *is separable over* $\mathbf{F}$ *and* α *separable over* $\mathbf{E}$, *then* α *is separable over* $\mathbf{F}$.

Proof. By Theorem 6.2.6, α is algebraic over $\mathbf{F}$. Since α is separable over $\mathbf{E}$, it satisfies an equation

$$f(\alpha) = \alpha^t + \gamma_1 \alpha^{t-1} + \ldots + \gamma_t = 0$$

where $f(x)$ is a separable polynomial of $\mathbf{E}[x]$. Due to the separability of $\mathbf{E}$ over $\mathbf{F}$, all the coefficients $\gamma_1, \ldots, \gamma_t$ are separable over $\mathbf{F}$. This gives

$$[\mathbf{F}(\gamma_1, \ldots, \gamma_i) : \mathbf{F}(\gamma_1, \ldots, \gamma_{i-1})] = [\mathbf{F}(\gamma_1, \ldots, \gamma_i) : \mathbf{F}(\gamma_1, \ldots, \gamma_{i-1})]_s \quad (7)$$

for $i = 1, \ldots, t$. Moreover, α is separable over $\mathbf{F}(\gamma_1, \ldots, \gamma_t)$ so that

$$[\mathbf{F}(\gamma_1, \ldots, \gamma_t, \alpha) : \mathbf{F}(\gamma_1, \ldots, \gamma_t)] = [\mathbf{F}(\gamma_1, \ldots, \gamma_t, \alpha) : \mathbf{F}(\gamma_1, \ldots, \gamma_t)]_s.$$

From the last equation and from the equations (7) we conclude that

$$[\mathbf{F}(\gamma_1, \ldots, \gamma_t, \alpha) : \mathbf{F}] = [\mathbf{F}(\gamma_1, \ldots, \gamma_t, \alpha) : \mathbf{F}]_s.$$

Therefore $\mathbf{F}(\gamma_1, \ldots, \gamma_t, \alpha)$ is separable over $\mathbf{F}$ so that, in particular, α is separable over $\mathbf{F}$ as was to be shown.

From Corollary 6.5.7, we deduce that separability is preserved under addition, substraction, multiplication, and division in the following sense:

6.5.9. Corollary. *If α and β are separable over $\mathbf{F}$, then so are $\alpha \pm \beta$, $\alpha\beta$, and α/β if $\beta \neq 0$.*

Proof. By Corollary 6.5.7, $\mathbf{F}(\alpha, \beta)$ is separable over $\mathbf{F}$. Since $\alpha \pm \beta$, $\alpha\beta$, and, for $\beta \neq 0$, also α/β are contained in $\mathbf{F}(\alpha, \beta)$, it follows that they are separable over $\mathbf{F}$.

Now let $\mathbf{K}$ be an arbitrary algebraic extension of $\mathbf{F}$. From the last corollary it follows that the elements of $\mathbf{K}$ which are separable over $\mathbf{F}$ form a subfield $\mathbf{S}$ of $\mathbf{K}$. By Theorem 6.5.8, every element of $\mathbf{K}$ that is separable over $\mathbf{S}$ is contained in $\mathbf{S}$. We refer to $\mathbf{S}$ as the maximal separable subfield of $\mathbf{K}$.

An algebraic extension $\mathbf{E}$ of $\mathbf{L}$ is called *purely inseparable* if every element of $\mathbf{E}$ that is separable over $\mathbf{L}$ is contained in $\mathbf{L}$. So the field $\mathbf{K}$ is purely inseparable over its maximal separable subfield.

We conclude this section by a few remarks about purely inseparable extensions. Let $\mathbf{K}$ be such an extension of $\mathbf{S}$. As we observed earlier in this section, for any element ϑ of $\mathbf{K}$ a certain power ϑ^{p^e} is separable over $\mathbf{S}$. Since $\mathbf{K}$ is purely inseparable over $\mathbf{S}$ it follows that ϑ^{p^e} is contained in $\mathbf{S}$. Therefore any element ϑ of $\mathbf{K}$ is a root of a polynomial

$$x^{p^e} - a \quad (8)$$

with $a \in \mathbf{S}$. Without loss of generality we may assume that a is not a p-th power of some element of $\mathbf{S}$. It is easy to see that under this assumption (8) is the minimal polynomial of ϑ over $\mathbf{S}$. From $\vartheta^{p^e} - a = 0$ it follows that

$$x^{p^e} - a = x^{p^e} - \vartheta^{p^e} = (x - \vartheta)^{p^e}.$$

Therefore ϑ coincides with all its conjugates over $\mathbf{S}$.

Since in the case of characteristic p the p-th root is uniquely determined we also write $\vartheta = a^{p^{-e}}$.

If $\mathbf{K}$ is purely inseparable and of finite degree over $\mathbf{S}$, then $\mathbf{K}$ can be obtained by adjoining finitely many algebraic elements to $\mathbf{S}$. From our last result it follows that

$$\mathbf{K} = \mathbf{S}(a_1^{p^{-e_1}}, a_2^{p^{-e_2}}, \ldots, a_r^{p^{-e_r}})$$

with suitable elements $a_1, a_2, \ldots, a_r$ of $\mathbf{S}$. The extraction of the p^e-th root can be performed in e steps each consisting of the extraction of a p-th root. Therefore $\mathbf{K}$ can be obtained from $\mathbf{S}$ by a finite number of successive extensions of degree p. This shows again that the degree of $\mathbf{K}$ over $\mathbf{S}$ is a power of p.

6.6. Galois theory

The Galois theory establishes a one-to-one correspondence between the subfields of separable normal extensions of finite degree and the subgroups of their automorphism groups. In Chapter 7 we shall apply this theory to the study of solutions of algebraic equations by radicals. The present approach to the Galois theory is due to E. Artin.

6.6.1. THEOREM. *Let $\mathbf{E}$ be a field and $\sigma_1, \sigma_2, \ldots, \sigma_n$ distinct automorphisms of $\mathbf{E}$. Then $\sigma_1, \sigma_2, \ldots, \sigma_n$ are linearly independent in the following sense: If*

$$c_1 a^{\sigma_1} + c_2 a^{\sigma_2} + \ldots + c_n a^{\sigma_n} = 0$$

for every $a \in \mathbf{E}$ with the same coefficients $c_1, c_2, \ldots, c_n$ of $\mathbf{E}$ for all $a \in \mathbf{E}$, then $c_1 = c_2 = \ldots = c_n = 0$.

Proof. We proceed by induction on n. For $n = 1$, we have $c_1 a^{\sigma_1} = 0$ for every $a \in \mathbf{E}$. Taking $a = 1$, the unit element of $\mathbf{E}$, we obtain $1^{\sigma_1} = 1$ and hence $c_1 1^{\sigma_1} = c_1 = 0$. We assume that the theorem holds for fewer than n automorphisms of $\mathbf{E}$. Suppose that the theorem is false for n automorphisms. Then we can find elements $c_1, c_2, \ldots, c_n$, not all zero, in $\mathbf{E}$ such that

$$c_1 a^{\sigma_1} + c_2 a^{\sigma_2} + \ldots + c_n a^{\sigma_n} = 0 \tag{1}$$

for every $a \in \mathbf{E}$. From the induction assumption we conclude that all c_i are distinct from zero for, otherwise, fewer than n automorphisms would be linearly dependent. Since σ_1 and σ_n are distinct, we can choose an element $b \in \mathbf{E}$ such that $b^{\sigma_1} \neq b^{\sigma_n}$. It is evident that $b \neq 0$. Using (1) for ab instead of a we obtain

$$c_1 a^{\sigma_1} b^{\sigma_1} + c_2 a^{\sigma_2} b^{\sigma_2} + \ldots + c_n a^{\sigma_n} b^{\sigma_n} = 0.$$

Since $b^{\sigma_n} \neq 0$ we may divide by b^{σ_n} to get

$$c_1' a^{\sigma_1} + c_2' a^{\sigma_2} + \ldots + c_n' a^{\sigma_n} = 0 \tag{2}$$

where $c_i' = c_i b^{\sigma_i} (b^{\sigma_n})^{-1}$ and, in particular, $c_n' = c_n$. Subtracting (2) from (1) we find

$$(c_1 - c_1')\, a^{\sigma_1} + (c_2 - c_2')\, a^{\sigma_2} + \ldots + (c_{n-1} - c_{n-1}')\, a^{\sigma_{n-1}} = 0.$$

This equation holds for every $a \in \mathbf{E}$ and

$$c_1 - c_1' = c_1 (1 - b^{\sigma_1} (b^{\sigma_n})^{-1}) \neq 0$$

which contradicts the induction assumption. Hence the theorem is proved for all n.

If $\sigma_1, \sigma_2, \ldots, \sigma_n$ are automorphisms of $\mathbf{E}$, then an element a of $\mathbf{E}$ is said to be *fixed* under $\sigma_1, \sigma_2, \ldots, \sigma_n$ if $a^{\sigma_1} = a^{\sigma_2} = \ldots = a^{\sigma_n}$. The name "fixed element" refers to a case with which we shall deal later, namely that one of the σ_i is the identity automorphism. In this case a is fixed, i.e. mapped onto itself, in the usual sense.

6.6.2. *The set of all fixed elements under $\sigma_1, \sigma_2, \ldots, \sigma_n$ forms a subfield of* $\mathbf{E}$.

Proof. Let a and b be fixed elements and $b \neq 0$. Then we have

$$(a - b)^{\sigma_i} = a^{\sigma_i} - b^{\sigma_i} = a^{\sigma_j} - b^{\sigma_j} = (a - b)^{\sigma_j},$$

$$(ab)^{\sigma_i} = a^{\sigma_i} b^{\sigma_i} = a^{\sigma_j} b^{\sigma_j} = (ab)^{\sigma_j},$$

$$(b^{-1})^{\sigma_i} = (b^{\sigma_i})^{-1} = (b^{\sigma_j})^{-1} = (b^{-1})^{\sigma_j}$$

for all pairs $i, j = 1, \ldots, n$. This shows that $a - b$, ab, and b^{-1} are fixed elements from which our proposition follows.

The field which consists of the fixed elements under $\sigma_1, \sigma_2, \ldots, \sigma_n$ is called the *fixed field* of $\sigma_1, \sigma_2, \ldots, \sigma_n$.

6.6.3. THEOREM. *If $\sigma_1, \sigma_2, \ldots, \sigma_n$ are distinct automorphisms of* $\mathbf{E}$ *and if* $\mathbf{F}$ *is the fixed field of $\sigma_1, \sigma_2, \ldots, \sigma_n$, then* $[\mathbf{E}:\mathbf{F}] \geqq n$.

Proof. Suppose that, on the contrary, $[\mathbf{E}:\mathbf{F}] = r < n$. Let $w_1, w_2, \ldots, w_r$ be a basis of $\mathbf{E}$ over $\mathbf{F}$. It follows from Corollary 3.13.5 that there are elements $c_1, c_2, \ldots, c_n$, not all zero, in $\mathbf{E}$ such that

$$\begin{aligned} c_1 w_1^{\sigma_1} + c_2 w_1^{\sigma_2} + \ldots + c_n w_1^{\sigma_n} &= 0, \\ c_1 w_2^{\sigma_1} + c_2 w_2^{\sigma_2} + \ldots + c_n w_2^{\sigma_n} &= 0, \\ \cdots\cdots\cdots\cdots\cdots\cdots\cdots\cdots& \\ c_1 w_r^{\sigma_1} + c_2 w_r^{\sigma_2} + \ldots + c_n w_r^{\sigma_n} &= 0. \end{aligned} \tag{3}$$

Since $w_1, w_2, \ldots, w_r$ form a basis of $\mathbf{E}$ over $\mathbf{F}$, any element $a \in \mathbf{E}$ has a unique representation

$$a = a_1 w_1 + a_2 w_2 + \ldots + a_r w_r \tag{4}$$

where $a_i \in \mathbf{F}$ for $i = 1, 2, \ldots, r$. We multiply the first equation (3) by $a_1^{\sigma_1}$, the second by $a_2^{\sigma_1}, \ldots$, finally the r-th equation by $a_r^{\sigma_1}$. Due to $a_i \in \mathbf{F}$ we have $a_i^{\sigma_1} = a_i^{\sigma_j}$ for $j = 2, \ldots, n$. This gives

$$\begin{aligned} c_1(a_1 w_1)^{\sigma_1} + c_2(a_1 w_1)^{\sigma_2} + \ldots + c_n(a_1 w_1)^{\sigma_n} &= 0, \\ c_1(a_2 w_2)^{\sigma_1} + c_2(a_2 w_2)^{\sigma_2} + \ldots + c_n(a_2 w_2)^{\sigma_n} &= 0, \\ \cdots\cdots\cdots\cdots\cdots\cdots\cdots\cdots\cdots\cdots& \\ c_1(a_r w_r)^{\sigma_1} + c_2(a_r w_r)^{\sigma_2} + \ldots + c_n(a_r w_r)^{\sigma_n} &= 0. \end{aligned}$$

Adding all these equations and using (4) we obtain

$$c_1 a^{\sigma_1} + c_2 a^{\sigma_2} + \ldots + c_n a^{\sigma_n} = 0.$$

Since a is an arbitrary element of $\mathbf{E}$ this contradicts Theorem 6.6.1. Thus, our theorem is proved.

In case the automorphisms $\sigma_1, \sigma_2, \ldots, \sigma_n$ form a group, the last theorem can be sharpened as follows.

6.6.4. THEOREM. *If the automorphisms $\sigma_1, \sigma_2, \ldots, \sigma_n$ of* $\mathbf{E}$ *form a group and if* $\mathbf{F}$ *is their fixed field, then* $[\mathbf{E}:\mathbf{F}] = n$.

Proof. Since $\sigma_1, \sigma_2, \ldots, \sigma_n$ form a group, one of these automorphisms, σ_1 say, is the identity. Assume that the theorem is false so that $[\mathbf{E}: \mathbf{F}] > n$ by Theorem 6.6.3. Then there exist $n + 1$ elements $a_1, a_2, \ldots, a_{n+1}$ of $\mathbf{E}$ that are linearly independent over $\mathbf{F}$. By Corollary 3.13.5, the system of linear equations

$$\begin{aligned} a_1^{\sigma_1}x_1 + a_2^{\sigma_1}x_2 + \ldots + a_{n+1}^{\sigma_1}x_{n+1} &= 0, \\ a_1^{\sigma_2}x_1 + a_2^{\sigma_2}x_2 + \ldots + a_{n+1}^{\sigma_2}x_{n+1} &= 0, \\ \ldots\ldots\ldots\ldots\ldots\ldots\ldots\ldots\ldots\ldots\ldots \\ a_1^{\sigma_n}x_1 + a_2^{\sigma_n}x_2 + \ldots + a_{n+1}^{\sigma_n}x_{n+1} &= 0 \end{aligned} \tag{5}$$

has a non-trivial solution in $\mathbf{E}$. We observe that not all the values of the unknowns $x_1, x_2, \ldots, x_{n+1}$ in such a solution can belong to $\mathbf{F}$ for, otherwise, since σ_1 is the identity, the first equation would mean that $a_1, a_2, \ldots, a_{n+1}$ are linearly dependent over $\mathbf{F}$.

Among all non-trivial solutions of (5) we choose one which has the least number of elements distinct from zero. By renumbering the unknowns, if necessary, we may assume that

$$x_1 = c_1, \quad \ldots, \quad x_r = c_r, \quad x_{r+1} = \ldots = x_{n+1} = 0 \tag{6}$$

is such a solution where $c_1, \ldots, c_r$ are distinct from zero. Moreover, $r > 1$, because $a_1^{\sigma_1}c_1 = a_1c_1 = 0$ implies that $c_1 = 0$ since $a_1 \neq 0$. We may also assume that $c_r = 1$ since, if we multiply the solution (6) by c_r^{-1}, we obtain a new solution with $x_r = 1$. Thus we have

$$a_1^{\sigma_i}c_1 + a_2^{\sigma_i}c_2 + \ldots + a_{r-1}^{\sigma_i}c_{r-1} + a_r^{\sigma_i} = 0 \quad (i = 1, 2, \ldots, n). \tag{7}$$

Since $c_1, \ldots, c_{r-1}$ cannot all belong to $\mathbf{F}$, one of them, say c_1, belongs, to $\mathbf{E}$ but not to $\mathbf{F}$. It follows that there is an automorphism σ_k such that $c_1^{\sigma_k} \neq c_1$. We now use the hypothesis that $\sigma_1, \sigma_2, \ldots, \sigma_n$ form a group. This implies that $\sigma_1\sigma_k, \sigma_2\sigma_k, \ldots, \sigma_n\sigma_k$ coincide with $\sigma_1, \sigma_2, \ldots, \sigma_n$ except for the order. Applying σ_k to (7) we obtain

$$a_1^{\sigma_i\sigma_k}c_1^{\sigma_k} + a_2^{\sigma_i\sigma_k}c_2^{\sigma_k} + \ldots + a_{r-1}^{\sigma_i\sigma_k}c_{r-1}^{\sigma_k} + a_r^{\sigma_i\sigma_k} = 0 \quad (i = 1, 2, \ldots, n). \tag{8}$$

Since $\sigma_1\sigma_k, \sigma_2\sigma_k, \ldots, \sigma_n\sigma_k$ is a permutation of $\sigma_1, \sigma_2, \ldots, \sigma_n$ the equations (8) show that

$$x_1 = c_1^{\sigma_k}, \quad \ldots, \quad x_{r-1} = c_{r-1}^{\sigma_k}, \quad x_r = 1, \quad x_{r+1} = \ldots = x_{n+1} = 0 \tag{9}$$

is another solution of (5). Therefore the difference of the solutions (6) and (9) is also a solution of (5), i.e.

$$x_1 = c_1 - c_1^{\sigma_k}, \ldots, x_{r-1} = c_{r-1} - c_{r-1}^{\sigma_k}, \; x_r = 0, \; x_{r+1} = \ldots = x_{n+1} = 0$$

is a solution of (5). This is not the trivial solution since $x_1 = c_1 - c_1^{\sigma_k} \neq 0$. But this solution has fewer than r non-zero elements which contradicts the choice of the solution (6). This completes the proof.

If $\sigma_1, \sigma_2, \ldots, \sigma_n$ form a group, then the fixed field F remains elementwise fixed so that the σ_i are automorphisms of E over F in the sense of § 6.1.

6.6.5. Corollary. *If H is a finite group of automorphisms of the field E and F the fixed field of H, then every automorphism of E over F belongs to H.*

Proof. Let $[\mathsf{E}:\mathsf{F}] = n = |H|$. Suppose that the automorphism σ of E leaves F elementwise fixed and does not belong to H. Then F is fixed under $n + 1$ automorphisms which contradicts Theorem 6.6.3.

An immediate consequence of Corollary 6.6.5 is the following.

6.6.6. Corollary. *Let H_1 and H_2 be two distinct finite groups of automorphisms of the field E. Then the fixed fields of H_1 and H_2 are distinct.*

If E is an extension field of F, then the automorphisms of E over F form a group which will be denoted by $\Gamma(\mathsf{E}/\mathsf{F})$. On the other hand, if a group G of automorphisms of E is given, then the elements that are fixed under G form a subfield $\Phi(G)$ of E.

For a given subfield F of E we consider the fixed field of $\Gamma(\mathsf{E}/\mathsf{F})$, i.e. the field $\Phi(\Gamma(\mathsf{E}/\mathsf{F}))$. It is clear that $\mathsf{F} \subseteq \Phi(\Gamma(\mathsf{E}/\mathsf{F}))$ but it may well happen that F is a proper subfield of $\Phi(\Gamma(\mathsf{E}/\mathsf{F}))$. For instance, let $\mathsf{F} = \mathbf{Q}$ and $\mathsf{E} = \mathbf{Q}(\sqrt[3]{2})$ where $\sqrt[3]{2}$ denotes the real cube root of 2. Every automorphism of $\mathbf{Q}(\sqrt[3]{2})$ over $\mathbf{Q}$ carries $\sqrt[3]{2}$ into some other root of $x^3 - 2$ because the coefficients of this polynomial belong to $\mathbf{Q}$ and are therefore fixed under such an automorphism. But the two other roots of $x^3 - 2$ are complex numbers and are therefore not contained in $\mathbf{Q}(\sqrt[3]{2})$. Hence the only automorphism of $\mathbf{Q}(\sqrt[3]{2})$

over Q is the identity, i.e. $\Gamma\left(Q\left(\sqrt[3]{2}\right)/Q\right)$ is the identity. Hence the fixed field $\Phi\left(\Gamma\left(Q\left(\sqrt[3]{2}\right)/Q\right)\right)$ is the entire field $Q\left(\sqrt[3]{2}\right)$.

An extension $\mathbf{E}$ of $\mathbf{F}$ is said to be a *Galois extension* of $\mathbf{F}$ if $[\mathbf{E}:\mathbf{F}]$ is finite and $\mathbf{F} = \Phi(\Gamma(\mathbf{E}/\mathbf{F}))$. The group $\Gamma(\mathbf{E}/\mathbf{F})$ is called the *Galois group* of $\mathbf{E}$ over $\mathbf{F}$.

The next theorem gives another characterization of Galois extensions.

6.6.7. THEOREM. *An extension field $\mathbf{E}$ of $\mathbf{F}$ of finite degree is a Galois extension of $\mathbf{F}$ if and only if $\mathbf{E}$ is normal and separable over $\mathbf{F}$.*

Proof. Let $\mathbf{E}$ be normal and separable over $\mathbf{F}$. It follows from Theorem 6.4.2 that $\mathbf{E}$ is a splitting field of a polynomial $f(x) \in \mathbf{F}[x]$. Since $\mathbf{E}$ is separable over $\mathbf{F}$ we conclude that the irreducible factors of $f(x)$ in $\mathbf{F}[x]$ have simple roots.

We have to show that $\mathbf{E}$ is a Galois extension of $\mathbf{F}$.

We proceed by induction on the number m of roots of $f(x)$ that are in $\mathbf{E}$ but not in $\mathbf{F}$.

The case $m = 0$ is trivial for then all the roots of $f(x)$ are in $\mathbf{F}$ so that $\mathbf{E} = \mathbf{F}$.

We now assume that our proposition holds for all pairs of fields with fewer than m roots of $f(x)$ outside the smaller field. It is obvious that the case $m = 1$ cannot occur.

Let

$$f(x) = f_1(x)\, f_2(x) \ldots f_r(x)$$

be the factorization of $f(x)$ into irreducible factors in $\mathbf{F}[x]$. We may assume that one of the factors is of degree greater than 1 for otherwise $\mathbf{F}$ were the splitting field of $f(x)$. Suppose that the degree s of $f_1(x)$ is greater than 1. Let α_1 be a root of $f_1(x)$. Then $[\mathbf{F}(\alpha_1):\mathbf{F}] = s$. Fewer roots of $f(x)$ than m are outside $\mathbf{F}(\alpha_1)$. Since $f(x)$ can be regarded as a polynomial with coefficients in $\mathbf{F}(\alpha_1)$ and $\mathbf{E}$ is the splitting field of $f(x)$ over $\mathbf{F}(\alpha_1)$, it follows from the induction assumption that $\mathbf{E}$ is a Galois extension of $\mathbf{F}(\alpha_1)$. This means that any element of $\mathbf{E}$ that does not belong to $\mathbf{F}(\alpha_1)$ cannot be fixed under all automorphisms of $\mathbf{E}$ over $\mathbf{F}(\alpha_1)$.

Since $\mathbf{E}$ is separable over $\mathbf{F}$, the roots $\alpha_1, \alpha_2, \ldots, \alpha_s$ of $f_1(x)$ are distinct. There exist isomorphisms $\sigma_1, \sigma_2, \ldots, \sigma_s$ mapping $\mathbf{F}(\alpha_1)$

onto $\mathbf{F}(\alpha_1)$, $\mathbf{F}(\alpha_2)$, ..., $\mathbf{F}(\alpha_s)$, respectively, such that σ_i leaves $\mathbf{F}$ elementwise fixed and carries α_1 into α_i. $\mathbf{E}$ is the splitting field of $f(x)$ over $\mathbf{F}(\alpha_1)$ and is also the splitting field of $f(x)$ over each $\mathbf{F}(\alpha_i)$. It follows from the proof of Theorem 6.3.4 that each σ_i can be extended to an automorphism τ_i of $\mathbf{E}$. So, $\tau_1, \tau_2, \ldots, \tau_s$ are automorphisms of $\mathbf{E}$ over $\mathbf{F}$ that carry α_1 into $\alpha_1, \alpha_2, \ldots, \alpha_s$, respectively.

Now let γ be an element of $\mathbf{E}$ that is fixed under all automorphisms of $\mathbf{E}/\mathbf{F}$. We know already that $\gamma \in \mathbf{F}(\alpha_1)$. Therefore γ has a representation of the form

$$\gamma = c_0 + c_1\alpha_1 + c_2\alpha_1^2 + \ldots + c_{s-1}\alpha_1^{s-1}$$

with coefficients $c_j \in \mathbf{F}$. By applying σ_i to this equation we obtain

$$\gamma = \gamma^{\sigma_i} = c_0 + c_1\alpha_i + c_2\alpha_i^2 + \ldots + c_{s-1}\alpha_i^{s-1}.$$

The polynomial

$$c_{s-1}x^{s-1} + c_{s-2}x^{s-2} + \ldots + c_1x + (c_0 - \gamma)$$

has therefore s distinct roots $\alpha_1, \alpha_2, \ldots, \alpha_s$. Therefore all the coefficients must vanish, in particular, $c_0 - \gamma = 0$ which means that γ belongs to $\mathbf{F}$.

Conversely, suppose that $\mathbf{E}$ is a Galois extension of $\mathbf{F}$ and let $\sigma_1, \sigma_2, \ldots, \sigma_n$ denote the elements of the Galois group G of $\mathbf{E}$ over $\mathbf{F}$. Let α be any element of $\mathbf{E}$ and let $\{\alpha, \alpha_2, \ldots, \alpha_r\}$ be the set of distinct elements in the sequence $\alpha^{\sigma_1}, \alpha^{\sigma_2}, \ldots, \alpha^{\sigma_n}$. Since G is a group we have

$$\alpha_i^{\sigma_j} = (\alpha^{\sigma_i})^{\sigma_j} = \alpha^{\sigma_i\sigma_j} = \alpha^{\sigma_k} = \alpha_k$$

when $\sigma_i\sigma_j = \sigma_k$ .This shows that $\alpha, \alpha_2, \ldots, \alpha_r$ are permuted under G. Therefore the coefficients of the polynomial

$$g(x) = (x - \alpha)(x - \alpha_2) \ldots (x - \alpha_r)$$

are fixed under G. Since $\mathbf{E}$ is a Galois extension of $\mathbf{F}$ the only elements of $\mathbf{E}$ that are fixed under the group G belong to $\mathbf{F}$. Therefore the coefficients of $g(x)$ belong to $\mathbf{F}$. If $h(x)$ is any polynomial of $\mathbf{F}[x]$ which has α as a root then, applying $\sigma_1, \sigma_2, \ldots, \sigma_n$ to the equation $h(\alpha) = 0$, we obtain $h(\alpha_i) = 0$ for $i = 2, \ldots, r$ so that the degree if $h(x)$ is at least equal to r. This shows that $g(x)$ is irreducible in $\mathbf{F}[x]$.

Thus we have shown that **E** is separable over **F** and that, moreover, every element of **E** is a root of an irreducible polynomial of **F**$[x]$ which splits into linear factors in **E**$[x]$.

Now let $w_1, \ldots, w_n$ be a basis of **E** over **F** and let $p_i(x)$ denote the minimal polynomial of w_i in **F**$[x]$. As we just observed, each $p_i(x)$ splits into linear factors in **E**$[x]$. Therefore **E** is the splitting field of

$$\prod_{i=1}^{n} p_i(x)$$

over **F**. Hence, it follows from Theorem 6.4.2 that **E** is normal over **F**. This completes the proof.

6.6.8. Corollary. *If* **E** *is a Galois extension of* **F** *and if* **K** *is any field such that* $\mathbf{F} \subseteq \mathbf{K} \subseteq \mathbf{E}$, *then* **E** *is also a Galois extension of* **K**.

Proof. This follows from Theorem 6.6.7 since, by the Lemmas 6.4.1 and 6.5.1, **E** is normal and separable over **K**.

We now turn to the relationship between the subfields of a Galois extension and the subgroups of its Galois group.

6.6.9. Principal Theorem of the Galois Theory. *Let* **E** *be a Galois extension of* **F** *and let* G *be the Galois group of* **E** *over* **F**.

(i) *There is a one-to-one correspondence between the subfields of* **E** *that contain* **F** *and the subgroups of* G. *If the subfield* **K** *and the subgroup* H *correspond to each other, then*

K *is the fixed field of* H,

H *is the Galois group of* **E** *over* **K**.

We say that **K** *and* H *belong to each other.*

(ii) $[\mathbf{E}:\mathbf{K}] = |H|$, $[\mathbf{K}:\mathbf{F}] = [G:H]$.

(iii) **K** *is normal over* **F** *if and only if* H *is a normal subgroup of* G. *In this case the Galois group of* **K** *over* **F** *is isomorphic to the factor group* G/H.

The following diagram illustrates the propositions of our theorem. ε denotes the identity of G.

Proof. Let **K** be any field such that $\mathbf{F} \subseteq \mathbf{K} \subseteq \mathbf{E}$. To **K** we assign the subgroup $H = \Gamma(\mathbf{E}/\mathbf{K})$ of G that leaves **K** fixed. By Corollary 6.6.8.

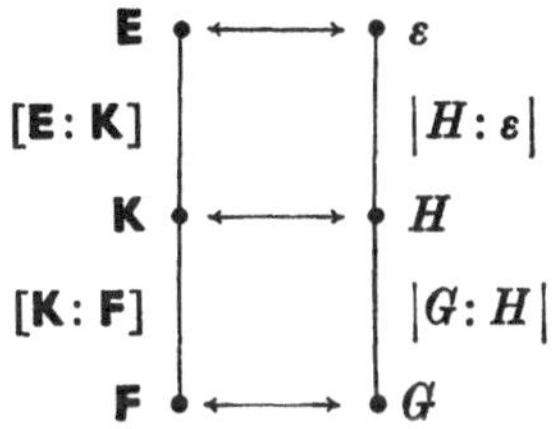

$\mathbf{E}$ is a Galois extension of $\mathbf{K}$ such that $\mathbf{K}$ is the fixed field of H or, in our previous notation, $\mathbf{K} = \Phi(H) = \Phi(\Gamma(\mathbf{E}/\mathbf{K}))$. To complete the proof of (i) we have to show that any given subgroup H_0 occurs as the Galois group of $\mathbf{E}$ over a suitable subfield. Let $\mathbf{K}_0$ denote the fixed field of H_0, i.e. $\mathbf{K}_0 = \Phi(H_0)$. On the other hand, we have $\mathbf{K}_0 = \Phi(\Gamma(\mathbf{E}/\mathbf{K}_0))$. Due to Corollary 6.6.6, $\Phi(H_0) = \Phi(\Gamma(\mathbf{E}/\mathbf{K}_0))$ implies that $H_0 = \Gamma(\mathbf{E}/\mathbf{K}_0)$. Thus, (i) is proved.

By Theorem 6.6.4, we have $[\mathbf{E}:\mathbf{K}] = |H|$ and, in particular, $[\mathbf{E}:\mathbf{F}] = |G|$. The other equality (ii) now follows from

$$|G:H|\,|H| = |G| = [\mathbf{E}:\mathbf{F}] = [\mathbf{E}:\mathbf{K}]\,[\mathbf{K}:\mathbf{F}].$$

To prove (iii) we consider the decomposition

$$G = \bigcup_{i=1}^{r} H\varrho_i$$

of G into right cosets of H. Each ϱ_i is an isomorphism of $\mathbf{K}$ onto a subfield $\mathbf{K}^{\varrho_i}$ of $\mathbf{E}$. If H belongs to $\mathbf{K}$, then the subgroup $\varrho_i^{-1}H\varrho_i$ belongs to $\mathbf{K}^{\varrho_i}$. Indeed, for any $\tau \in H$ we have $b^{\varrho_i\varrho_i^{-1}\tau\varrho_i} = b^{\varrho_i}$ for any $b \in \mathbf{K}$ since $b^\tau = b$; on the other hand, $b^{\varrho_i\lambda} = b^{\varrho_i}$ for any $b^{\varrho_i} \in \mathbf{K}^{\varrho_i}$ implies that $b^{\varrho_i\lambda\varrho_i^{-1}} = b$ and, hence, $\varrho_i\lambda\varrho_i^{-1} \in H$, i.e. $\lambda \in \varrho_i^{-1}H\varrho_i$.

Now suppose that $\mathbf{K}$ is normal over $\mathbf{F}$. By Theorem 6.4.2, this implies that $\varrho_1, \varrho_2, \ldots, \varrho_r$ are automorphisms of $\mathbf{K}$, i.e.

$$\mathbf{K} = \mathbf{K}^{\varrho_1} = \mathbf{K}^{\varrho_2} = \ldots = \mathbf{K}^{\varrho_r}.$$

From part (i) of our theorem it follows that $H = \varrho_i^{-1}H\varrho_i$ for $i = 1, 2, \ldots, r$. In other words, H is a normal subgroup of G.

Conversely, suppose that H is a normal subgroup of G i.e. $H = \varrho_i^{-1}H\varrho_i$ for $i = 1, 2, \ldots, r$. Then all the fields $\mathbf{K}^{\varrho_i}$ coincide with $\mathbf{K}$. Therfeore every element of G induces an automorphism of $\mathbf{K}/\mathbf{F}$

and two elements of G induce the same automorphism of $\mathbf{K}/\mathbf{F}$ if and only if they are contained in the same coset of H. Thus G induces a group T of automorphisms of $\mathbf{K}/\mathbf{F}$ which is isomorphic to the factor group G/H. Let $\mathbf{L}$ be the fixed field of T. By Theorem 6.6.4 we have $|T| = [\mathbf{K}:\mathbf{L}]$. But on the other hand,

$$|T| = |G:H| = [\mathbf{K}:\mathbf{F}]$$

so that $[\mathbf{K}:\mathbf{L}] = [\mathbf{K}:\mathbf{F}]$. Since obviously $\mathbf{F} \subseteq \mathbf{L}$ we conclude that $\mathbf{F} = \mathbf{L}$. Therefore $\mathbf{K}$ is normal over $\mathbf{F}$. Moreover, we see that the Galois group of $\mathbf{K}$ over $\mathbf{F}$ is isomorphic to G/H. This completes the proof.

Since any Galois extension $\mathbf{E}$ of $\mathbf{F}$ is separable over $\mathbf{F}$ it follows from Theorem 6.5.2 that it is a simple extension, $\mathbf{E} = \mathbf{F}(\vartheta)$. If $[\mathbf{E}:\mathbf{F}] = n$, then the minimal polynomial of ϑ in $\mathbf{F}[x]$ is of degree n and has n distinct roots $\vartheta_1 = \vartheta, \vartheta_2, \ldots, \vartheta_n$. Clearly, any element σ_i of the Galois group of $\mathbf{E}$ over $\mathbf{F}$ carries ϑ_1 into some conjugate of ϑ_1, say $\vartheta_1^{\sigma_i} = \vartheta_i$. Representing the elements of $\mathbf{F}(\vartheta_1)$ in the form $c_0 + c_1\vartheta_1 + \ldots + c_{n-1}\vartheta_1^{n-1}$ with coefficients c_j of $\mathbf{F}$ we obtain

$$(c_0 + c_1\vartheta_1 + \ldots + c_{n-1}\vartheta_1^{n-1})^{\sigma_i} = c_0 + c_1\vartheta_i + \ldots + c_{n-1}\vartheta_i^{n-1}.$$

Conversely, these n mappings σ_i are automorphisms of $\mathbf{E}/\mathbf{F}$. Thus a primitive element and its conjugates can occasionally be used to compute the Galois group. (Cf. 6.7.2.)

We shall now consider two examples. In both of them we shall deal with extensions of the field $\boldsymbol{Q}$ of the rational numbers. By 5.4.7, all algebraic extensions of $\boldsymbol{Q}$ are separable.

Example 1. Let $\mathbf{E}$ denote the splitting field of the polynomial $(x^2 - 2)(x^2 - 3)$ over $\boldsymbol{Q}$, i.e. $\mathbf{E} = \boldsymbol{Q}(\sqrt{2}, \sqrt{3})$, Since $x^2 - 2$ is irreducible in $\boldsymbol{Q}[x]$ we have $[\boldsymbol{Q}(\sqrt{2}):\boldsymbol{Q}] = 2$. Clearly, $x^2 - 3$ is irreducible over $\boldsymbol{Q}(\sqrt{2})$ so that $[\boldsymbol{Q}(\sqrt{2}, \sqrt{3}):\boldsymbol{Q}(\sqrt{2})] = 2$. This gives $[\boldsymbol{Q}(\sqrt{2}, \sqrt{3}):\boldsymbol{Q}] = 4$. Hence the Galois group G of $\boldsymbol{Q}(\sqrt{2}, \sqrt{3})$ over $\boldsymbol{Q}$ is of order 4. There are two groups of order 4, the cyclic group and the four-group. The group G cannot be cyclic since, otherwise, G would contain a single subgroup of index 2 whereas $\boldsymbol{Q}(\sqrt{2}, \sqrt{3})$ contains three subfields of degree 2 over $\boldsymbol{Q}$, namely $\boldsymbol{Q}(\sqrt{2})$, $\boldsymbol{Q}(\sqrt{3})$, and $\boldsymbol{Q}(\sqrt{2}\sqrt{3})$. Therefore G is isomorphic to the four-group.

Every element of G permutes the roots $\sqrt{2}$, $-\sqrt{2}$, $\sqrt{3}$, $-\sqrt{3}$ of $(x^2 - 2)(x^2 - 3)$ and distinct elements induce distinct permutations. Every element of G leaves $\sqrt{2}$ fixed or carries $\sqrt{2}$ into the other root $-\sqrt{2}$ of $x^2 - 2$. The same holds for the two roots $\sqrt{3}$ and $-\sqrt{3}$ of $x^2 - 3$. Thus the permutations corresponding to the elements of G are the identity P_ε and

$$P_\alpha = \begin{pmatrix} \sqrt{2} & -\sqrt{2} & \sqrt{3} & -\sqrt{3} \\ -\sqrt{2} & \sqrt{2} & \sqrt{3} & -\sqrt{3} \end{pmatrix}, \quad P_\beta = \begin{pmatrix} \sqrt{2} & -\sqrt{2} & \sqrt{3} & -\sqrt{3} \\ \sqrt{2} & -\sqrt{2} & -\sqrt{3} & \sqrt{3} \end{pmatrix},$$

$$P_\gamma = \begin{pmatrix} \sqrt{2} & -\sqrt{2} & \sqrt{3} & -\sqrt{3} \\ -\sqrt{2} & \sqrt{2} & -\sqrt{3} & \sqrt{3} \end{pmatrix}.$$

We have $P_\alpha^2 = P_\beta^2 = P_\varepsilon$ and $P_\alpha P_\beta = P_\beta P_\alpha$ which shows once more that G is isomorphic to the four-group.

There are three subgroups of index 2, namely $\langle P_\alpha \rangle$, $\langle P_\beta \rangle$, and $\langle P_\gamma \rangle$. The subfields, belonging to these subgroups are $\boldsymbol{Q}(\sqrt{3})$, $\boldsymbol{Q}(\sqrt{2})$, and $\boldsymbol{Q}(\sqrt{2}\sqrt{3})$, respectively.

Example 2. As we observed in § 6.3, the splitting field $\mathbf{E} = \boldsymbol{Q}\left(\sqrt[3]{2}, \varrho\sqrt[3]{2}\right)$ of $x^3 - 2$ over $\boldsymbol{Q}$ is of degree 6. Here we may take $\sqrt[3]{2}$ as the real cube root and $\varrho = \frac{1}{2}(-1 + \sqrt{-3})$. The Galois group G of $\mathbf{E}$ over $\boldsymbol{Q}$ is of order 6. It is evident that the splitting field can also be obtained by adjoining $\sqrt[3]{2}$ and ϱ to $\boldsymbol{Q}$, i.e. $\mathbf{E} = \boldsymbol{Q}\left(\sqrt[3]{2}, \varrho\right)$. Since $\sqrt[3]{2}$ is a root of $x^3 - 2$, every element of G leaves $\sqrt[3]{2}$ fixed or carries it into one of the other roots $\varrho\sqrt[3]{2}$, $\bar{\varrho}\sqrt[3]{2}$ of $x^3 - 2$. Similarly, ϱ and $\bar{\varrho} = \varrho^2$ are the roots of $x^2 + x + 1$ so that every element of G leaves both ϱ and $\bar{\varrho}$ fixed or interchanges ϱ and $\bar{\varrho}$. Since every element of $\mathbf{E}$ is a rational expression of $\sqrt[3]{2}$ and ϱ with rational coefficients it is clear that every element of G is uniquely determined by its effect on $\sqrt[3]{2}$ and ϱ. Using an obvious notation we put

$$\alpha = \left(\sqrt[3]{2} \to \varrho\sqrt[3]{2}, \varrho \to \varrho\right), \quad \beta = \left(\sqrt[3]{2} \to \sqrt[3]{2}, \varrho \to \bar{\varrho}\right).$$

Due to $\varrho^2 = \bar{\varrho}$ and $\varrho\bar{\varrho} = 1$ we obtain

$$\alpha^3 = \beta^2 = \varepsilon, \quad \beta\alpha = \alpha^2\beta$$

where ε denotes the identity automorphism. This shows that G is isomorphic to the symmetric group S_3.

The cyclic group $\langle\alpha\rangle$ is normal in G. To $\langle\alpha\rangle$ there belongs the subfield $\mathbf{Q}(\varrho)$ of E which is normal over $\mathbf{Q}$ since it is the splitting field of $x^2 + x + 1$. The only non-identical automorphism of $\mathbf{Q}(\varrho)/\mathbf{Q}$ interchanges ϱ and $\bar{\varrho}$ so that the Galois group of $\mathbf{Q}(\varrho)$ over $\mathbf{Q}$ is of order 2 and hence isomorphic to the factor group $G/\langle\alpha\rangle$ in accordance with (iii) of Theorem 6.6.9.

The group G contains three subgroups of index 3, namely

$$\langle\beta\rangle, \langle\alpha\beta\rangle, \langle\alpha^2\beta\rangle.$$

To these subgroups there belong the subfields

$$\mathbf{Q}\left(\sqrt[3]{2}\right), \quad \mathbf{Q}\left(\varrho\sqrt[3]{2}\right), \quad \mathbf{Q}\left(\bar{\varrho}\sqrt[3]{2}\right),$$

respectively.

The next two paragraphs will provide further examples.

Let K be a separable extension of F of finite degree $[\mathsf{K}:\mathsf{F}] = r$. By Theorem 6.5.2, K is a simple extension of F, say $\mathsf{K} = \mathsf{F}(\gamma)$. By adjoining all the r conjugates of γ to F we obtain a Galois extension E of F. Let G be the Galois group of E over F. To $\mathsf{K} = \mathsf{F}(\gamma)$ there belongs a subgroup H of G. This means that an element of G leaves γ fixed if and only if it is contained in H. We decompose G into right cosets of H,

$$G = \bigcup_{i=1}^{r} H\varrho_i$$

where $H\varrho_1 = H$. Since γ is fixed under H, all elements of a coset $H\varrho_i$ carry γ into one and the same element γ^{ϱ_i}. Moreover, $\gamma^{\varrho_i} = \gamma^{\varrho_j}$ implies that $i = j$; indeed, from $\gamma^{\varrho_i} = \gamma^{\varrho_j}$ it follows that $\gamma^{\varrho_i\varrho_j^{-1}} = \gamma$ and hence $\varrho_i\varrho_j^{-1} \in H$ or $\varrho_i \in H\varrho_j$, i.e. $i = j$. This shows that the set $\{\gamma^\sigma : \sigma \in G\}$ consists precisely of r elements, namely

$$\gamma^{\varrho_1} = \gamma, \gamma^{\varrho_2}, \ldots, \gamma^{\varrho_r}.$$

Every element of G permutes these elements. Therefore the coefficients of the polynomial

$$(x - \gamma^{\varrho_1})(x - \gamma^{\varrho_2}) \ldots (x - \gamma^{\varrho_r})$$

are fixed under G so that they belong to $\mathbf{F}$. Since the degree of this polynomial is equal to the degree $[\mathbf{F}(\gamma):\mathbf{F}]$ we conclude that it is the minimal polynomial of γ in $\mathbf{F}[x]$.

Now let α be an arbitrary element of $\mathbf{K} = \mathbf{F}(\gamma)$. Then α is certainly fixed under H but in general, namely unless α is a primitive element of $\mathbf{F}(\gamma)$, the precise subgroup U of G that leaves α fixed is larger than H. The following diagram shows the correspondence between the fields in question and the subgroups of G:

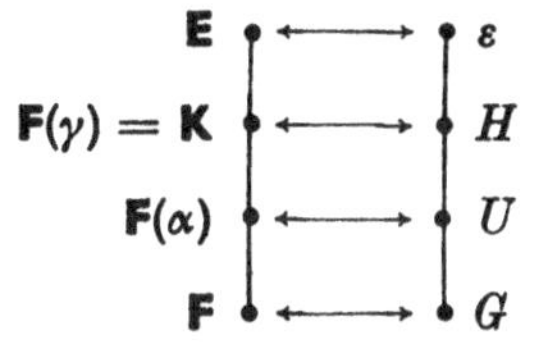

Let

$$G = \bigcup_{j=1}^{t} U\tau_j$$

be the partition of G into right cosets of U where $U\tau_1 = U$. As above we conclude that

$$g(x) = (x - \alpha^{\tau_1})(x - \alpha^{\tau_2}) \dots (x - \alpha^{\tau_t})$$

is the minimal polynomial of α in $\mathbf{F}[x]$. Each coset of U splits into $m = |U:H|$ cosets of H, and ϱ_i and ϱ_j carry α into the same element if and only if they belong to the same coset of U. This shows that among

$$\alpha^{\varrho_1}, \alpha^{\varrho_2}, \dots, \alpha^{\varrho_r}$$

each α^{τ_j} occurs precisely m times. Therefore we obtain

$$\prod_{i=1}^{r} (x - \alpha^{\varrho_i}) = (g(x))^m .$$

Two coefficients of this polynomial occasionally play an important role, namely

$$\sum_{i=1}^{r} \alpha^{\varrho_i} = T_{\mathbf{K}/\mathbf{F}}(\alpha), \quad \prod_{i=1}^{r} \alpha^{\varrho_i} = N_{\mathbf{K}/\mathbf{F}}(\alpha).$$

They are respectively called the *trace* and the *norm* of α from **K** to **F**. Since the ϱ_i are automorphisms it follows that for $\alpha, \beta \in \mathbf{K}$

$$T_{\mathbf{K}/\mathbf{F}}(\alpha + \beta) = T_{\mathbf{K}/\mathbf{F}}(\alpha) + T_{\mathbf{K}/\mathbf{F}}(\beta),$$
$$N_{\mathbf{K}/\mathbf{F}}(\alpha\beta) = N_{\mathbf{K}/\mathbf{F}}(\alpha)\, N_{\mathbf{K}/\mathbf{F}}(\beta).$$

The trace and the norm can also be defined for inseparable extensions but we shall not deal with this case.

Later we shall have to consider the following situation. Let **F** be a field and **E** a Galois extension of **F**. Then **E** is the splitting field of a separable polynomial $f(x) \in \mathbf{F}[x]$. Now let $\mathbf{F}^*$ be any extension field of **F**. Then $f(x)$ can be regarded as a polynomial in $\mathbf{F}^*[x]$ and we can form the splitting field $\mathbf{E}^*$ of $f(x)$ over $\mathbf{F}^*$. If $\alpha_1, \alpha_2, \ldots, \alpha_r$ are the roots of $f(x)$ in $\mathbf{E}^*$, then we have $\mathbf{E}^* = \mathbf{F}^*(\alpha_1, \alpha_2, \ldots, \alpha_r)$. The field $\mathbf{E}^*$ contains the subfield $\mathbf{F}(\alpha_1, \alpha_2, \ldots, \alpha_r)$ which is isomorphic to **E**. Therefore it is no loss of generality to assume that **E** is a subfield of $\mathbf{E}^*$. From Corollary 6.5.7 it follows that $\mathbf{E}^*$ is separable over $\mathbf{F}^*$ and hence a Galois extension of $\mathbf{F}^*$. Our aim is to determine the Galois group of $\mathbf{E}^*$ over $\mathbf{F}^*$.

6.6.10. THEOREM. *If G is the Galois group of* **E** *over* **F**, *then the Galois group G^* of* $\mathbf{E}^*$ *over* $\mathbf{F}^*$ *is isomorphic to the subgroup H of G having* $\mathbf{E} \cap \mathbf{F}^*$ *as its fixed field.*

Proof. Every automorphism of $\mathbf{E}^*$ over $\mathbf{F}^*$ permutes $\alpha_1, \alpha_2, \ldots, \alpha_r$ and leaves $\mathbf{F}^*$, and hence also **F**, fixed. Since the elements of $\mathbf{E}^*$ are rational expressions of $\alpha_1, \alpha_2, \ldots, \alpha_r$ with coefficients in $\mathbf{F}^*$, each automorphism of $\mathbf{E}^*/\mathbf{F}^*$ is uniquely determined by the corresponding permutation of the α_i. The elements of **E** are rational expressions of $\alpha_1, \alpha_2, \ldots, \alpha_r$ with coefficients in **F**, and **F** is fixed under all automorphisms of $\mathbf{E}^*/\mathbf{F}^*$. Therefore every automorphism of $\mathbf{E}^*/\mathbf{F}^*$ induces an automorphism of $\mathbf{E}/\mathbf{F}$, and distinct automorphisms of $\mathbf{E}^*/\mathbf{F}^*$ induce distinct automorphisms of $\mathbf{E}/\mathbf{F}$ because the elements $\alpha_1, \alpha_2, \ldots, \alpha_r$ of **E** are permuted in distinct ways. Denoting the permutation groups on the α_i corresponding to G and G^* by G_P and G_P^*, respectively, we see that G_P^* is a subgroup of G_P. It is clear that G_P^* leaves $\mathbf{E} \cap \mathbf{F}^*$ fixed because $\mathbf{F}^*$ is fixed under G^*. But if a is any element of **E** that does not belong to $\mathbf{E} \cap \mathbf{F}^*$, then a is not contained in $\mathbf{F}^*$ so that a is not fixed under all permutations of G_P^*. This shows that $\mathbf{E} \cap \mathbf{F}^*$ is the fixed field of G_P^* as was to be proved.

If **E** is a Galois extension of **F**, then **E** is separable over **F** and of finite degree $[\mathbf{E} : \mathbf{F}] = n$. Therefore **E** is a simple extension of **F**, say $\mathbf{E} = \mathbf{F}(\vartheta)$. The minimal polynomial $p(x)$ of ϑ over **F**, is of degree n and has n distinct roots $\vartheta_1 = \vartheta, \vartheta_2, \ldots, \vartheta_n$. Since **E** is normal over **F**, all these roots are contained in **E** and we have

$$\mathbf{E} = \mathbf{F}(\vartheta_1) = \mathbf{F}(\vartheta_2) = \ldots = \mathbf{F}(\vartheta_n)$$

because $[\mathbf{E} : \mathbf{F}(\vartheta_i)] = n$ for $i = 1, 2, \ldots, n$. It follows that for each pair ϑ_i, ϑ_k there exists a unique polynomial $g_{ik}(x)$ in $\mathbf{F}[x]$ whose degree does not exceed $n - 1$ such that

$$\vartheta_i = g_{ik}(\vartheta_k) \quad (i, k = 1, 2, \ldots, n).$$

In short, every root of $p(x)$ can be expressed as a polynomial in every other root. Due to this property $p(x)$ is called a *normal polynomial*. Conversely, if $p(x)$ is a separable normal polynomial in $\mathbf{F}[x]$ and if ϑ is a root of $p(x)$, then $\mathbf{F}(\vartheta)$ is a Galois extension of **F**.

Let G denote the Galois group of **E** over **F**. We choose a fixed root of $p(x)$, say ϑ. If σ is any element of G, then ϑ^σ is also a root of $p(x)$ so that as we observed above

$$\vartheta^\sigma = g_\sigma(\vartheta) \quad (\sigma \in G) \tag{10}$$

where $g_\sigma(x)$ is a polynomial in $\mathbf{F}[x]$ whose degree is at most equal to $n - 1$. (Clearly, $g_\sigma(x)$ is one of the $g_{ik}(x)$.) Since the coefficients of $g_\sigma(x)$ belong to **F** the equation (10) gives $\vartheta^{\sigma\tau} = g_\sigma(\vartheta^\tau)$ for any $\tau \in G$. On the other hand, we have $\vartheta^{\sigma\tau} = g_{\sigma\tau}(\vartheta)$ and $\vartheta^\tau = g_\tau(\vartheta)$. So we obtain

$$g_{\sigma\tau}(\vartheta) = g_\sigma(g_\tau(\vartheta)).$$

Because $p(x)$ is irreducible in $\mathbf{F}[x]$ this yields the following rule for combining the polynomials $g_\sigma(x)$:

$$g_{\sigma\tau}(x) \equiv g_\sigma(g_\tau(x)) \pmod{p(x)}.$$

6.7. Cyclotomic fields

Let **P** be any prime field and n a natural number such that the characteristic of **P** does not divide n. By the n-th *roots of unity* we understand the roots of the polynomial $x^n - 1$ in some extension field of P. The splitting field K of $x^n - 1$ over P is called the n-th *cyclotomic field* over P. In case $\mathsf{P} = \mathbb{Q}$ the roots of $x^n - 1$ form the vertices of a regular n-gon inscribed in the unit circle of the complex plane. This accounts for the name cyclotomic field.

Due to the condition that n is not divisible by the characteristic of P the derivative nx^{n-1} is obviously relatively prime to $x^n - 1$ so that all the roots of $x^n - 1$ are distinct.

6.7.1. LEMMA. *If C is a finite group whose elements belong to a field F and whose operation is the multiplication in F, then C is cyclic.*

Proof. Clearly, C is abelian. Let C be of order $n > 1$ and let

$$n = \prod_{i=1}^{r} p_i^{\nu_i}$$

be the factorization of n into powers of distinct prime numbers p_i. In C there are at most n/p_i elements y for which

$$y^{n/p_i} = 1$$

because the polynomial $x^{n/p_i} - 1$ cannot have more than n/p_i roots in F. So, for $i = 1, 2, \ldots, r$, there exist elements y_i in C such that

$$y_i^{n/p_i} \neq 1 \quad (i = 1, 2, \ldots, r).$$

Therefore

$$z_i = y_i^{n/p_i^{\nu_i}} \quad (i = 1, 2, \ldots, r)$$

is of order $p_i^{\nu_i}$. Hence the product $\prod_{i=1}^{r} z_i$ is of order n which shows that C is cyclic.

If α and β are n-th roots of unity, i.e. $\alpha^n = \beta^n = 1$, then $(\alpha\beta)^n = 1$ and $(\alpha^{-1})^n = 1$. Therefore the n-th roots of unity form a multiplicative group of order n. By Lemma 6.7.1, this group is cyclic. So we have:

6.7.2. *The n-th roots of unity form a cyclic group of order n.*

Applying Corollary 3.6.5 to the group of the n-th roots of unity we obtain the following result: There are $\varphi(n)$ generating elements of the group of the n-th roots of unity. They are called the *primitive* n-th roots of unity. If ζ is any primitive n-th root of unity, then all the primitive roots of unity are the powers ζ^k with $1 \leqq k \leqq n$ and $(k, n) = 1$.

The polynomial

$$\Phi_n(x) = \prod_{\substack{k=1 \\ (k,n)=1}}^{n} (x - \zeta^k)$$

whose roots are the primitive n-th roots of unity is called the n-th *cyclotomic polynomial* over P. In view of the definition of Euler's function in § 2.7, the degree of $\Phi_n(x)$ is equal to $\varphi(n)$.

6.7.3. Theorem. *The coefficients of $\Phi_n(x)$ belong to P. In case P is the field of the rational numbers, the coefficients of $\Phi_n(x)$ are integers.*

Proof. Any non-primitive n-th root of unity α is a primitive d-th root of unity for some proper divisor d of n. Here d is the order of α as an element of the group of the n-th roots of unity. Conversely, if d is a divisor of n, then every primitive d-th root of unity is also an n-th root of unity. This gives

$$x^n - 1 = \prod_{d \mid n} \Phi_d(x) \tag{1}$$

where d ranges over all positive divisors of n.

We now prove our theorem by induction on n. For $n = 1$, the cyclotomic polynomial is $x - 1$ so that the theorem holds. Suppose that the theorem is true for all m-th cyclotomic polynomials with $m < n$. Consequently all the $\Phi_d(x)$ for $d \mid n$, $d < n$, have coefficients in P. The polynomial $\Phi_n(x)$ is obtained by dividing $x^n - 1$ by the product $f(x)$ of the $\Phi_d(x)$ with $d \mid n$, $d < n$. Since both $x^n - 1$ and $f(x)$ have coefficients in P, the same holds for their quotient. In case $\mathsf{P} = Q$ we have to divide $x^n - 1$ by the *monic* polynomial $f(x)$ with integral coefficients so that the coefficients of the quotient are also integers. This completes the proof.

To derive an explicit formula for $\Phi_n(x)$ we define the so-called *Möbius function* $\mu(m)$ for all natural numbers m as follows:

$$\mu(m) = \begin{cases} 1 & \text{for } m = 1, \\ 0 & \text{if } m \text{ is divisible by the square of a prime number,} \\ (-1)^r & \text{if } m = p_1 p_2 \ldots p_r \text{ where the } p_i \text{ are distinct prime numbers.} \end{cases}$$

It is easy to verify that, for distinct prime numbers $p_1, p_2, \ldots, p_r$,

$$\mu(p_1^{a_1} p_2^{a_2} \ldots p_r^{a_r}) = \mu(p_1^{a_1})\, \mu(p_2^{a_2}) \ldots \mu(p_r^{a_r}).$$

Moreover, we have

$$\sum_{d|m} \mu(d) = \begin{cases} 1 \text{ for } m = 1, \\ 0 \text{ for } m > 1 \end{cases} \tag{2}$$

where d ranges over all positive divisors of m. To prove this formula we consider the prime factorization

$$m = p_1^{a_1} p_2^{a_2} \ldots p_r^{a_r}.$$

The positive divisors of m are precisely the numbers

$$d = p_1^{b_1} p_2^{b_2} \ldots p_r^{b_r} \text{ where } 0 \leqq b_i \leqq a_i.$$

For $m > 1$ we thus obtain

$$\begin{aligned} \sum_{d|m} \mu(d) &= \sum_{b_1,\ldots,b_r} \mu(p_1^{b_1} \ldots p_r^{b_r}) = \sum_{b_1,\ldots,b_r} \mu(p_1^{b_1}) \ldots \mu(p_r^{b_r}) \\ &= \Big(\sum_{b_1} \mu(p_1^{b_1})\Big) \cdots \Big(\sum_{b_r} \mu(p_r^{b_r})\Big) \\ &= (1 + \mu(p_1) + \mu(p_1^2) + \ldots) \ldots (1 + \mu(p_r) + \mu(p_r^2) + \ldots) \\ &= (1-1) \ldots (1-1) = 0. \end{aligned}$$

We are now ready to verify the following explicit formula:

$$\Phi_n(x) = \prod_{d|n} (x^d - 1)^{\mu(n/d)}. \tag{3}$$

It is evident that this formula holds for $n = 1$. We assume that

$$\Phi_d(x) = \prod_{t|d} (x^t - 1)^{\mu(d/t)} \tag{4}$$

for any proper divisor d of n. To verify (3) we show that (1) is satisfied if we substitute the right-hand sides of (4) and (3) for

$\Phi_d(x)$ and $\Phi_n(x)$, respectively. We obtain

$$\prod_{d|n} \Phi_d(x) = \prod_{d|n} \prod_{t|d} (x^t - 1)^{\mu(d/t)}. \tag{5}$$

The exponent of a fixed factor $x^t - 1$ on the right-hand side of (5) is the sum of all $\mu(d/t)$ where $t \mid d$ and $d \mid n$. This exponent is therefore equal to $\Sigma\mu(f)$ where f ranges over all divisors of n/t. By (2), this sum is equal to zero unless $n/t = 1$ in which case the sum in question is equal to 1. This shows that the product (5) is equal to $x^n - 1$ which proves our theorem.

We now restrict ourselves to the cyclotomic polynomials over the field $\boldsymbol{Q}$ of the rational numbers. In § 5.3 we proved that, for any prime number p, $\Phi_p(x)$ is irreducible in $\boldsymbol{Q}[x]$. We now generalize this result for arbitrary $\Phi_n(x)$.

6.7.4. Theorem. *For any natural number n, the cyclotomic polynomial $\Phi_n(x)$ is irreducible in $\boldsymbol{Q}[x]$.*

Proof. Let ζ be a primitive n-th root of unity and let $g(x)$ denote the minimal polynomial of ζ in $\boldsymbol{Q}[x]$. The polynomial $g(x)$ is monic and divides $\Phi_n(x)$. Since $\Phi_n(x)$ has integral coefficients it follows from Theorem 5.3.4 that the coefficients of $g(x)$ are integers. To prove the irreducibility of $\Phi_n(x)$ we show that all the primitive n-th roots of unity are also roots of $g(x)$, i.e. $g(\zeta^k) = 0$ for all k such that $1 \leqq k \leqq n$ and $(k, n) = 1$.

Let m denote the degree of $g(x)$. Then every element of $\boldsymbol{Q}(\zeta)$ has a unique representation

$$a_0 + a_1\zeta + \dots + a_{m-1}\zeta^{m-1}$$

with coefficients $a_i \in \boldsymbol{Q}$. If $h(x)$ is a polynomial with integral coefficients, then

$$h(\zeta) = b_0 + b_1\zeta + \dots + b_{m-1}\zeta^{m-1} \tag{6}$$

where the b_i are integers. Indeed, if we divide $h(x)$ by $g(x)$,

$$h(x) = g(x)\,q(x) + r(x),$$

then the degree of the remainder $r(x)$ does not exceed $m - 1$, and the coefficients of $r(x)$ are integers since both $h(x)$ and $g(x)$ have integral coefficients and $g(x)$ is monic. Substituting ζ for x we thus obtain a representation (6) with integral coefficients. In particular, for any natural number k such that $1 \leqq k \leqq n$ and $(k, n) = 1$ we obtain

$$g(\zeta^k) = c_{k,0} + c_{k,1}\zeta + \ldots + c_{k,m-1}\zeta^{m-1} \tag{7}$$

with integers $c_{k,l}$. Let c denote the maximum of the absolute values $|c_{k,l}|$ of the $\varphi(n)\,m$ integers $c_{k,l}$.

Now let p be a prime number that does not divide n. From

$$g(x)^p \equiv g(x^p) \pmod p$$

we conclude that

$$g(x^p) = g(x)^p + pG(x)$$

where $G(x)$ has integral coefficients. Substituting ζ for x we get

$$g(\zeta^p) = pG(\zeta).$$

Representing $G(\zeta)$ in the form (6) we find that in the expression (7) for $g(\zeta^p)$ all the coefficients on the right-hand side are divisible by p. If, moreover, $p > c$, then $|c_{k,l}| \leqq c$ implies that all the coefficients $c_{k,l}$ are equal to zero, i.e. $g(\zeta^p) = 0$. Thus we have the following result: If p is any prime number such that $(p, n) = 1$ and $p > c$, then $g(\zeta^p) = 0$.

Now let $p_1, \ldots, p_s$ be prime numbers, not necessarily distinct, such that $(p_i, n) = 1$ and $p_i > c$ for $i = 1, \ldots, s$. Applying the last result s times we find that $g(\zeta^{p_1 \cdots p_s}) = 0$.

To complete the proof we show that for any integer k with $1 \leqq k \leqq n$ and $(k, n) = 1$ there exist prime numbers $p_1, \ldots, p_s$ such that $(p_i, n) = 1$, $p_i > c$, and

$$k \equiv p_1 \ldots p_s \pmod n.$$

By our previous results this gives $g(\zeta^k) = g(\zeta^{p_1 \cdots p_s}) = 0$.

The existence of such prime numbers can be established as follows: Let q denote the product of all prime numbers q_j such that $(q_j, n) = 1$ and $q_j \leqq c$. By Theorem 2.7.1, there is a natural number l which satisfies the congruences

$$l \equiv k \pmod n, \quad l \equiv 1 \pmod q.$$

Due to the second congruence all the prime factors of $l = p_1 \dots p_s$ are greater than c. So we obtain the desired result

$$k \equiv p_1 \dots p_s \pmod n, \ p_i > c.$$

This completes the proof.

Since all the n-th roots of unity are powers of a primitive one, the splitting field of $x^n - 1$ over $\mathbf{Q}$ is the field $\mathbf{Q}(\zeta)$ where ζ is an arbitrary primitive n-th root of unity. Due to the irreducibility of $\Phi_n(x)$ we have

$$[\mathbf{Q}(\zeta) : \mathbf{Q}] = \varphi(n).$$

Any element of the Galois group of $\mathbf{Q}(\zeta)$ over $\mathbf{Q}$ carries ζ into some power ζ^k with $1 \leqq k \leqq n$ and $(k, n) = 1$; conversely, if k satisfies these conditions, then the mapping $\zeta \to \zeta^k$ gives rise to an automorphism of $\mathbf{Q}(\zeta)/\mathbf{Q}$. Since only the residue classes mod n of the exponents of ζ are relevant we can state the last result as follows: There is a one-to-one correspondence between the elements of the Galois group of $\mathbf{Q}(\zeta)$ over $\mathbf{Q}$ and the prime residue classes mod n. Let σ_k denote the element of the Galois group corresponding to the residue class of k, i.e. $\zeta^{\sigma_k} = \zeta^k$. If i, j, k are representatives of prime residue classes mod n such that

$$ij \equiv k \pmod n,$$

then we have

$$\zeta^{\sigma_i\sigma_j} = (\zeta^{\sigma_i})^{\sigma_j} = (\zeta^i)^{\sigma_j} = (\zeta^i)^j = \zeta^{ij} = \zeta^k = \zeta^{\sigma_k}$$

so that $\sigma_i\sigma_j = \sigma_k$. This gives the following result:

6.7.5. Theorem. *The Galois group of the n-th cyclotomic field over $\mathbf{Q}$ is isomorphic to the multiplicative group of the prime residue classes mod n.*

We shall now turn to a more detailed study of the p-th cyclotomic fields for a prime number p. Let ζ denote a primitive p-th root of unity and $\mathbf{E} = \mathbf{Q}(\zeta)$. We have $[\mathbf{E} : \mathbf{Q}] = p - 1$. To prove this we need not the general Theorem 6.7.4 but only the irreducibility of $\Phi_p(x)$ in $\mathbf{Q}[x]$ which was proved in § 5.3 as a simple consequence of Eisenstein's Theorem.

Since the residue class ring $\mathbf{Z}/(p)$ is a field it follows from Lemma 6.7.1 that the group of the prime residue classes mod p is cyclic. Let g be an integer that belongs to one of the generating elements of the group of the prime residue classes mod p. Such an integer is called a *primitive root* mod p. All the prime residue classes mod p are represented by the powers

$$1, g, g^2, \ldots, g^{p-2}.$$

Therefore

$$\zeta, \zeta^g, \zeta^{g^2}, \ldots, \zeta^{g^{p-2}}$$

are all the primitive p-th roots of unity. The automorphism σ defined by $\zeta^\sigma = \zeta^g$ is of order $p - 1$ and generates the Galois group of $\mathbf{E}$ over $\boldsymbol{Q}$

To simplify the notation we write

$$\zeta^{\sigma^i} = \zeta^{g^i} = \zeta_i.$$

In view of $\sigma^i = \sigma^k$ if $i \equiv k \pmod{p-1}$ we admit arbitrary subscripts of the ζ_i on the understanding that $\zeta_i = \zeta_k$ if $i \equiv k \pmod{p-1}$. Then we have the rule

$$\zeta_i^{\sigma^j} = \zeta_{i+j}. \tag{8}$$

The elements $\zeta_0, \zeta_1, \ldots, \zeta_{p-2}$ form a basis of $\mathbf{E}$ over $\boldsymbol{Q}$. Indeed, except for the order they coincide with the powers $\zeta, \zeta^2, \ldots, \zeta^{p-1}$ which are linearly independent over $\boldsymbol{Q}$; for $a_0\zeta + a_1\zeta^2 + \ldots + a_{p-2}\zeta^{p-1} = 0$ gives

$$a_0 + a_1\zeta + \ldots + a_{p-2}\zeta^{p-2} = 0$$

so that $a_0 = a_1 = \ldots = a_{p-2} = 0$ because the minimal polynomial $\Phi_p(x)$ of ζ in $\boldsymbol{Q}[x]$ is of degree $p - 1$.

Let f be a positive divisor of $p - 1$. By Theorem 3.6.1, the cyclic group $\langle\sigma\rangle$ of order $p - 1$ contains a unique subgroup H of order f. If $p - 1 = ef$, then H consists of the identity and the powers $\sigma^e, \sigma^{2e}, \ldots, \sigma^{(f-1)e}$. Our aim is to find a basis of the fixed field $\mathbf{K}$ of H. As we observed above, any element α of $\mathbf{E}$ has a unique representation

$$\alpha = a_0\zeta_0 + a_1\zeta_1 + \ldots + a_{p-2}\zeta_{p-2} \tag{9}$$

with coefficients $a_i \in \boldsymbol{Q}$. Evidently, α belongs to $\mathbf{K}$ if and only if $\alpha^{\sigma^e} = \alpha$. In view of (8) this means

$$\alpha^{\sigma^e} = a_0\zeta_e + a_1\zeta_{e+1} + \ldots + a_{p-2}\zeta_{e+p-2} = \alpha$$

or

$$a_0 = a_e, a_1 = a_{e+1}, \ldots, a_{p-2} = a_{e+p-2}$$

on the understanding that $a_i = a_k$ if $i \equiv k \pmod{p-1}$. Collecting the terms with the same coefficients a_i in (9) we obtain

$$\alpha = a_0\eta_0 + a_1\eta_1 + \ldots + a_{e-1}\eta_{e-1}$$

where

$$\eta_i = \zeta_i + \zeta_{e+i} + \zeta_{2e+i} + \ldots + \zeta_{(f-1)e+i} \quad (i = 0, 1, \ldots, e-1). \quad (10)$$

The linear independence of $\zeta_0, \zeta_1, \ldots, \zeta_{p-2}$ over $\mathbf{Q}$ implies that $\eta_0, \eta_1, \ldots, \eta_{e-1}$ are linearly independent over $\mathbf{Q}$. Therefore the η_i form a basis of K over $\mathbf{Q}$. They are called the periods of f terms.

From (8) and (10) we conclude that

$$\eta_i^{\sigma_j} = \eta_{i+j}$$

on the understanding that $\eta_i = \eta_k$ if $i \equiv k \pmod{e}$. Since $\langle\sigma\rangle$ is abelian, the subgroup H is normal and, hence, K is a normal extension of $\mathbf{Q}$. This gives

$$\mathsf{K} = \mathbf{Q}(\eta_0, \ldots, \eta_{e-1}) = \mathbf{Q}(\eta_i)$$

for an arbitrary subscript i. The coefficients of the polynomial

$$(x - \eta_0)(x - \eta_1) \ldots (x - \eta_{e-1}) \quad (11)$$

are fixed under σ so that they belong to $\mathbf{Q}$. Due to $[\mathsf{K} : \mathbf{Q}] = e$ this polynomial is irreducible in $\mathbf{Q}[x]$.

For any i with $0 \leqq i \leqq e-1$ the elements

$$\zeta_i, \zeta_{e+i}, \zeta_{2e+i}, \ldots, \zeta_{(f-1)e+i}$$

undergo a cyclic permutation under σ^e. Therefore the coefficients of each polynomial

$$(x - \zeta_i)(x - \zeta_{e+i}) \ldots (x - \zeta_{(f-1)e+i}) \quad (i = 0, 1, \ldots, e-1) \quad (12)$$

belong to K, and due to $[\mathsf{E} : \mathsf{K}] = [\mathbf{Q}(\zeta_i) : \mathsf{K}] = f$ these polynomials are irreducible in $\mathsf{K}[x]$.

We consider the example $p = 13$. One easily verifies that 2 is a primitive root mod 13. The following table gives the values of i and k that correspond to each other by the congruence

$$2^i \equiv k \pmod{13}:$$

i	0	1	2	3	4	5	6	7	8	9	10	11
k	1	2	4	8	3	6	12	11	9	5	10	7

Let ζ denote any primitive 13th root of unity. In the decomposition $13 - 1 = ef$ we take $e = 3$, $f = 4$. The periods of four terms are

$$\eta_0 = \zeta_0 + \zeta_3 + \zeta_6 + \zeta_9 = \zeta + \zeta^8 + \zeta^{12} + \zeta^5,$$

$$\eta_1 = \zeta_1 + \zeta_4 + \zeta_7 + \zeta_{10} = \zeta^2 + \zeta^3 + \zeta^{11} + \zeta^{10},$$

$$\eta_2 = \zeta_2 + \zeta_5 + \zeta_8 + \zeta_{11} = \zeta^4 + \zeta^6 + \zeta^9 + \zeta^7.$$

A computation shows that

$$\eta_0 + \eta_1 + \eta_2 = -1,$$

$$\eta_0\eta_1 + \eta_1\eta_2 + \eta_2\eta_0 = -4,$$

$$\eta_0\eta_1\eta_2 = -1.$$

Therefore the polynomial (11) becomes

$$(x - \eta_0)(x - \eta_1)(x - \eta_2) = x^3 + x^2 - 4x + 1.$$

Computing the polynomials (12) we find

$$(x - \zeta_0)(x - \zeta_3)(x - \zeta_6)(x - \zeta_9)$$

$$= x^4 - \eta_0 x^3 + (2 + \eta_2)x^2 - \eta_0 x + 1,$$

$$(x - \zeta_1)(x - \zeta_4)(x - \zeta_7)(x - \zeta_{10})$$

$$= x^4 - \eta_1 x^3 + (2 + \eta_0)x^2 - \eta_1 x + 1,$$

$$(x - \zeta_2)(x - \zeta_5)(x - \zeta_8)(x - \zeta_{11})$$

$$= x^4 - \eta_2 x^3 + (2 + \eta_1)x^2 - \eta_2 x + 1.$$

The group $H = \langle\sigma^3\rangle$ is cyclic of order 4. To its subgroup $\langle\sigma^6\rangle$ of order 2 there belongs a subfield $\mathbf{L}$ such that $[\mathbf{Q}(\zeta):\mathbf{L}] = 2$ and $[\mathbf{L}:\mathbf{Q}(\eta_0)] = 2$. There are six periods of two terms that are fixed under $\langle\sigma^6\rangle$, namely

$$\xi_0 = \zeta_0 + \zeta_6 = \zeta + \zeta^{12}, \quad \xi_3 = \zeta_3 + \zeta_9 = \zeta^8 + \zeta^5,$$

$$\xi_1 = \zeta_1 + \zeta_7 = \zeta^2 + \zeta^{11}, \quad \xi_4 = \zeta_4 + \zeta_{10} = \zeta^3 + \zeta^{10},$$

$$\xi_2 = \zeta_2 + \zeta_8 = \zeta^4 + \zeta^9, \quad \xi_5 = \zeta_5 + \zeta_{11} = \zeta^6 + \zeta^7.$$

These periods are roots of the polynomials

$$(x - \xi_0)(x - \xi_3) = x^2 - \eta_0 x + \eta_2,$$
$$(x - \xi_1)(x - \xi_4) = x^2 - \eta_1 x + \eta_0,$$
$$(x - \xi_2)(x - \xi_5) = x^2 - \eta_2 x + \eta_1.$$

Finally, the primitive 13th roots of unity are roots of the polynomial

$$(x - \zeta_0)(x - \zeta_6) = x^2 - \xi_0 x + 1$$

and its images under $\sigma, \sigma^2, \ldots, \sigma^5$.

6.8. Galois fields

A *Galois field* is a field with finitely many elements. For instance, for any prime number p the residue class field $\mathbf{Z}/(p)$ is a Galois field. The characteristic of every Galois field is a prime number for otherwise there would be infinitely many multiples of the unit element.

Let $\mathbf{K}$ be a Galois field of characteristic p and $\mathbf{P}$ the prime field of $\mathbf{K}$. By Theorem 4.7.2, $\mathbf{P}$ is isomorphic to $\mathbf{Z}/(p)$. It is evident that the degree of $\mathbf{K}$ over $\mathbf{P}$ is finite, say $[\mathbf{K}: \mathbf{P}] = n$. Let $\omega_1, \ldots, \omega_n$ be a basis of $\mathbf{K}$ over $\mathbf{P}$. Then every element of $\mathbf{K}$ has a unique representation of the form

$$a_1\omega_1 + a_2\omega_2 + \ldots + a_n\omega_n$$

with coefficients $a_i \in \mathbf{P}$. Since all the a_i range independently over the p elements of $\mathbf{P}$ it follows that $\mathbf{K}$ consists of p^n elements.

The non-zero elements of $\mathbf{K}$ form an abelian group of order $p^n - 1$ with respect to multiplication, the multiplicative group of $\mathbf{K}$. By Lemma 6.7.1 we obtain the following result:

6.8.1. *The multiplicative group of every Galois field is cyclic.*

By Theorem 3.5.8, any element α of $\mathbf{K}$ other than 0 satisfies the equation $\alpha^{p^n-1} = 1$. Hence we have $\alpha^{p^n} = \alpha$ for every $\alpha \in \mathbf{K}$. In other words, every element of $\mathbf{K}$ is a root of the polynomial $x^{p^n} - x$. Since the degree of this polynomial is equal to the number of elements in $\mathbf{K}$ it follows that $\mathbf{K}$ consists precisely of the roots of $x^{p^n} - x$.

This means that **K** is the splitting field of $x^{p^n} - x$ over **P**. Since the splitting field is unique up to isomorphism over **P** it follows that a Galois field is uniquely determined, up to isomorphism, by the number of its elements. So we introduce the notation $\mathrm{GF}(p^n)$ for the Galois field with p^n elements.

By 5.4.9, $\mathrm{GF}(p)$ is perfect. Therefore every Galois field is a separable extension of any subfield. Being a splitting field of a polynomial with coefficients in its prime field every Galois field is a normal extension of every subfield. Therefore every Galois field is a Galois extension of every subfield.

Theorem 6.5.2 has been proved under the additional assumption that the fields contain infinitely many elements. From 6.8.1 we conclude that Theorem 6.5.2 is also valid for Galois fields. Indeed, $\mathrm{GF}(p^n)$ can be obtained out of any subfield by adjoining a generating element of the multiplicative group of $\mathrm{GF}(p^n)$.

We now prove that for any given prime power p^n there exists a $\mathrm{GF}(p^n)$. Let **P** be the prime field of characteristic p and let **S** denote the splitting field of $x^{p^n} - x$ over **P**. The derivative

$$p^n x^{p^n - 1} - 1 = -1$$

is distinct from zero so that $x^{p^n} - x$ has p^n distinct roots in **S**. Let **K** be the set of all these roots. We show that **K** is a subfield of **S**. (By our previous results this implies that **K** = **S**.) If α and β belong to **K**, i.e. $\alpha^{p^n} = \alpha$, $\beta^{p^n} = \beta$, then $(\alpha\beta)^{p^n} = \alpha\beta$ and, if $\beta \neq 0$, $(\alpha/\beta)^{p^n} = \alpha/\beta$. Moreover, since **S** is of characteristic p, we have $(\alpha \pm \beta)^p = \alpha^p \pm \beta^p$ and hence $(\alpha \pm \beta)^{p^n} = \alpha^{p^n} \pm \beta^{p^n} = \alpha \pm \beta$. This shows that **K** is a field. So we have the following result:

6.8.2. Theorem. *For any prime power p^n there exists one and, up to isomorphism, only one* $\mathrm{GF}(p^n)$.

Next, we determine the subfields of $\mathrm{GF}(p^n)$.

6.8.3. Theorem. *For any positive divisor m of n,* $\mathrm{GF}(p^n)$ *contains precisely one* $\mathrm{GF}(p^m)$. *These* $\mathrm{GF}(p^m)$ *are the only subfields of* $\mathrm{GF}(p^n)$.

Proof. Clearly, every subfield of $\mathrm{GF}(p^n)$ is some $\mathrm{GF}(p^m)$. We have $[\mathrm{GF}(p^n):\mathrm{GF}(p)] = n$, $[\mathrm{GF}(p^m):\mathrm{GF}(p)] = m$. From Theorem

6.2.3 it follows that m is a divisor of n. The elements of $\mathrm{GF}(p^m)$ are the roots of the polynomial $x^{p^m} - x$. If m divides n, then $(p^m - 1) \mid (p^n - 1)$ so that $(x^{p^m-1} - 1) \mid (x^{p^n-1} - 1)$ and hence also

$$(x^{p^m} - x) \mid (x^{p^n} - x).$$

Therefore, the roots of $x^{p^m} - x$ form a subset of the roots of $x^{p^n} - x$. Hence, $\mathrm{GF}(p^n)$ contains precisely one $\mathrm{GF}(p^m)$ for any divisor m of n. This completes the proof.

We now consider the automorphisms of $\mathrm{GF}(p^n)$. By 4.7.4, every automorphism of $\mathrm{GF}(p^n)$ leaves the subfield $\mathrm{GF}(p)$ fixed. The mapping σ defined by

$$\alpha^\sigma = \alpha^p \text{ for every } \alpha \in \mathrm{GF}(p^n)$$

is an automorphism of $\mathrm{GF}(p^n)$ because

$$(\alpha + \beta)^p = \alpha^p + \beta^p, \quad (\alpha\beta)^p = \alpha^p\beta^p.$$

We have $\alpha^{\sigma^k} = \alpha^{p^k}$. Since not all the p^n elements of $\mathrm{GF}(p^n)$ are roots of the polynomial $x^{p^k} - x$ unless $k = n$ it follows that, for $0 < k < n$, σ^k is distinct from the identity automorphism ε. In other words, σ is of order n, and the automorphisms

$$\varepsilon, \sigma, \sigma^2, \ldots, \sigma^{n-1} \tag{1}$$

are distinct. In view of $[\mathrm{GF}(p^n):\mathrm{GF}(p)] = n$ it follows from Theorem 6.6.3 that (1) are the only automorphisms of $\mathrm{GF}(p^n)$. This gives the following result:

6.8.4. Theorem. *The Galois group of* $\mathrm{GF}(p^n)$ *over* $\mathrm{GF}(p)$ *is cyclic of order* n.

If m is a natural number dividing n, then, by Theorem 3.6.1, the cyclic group $G = \langle\sigma\rangle$ of order n contains precisely one subgroup H of index m, namely $H = \langle\sigma^m\rangle$. The fixed field of H consists of all $\alpha \in \mathrm{GF}(p^n)$ such that $\alpha^{\sigma^m} = \alpha^{p^m} = \alpha$. These elements are therefore the roots of $x^{p^m} - x$ and form the subfield $\mathrm{GF}(p^m)$ of $\mathrm{GF}(p^n)$. Moreover, the Galois group of $\mathrm{GF}(p^m)$ over $\mathrm{GF}(p)$ is cyclic of order m and, hence, isomorphic to the factor group G/H. So we have once more proved the Principal Theorem of the Galois Theory for Galois fields.

The fact that $\mathrm{GF}(p)$ is isomorphic to $\boldsymbol{Z}/(p)$ yields a number theoretical result.

6.8.5. WILSON'S THEOREM. *For any prime number p*

$$(p-1)! \equiv -1 \pmod{p}.$$

Proof. The elements of GF(p) are the roots of the polynomial $x^p - x$. Therefore the non-zero elements of GF(p) are the roots of $x^{p-1} - 1$.
Hence

$$x^{p-1} - 1 = \prod (x - \alpha)$$

where α ranges over all non-zero elements of GF(p). This shows that the product of all non-zero elements of GF(p) is equal to -1. Due to the isomorphism of GF(p) and $\mathbf{Z}/(p)$ the product of all non-zero residue classes mod p is equal to the residue class of -1. Since the non-zero residue classes mod p are represented by $1, 2, \ldots, p-1$ the theorem follows.

We conclude this section by a theorem which states that in the case of finite fields the commutative law of multiplication is a consequence of the other field axioms.

6.8.6. WEDDERBURN'S THEOREM. *Every finite skew field is a field.*

Proof. We consider the n-th roots of unity in the field of the complex numbers, i.e. the numbers

$$\varepsilon^r = \cos 2\pi r/n + i \sin 2\pi r/n \quad (r = 0, 1, \ldots, n-1).$$

The primitive roots of unity are the powers ε^k with $(k, n) = 1$. They are the roots of the cyclotomic polynomial

$$\Phi_n(x) = \prod_{\substack{k=1 \\ (k,n)=1}}^{n-1} (x - \varepsilon^k).$$

By Theorem 6.7.3, the coefficients of $\Phi_n(x)$ are integers. We first show that, for any natural divisor d of n with $d < n$, the quotient

$$(x^n - 1)/(x^d - 1) = x^{n-d} + x^{n-2d} + \ldots + x^d + 1 \qquad (2)$$

is divisible by $\Phi_n(x)$. Indeed, the numerator $x^n - 1$ is the product of all the linear factors $x - \varepsilon^r$ $(r = 0, 1, \ldots, n-1)$ whereas the denominator $x^d - 1$ contains only linear factors $x - \eta$ for which $\eta^d = 1$ so that η is not a primitive n-th root of unity. Therefore the linear factors of $x^n - 1$ corresponding to the primitive roots of unity are not factors of the denominator $x^d - 1$. This shows that the quotient (2) is divisible by $\Phi_n(x)$. Substituting any natural number q for x we find that $(q^n - 1)/(q^d - 1)$ is divisible by $\Phi_n(q)$.

Now let $\mathbf{S}$ be a skew field with finitely many elements. By $\mathbf{Z}$ we denote the centre of $\mathbf{S}$ (cf. § 4.1). It is evident that $\mathbf{Z}$ is a Galois field. Let q denote the number of elements in $\mathbf{Z}$. (We shall not need the fact that q is a prime power.)

Since $\mathbf{S}$ contains finitely many elements it is an algebra of finite rank n over $\mathbf{Z}$. Let $\omega_1, \ldots, \omega_n$ be a basis of $\mathbf{S}$ over $\mathbf{Z}$. Then any element of $\mathbf{S}$ has a unique representation of the form

$$a_1\omega_1 + \ldots + a_n\omega_n$$

with coefficients $a_k \in \mathbf{Z}$. Since the a_k range independently over the elements of $\mathbf{Z}$ it follows that $\mathbf{S}$ contains q^n elements.

We now consider a fixed element α of $\mathbf{S}$ and form the set $\mathbf{Z}_\alpha$ of all elements ζ of $\mathbf{S}$ such that $\alpha\zeta = \zeta\alpha$. It is easy to verify that $\mathbf{Z}_\alpha$ is a skew field contained in $\mathbf{S}$. As above, we conclude that the number of elements in $\mathbf{Z}_\alpha$ is some power q^d where the exponent d depends on α. For $\alpha = 0$ we obviously have $\mathbf{Z}_0 = \mathbf{S}$. For $\alpha \neq 0$, the multiplicative group $\mathbf{Z}_\alpha^\times$ of the non-zero elements of $\mathbf{Z}_\alpha$ is the normalizer of α in the multiplicative group $\mathbf{S}^\times$ of the non-zero elements of $\mathbf{S}$ (cf. § 3.11).

We decompose the group $\mathbf{S}^\times$ into its conjugacy classes. Precisely the elements of $\mathbf{Z}^\times$ are the only elements in their conjugacy classes. By Theorem 3.11.5, the numbers of elements in the other conjugacy classes are of the form $(q^n - 1)/(q^d - 1)$ where d is a divisor of n and $1 \leqq d < n$. Counting the elements of $\mathbf{S}^\times$ we obtain

$$q^n - 1 = q - 1 + \sum_{\substack{d|n \\ 1 \leqq d < n}} (q^n - 1)/(q^d - 1).$$

As we saw above, $q^n - 1$ and each summand $(q^n - 1)/(q^d - 1)$ is divisible by $\Phi_n(q)$. Therefore $q - 1$ is also divisible by $\Phi_n(q)$.

In case $n > 1$ we have

$$\Phi_n(q) = \prod_{\substack{k=1 \\ (k,n)=1}}^{n-1} (q - \varepsilon^k).$$

Regarding ε^k as a point of the complex plane one easily verifies that each factor $q - \varepsilon^k$ has an absolute value which is greater than $q - 1$. This gives $|\Phi_n(q)| > q - 1$ which contradicts the divisibility of $q - 1$ by $\Phi_n(q)$. It follows that $n = 1$, i.e. $\mathbf{S} = \mathbf{Z}$, which means that multiplication in $\mathbf{S}$ is commutative.

Exercises

1. Prove that every algebraically closed field contains infinitely many elements.

2. Let $\mathbf{F}$ be a field whose characteristic is distinct from 2 and let $\mathbf{E}$ be an extension field of $\mathbf{F}$ such that $[\mathbf{E}:\mathbf{F}] = 2$. Show that $\mathbf{E}$ is normal over $\mathbf{F}$.

3. Let $\mathbf{F}$ be any field, $\mathbf{F}(\alpha, \beta) = \mathbf{F}(\gamma)$ and $[\mathbf{F}(\alpha):\mathbf{F}] = a$, $[\mathbf{F}(\beta):\mathbf{F}] = b$, $[\mathbf{F}(\gamma):\mathbf{F}] = c$. Prove the following propositions:

   (i) $c \leqq ab$.

   (ii)   The degree a' of α over $\mathbf{F}(\beta)$ is equal to c/b.

   (iii) If β is of degree b' over $\mathbf{F}(\alpha)$, then $ab' = a'b$.

   (iv)   Let $\mathbf{F}(\alpha) \cap \mathbf{F}(\beta) = \mathbf{F}(\delta)$ and $[\mathbf{F}(\delta):\mathbf{F}] = d$. Then $c \leqq ab/d$ and $c = ab/d$ in case $(a/d, b/d) = 1$.

4, Prove that every algebraic extension of a perfect field is perfect.

5. Let $\mathbf{K} = \boldsymbol{Q}\!\left(\sqrt[4]{2}\right)$, $\mathbf{E} = \boldsymbol{Q}(\sqrt{2})$. Show that $\mathbf{K}$ is normal over $\mathbf{E}$ and $\mathbf{E}$ is normal over $\boldsymbol{Q}$ whereas $\mathbf{K}$ is not normal over $\boldsymbol{Q}$.

6. Find a primitive element of the splitting field of $x^3 - 2$ over $\boldsymbol{Q}$.

7. Determine the Galois group G of the splitting field $\mathbf{E}$ of $x^4 - 2$ over $\boldsymbol{Q}$. Find the subgroups of G and the corresponding subfields of $\mathbf{E}$.

8. Let $\mathbf{E}$ and $\mathbf{K}$ be separable extensions of finite degree over $\mathbf{F}$ such that $\mathbf{F} \subseteq \mathbf{E} \subseteq \mathbf{K}$. Show that for any $\alpha \in \mathbf{K}$.

$$T_{\mathbf{K}/\mathbf{F}}(\alpha) = T_{\mathbf{K}/\mathbf{E}}\big(T_{\mathbf{E}/\mathbf{F}}(\alpha)\big), \; N_{\mathbf{K}/\mathbf{F}}(\alpha) = N_{\mathbf{K}/\mathbf{E}}\big(N_{\mathbf{E}/\mathbf{F}}(\alpha)\big).$$

9. Let $\mathbf{F}$ be any field and $x_1, x_2, \ldots, x_n$ indeterminates. By $C_1, C_2, \ldots, C_n$ we denote the elementary symmetric polynomials in $x_1, x_2, \ldots, x_n$ (cf. § 5.6). Show that $\mathbf{F}(x_1, x_2, \ldots, x_n)$ is a Galois extension of $\mathbf{F}(C_1, C_2, \ldots, C_n)$ and determine the Galois group. Hence show that every symmetric polynomial in $x_1, x_2, \ldots, x_n$ can be expressed as a rational function of $C_1, C_2, \ldots, C_n$.

10. (i)   Using an irreducible quadratic polynomial with coefficients in GF(3) construct tables for addition and multiplication in GF(9).

    (ii) Find a primitive element of GF(9) which is not a generating element of the multiplicative group of the non-zero elements of GF(9).

11. Let the function f be defined for all natural numbers n and put

$$g(n) = \sum_{d|n} f(d)$$

where d ranges over all natural numbers that divide n. Prove that

$$f(n) = \sum_{d|n} \mu(d)\, g(n/d).$$

12. Let $\mathbf{F} = \mathrm{GF}(q)$ where $q = p^m$, p a prime number. Let $J(d)$ denote the number of irreducible monic polynomials of degree d in $\mathbf{F}[x]$, and let $P_d(x)$ be the product of these $J(d)$ polynomials
    (i) Show that for any natural number n

$$x^{q^n} - x = \prod_{d|n} P_d(x)$$

    where d ranges over all positive divisors of n.
    (ii) Use (i) to prove that

$$q^n = \sum_{d|n} dJ(d).$$

    Hence show that

$$J(n) = \frac{1}{n} \sum_{d|n} \mu(d)\, q^{n/d}.$$

13. (i) Let a be an element of $\mathrm{GF}(p^n)$ that generates the multiplicative group $\mathrm{GF}(p^n)^\times$ of all non-zero elements of $\mathrm{GF}(p^n)$. Prove that all conjugates of a over $\mathrm{GF}(p)$ are also generators of $\mathrm{GF}(p^n)^\times$.
    (ii) By (i), or otherwise, show that, for any prime number p, $\varphi(p^n - 1)$ is divisible by n.
14. Let $a \in \mathrm{GF}(p)$ and $a \neq 0$. Show that $x^p - x - a$ is irreducible in $\mathrm{GF}(p)\,[x]$.
15. Let $\mathbf{K}$ be any field of characteristic $p > 0$ and let $x^p - x - a$, $a \in \mathbf{K}$, be irreducible in $\mathbf{K}[x]$. Show that over any extension field of $\mathbf{K}$ the polynomial is either irreducible or splits into linear factors.
16. Show that the cyclotomic polynomial $\Phi_{12}(x)$ is reducible over GF(11).
17. Let $\mathbf{E}$ be an extension field of finite degree over $\mathbf{Q}$. Show that $\mathbf{E}$ contains only a finite number of roots of unity.

18. Study the field of the 17th roots of unity similarly to the example at the end of § 6.7. Show, in particular, that the 17th roots of unity can be obtained by solving three successive quadratic equations.
19. Let m and n be natural numbers such that $(m, n) = 1$ and let ξ and η denote primitive m-th and n-th roots of unity over $\mathbf{Q}$, respectively. Show that $\mathbf{Q}(\xi) \cap \mathbf{Q}(\eta) = \mathbf{Q}$.
20. Let $\Phi_n(x)$ denote the n-th cyclotomic polynomial in $\mathbf{Q}[x]$.
    (i) Let k denote the product of the distinct prime factors of n and put $n = km$. Show that $\Phi_n(x) = \Phi_k(x^m)$.
    (ii) If n is odd, then $\Phi_{2n}(x) = \Phi_n(-x)$.
    (iii) $\Phi_{np}(x) = \Phi_n(x^p)/\Phi_n(x)$ for any odd prime number p that does not divide n.
21. Let $\mathbf{F}$ be the Galois field with $q = p^r$ elements and let $\mathbf{K}$ denote the n-th cyclotomic field over $\mathbf{F}$. (We recall that this implies $(n, p) = 1$.) Prove that $[\mathbf{K}: \mathbf{F}]$ is equal to the least natural number k such that $q^k - 1$ is divisible by n.
22. Let $\mathbf{F} = \mathrm{GF}(p^n)$ and let $\mathbf{K}$ be an extension field of $\mathbf{F}$ of finite degree.
    (i) Prove that for any $a \in \mathbf{F}$ there exist elements $\gamma \in \mathbf{K}$ such that $a = T_{\mathbf{K}/\mathbf{F}}(\gamma)$.
    (ii) Prove the analogous proposition for the norm.
23. Let $\mathbf{K}$ be a separable extension field of $\mathbf{F}$ of finite degree $[\mathbf{K}: \mathbf{F}] = r$. If $\omega_1, \ldots, \omega_r$ is a basis of $\mathbf{K}$ over $\mathbf{F}$ and if α is any element of $\mathbf{K}$ there hold equations of the form

$$\omega_i\alpha = \sum_{k=1}^{r} a_{ik}\omega_k \qquad (i = 1, \ldots, r)$$

where $a_{ik} \in \mathbf{F}$. Consider the $r \times r$ matrix $A = [a_{ik}]$. Prove that

$$T_{\mathbf{K}/\mathbf{F}}(\alpha) = \text{trace of } A,$$
$$N_{\mathbf{K}/\mathbf{F}}(\alpha) = \text{determinant of } A.$$

24. Let $\mathbf{F}$ be any field and suppose that the polynomial $f(x) \in \mathbf{F}[x]$ of degree m is irreducible over $\mathbf{F}$. If $\mathbf{K}$ is an extension field of $\mathbf{F}$ such that $[\mathbf{K}: \mathbf{F}]$ is relatively prime to m show that $f(x)$ is irreducible in $\mathbf{K}[x]$.

7

Galois theory of equations

As is well known the roots of a quadratic equation can be computed from its coefficients by rational operations (i.e. addition, subtraction, multiplication, and division) and the extraction of a square root. In § 7.3 we shall derive similar formulae for the roots of cubic and quartic equations. Since these formulae involve only rational operations and root extractions we say that cubic and quartic equations are soluble by radicals. After vain attempts of numerous mathematicians to solve equations of degree higher than four by radicals Ruffini and Abel proved that equations of degree five cannot be solved by radicals. Later, Galois derived a necessary and sufficient condition under which an equation is soluble by radicals. The work of Galois is of great importance for the development of mathematical concepts. In this chapter we shall apply the results of the previous chapter to the study of algebraic equations and related topics.

7.1. The Galois group of a polynomial

Let $f(x)$ be a polynomial of degree $n > 0$ whose coefficients belong to a field $\mathbf{F}$. We assume that $f(x)$ has no multiple roots in the algebraic closure of $\mathbf{F}$. Adjoining the roots $\alpha_1, \alpha_2, \ldots, \alpha_n$ of $f(x)$ to $\mathbf{F}$ we obtain the splitting field $\mathbf{E} = \mathbf{F}(\alpha_1, \alpha_2, \ldots, \alpha_n)$ which is a normal extension of $\mathbf{F}$. Due to the assumption that $f(x)$ has no multiple roots, $\mathbf{E}$ is separable over $\mathbf{F}$. Let G be the Galois group of $\mathbf{E}$ over $\mathbf{F}$ and let g denote the order of G.

By Theorem 6.5.2, $\mathbf{E}$ is a simple extension of $\mathbf{F}$,

$$\mathbf{E} = \mathbf{F}(\alpha_1, \alpha_2, \ldots, \alpha_n) = \mathbf{F}(\vartheta).$$

Moreover, $[\mathbf{E}:\mathbf{F}] = g$ so that the minimal polynomial $R(x)$ of ϑ in $\mathbf{F}[x]$ is of degree g. The polynomial $R(x)$ is said to be a *Galois resolvent* of $f(x)$.

From our point of view, to solve the equation $f(x) = 0$ means to construct the splitting field $\mathbf{F}(\alpha_1, \alpha_2, \ldots, \alpha_n)$ of $f(x)$.[1]

Let σ be any element of G. Since the coefficients of $f(x)$ are fixed under G, $f(\alpha_i) = 0$ implies that $f(\alpha_i^\sigma) = 0$. Moreover, from $\alpha_i \neq \alpha_j$ for $i \neq j$ it follows that $\alpha_i^\sigma \neq \alpha_j^\sigma$. Therefore σ induces a permutation

$$P_\sigma = \begin{pmatrix} \alpha_1 \; \alpha_2 \ldots \alpha_n \\ \alpha_1^\sigma \; \alpha_2^\sigma \ldots \alpha_n^\sigma \end{pmatrix}$$

of $\alpha_1, \alpha_2, \ldots, \alpha_n$. For $\sigma, \tau \in G$ we have

$$P_\sigma P_\tau = \begin{pmatrix} \alpha_1 \ldots \alpha_n \\ \alpha_1^\sigma \ldots \alpha_n^\sigma \end{pmatrix} \begin{pmatrix} \alpha_1 \ldots \alpha_n \\ \alpha_1^\tau \ldots \alpha_n^\tau \end{pmatrix}$$

$$= \begin{pmatrix} \alpha_1 \ldots \alpha_n \\ \alpha_1^\sigma \ldots \alpha_n^\sigma \end{pmatrix} \begin{pmatrix} \alpha_1^\sigma \ldots \alpha_n^\sigma \\ \alpha_1^{\sigma\tau} \ldots \alpha_n^{\sigma\tau} \end{pmatrix} = \begin{pmatrix} \alpha_1 \ldots \alpha_n \\ \alpha_1^{\sigma\tau} \ldots \alpha_n^{\sigma\tau} \end{pmatrix} = P_{\sigma\tau}.$$

Therefore the mapping $\sigma \to P_\sigma$ is a homomorphism of G onto the group P_G of the P_σ. This mapping is even an isomorphism. For, if P_σ is the identity permutation, then we have $\alpha_i^\sigma = \alpha_i$, for $i = 1, \ldots, n$ so that σ leaves every element of $\mathbf{F}(\alpha_1, \ldots, \alpha_n)$ fixed, i.e. σ is the identity of G.

The permutation group P_G is called the *Galois group* of the polynomial $f(x)$ over $\mathbf{F}$ or the Galois group of the equation $f(x) = 0$ over $\mathbf{F}$.

---

[1]) Of course, $f(x) = 0$ does not mean that $f(x)$ is the zero polynomial. We rather adopt the following expedient expression: To solve the equation $f(x) = 0$ means to determine the roots of the polynomial $f(x)$. This can be understood in several ways, e.g. to find the numerical values of the roots or approximations to them in case $f(x) \in C[x]$. In our context, however, to solve the equation $f(x) = 0$ has always the meaning indicated in the text.

In Example 2 of § 6.6 we have $\mathsf{F} = \mathcal{Q}$, $f(x) = x^3 - 2$,

$$P_\alpha = \begin{pmatrix} \sqrt[3]{2} & \varrho\sqrt[3]{2} & \bar{\varrho}\sqrt[3]{2} \\ \varrho\sqrt[3]{2} & \bar{\varrho}\sqrt[3]{2} & \sqrt[3]{2} \end{pmatrix}, \quad P_\beta = \begin{pmatrix} \sqrt[3]{2} & \varrho\sqrt[3]{2} & \bar{\varrho}\sqrt[3]{2} \\ \sqrt[3]{2} & \bar{\varrho}\sqrt[3]{2} & \varrho\sqrt[3]{2} \end{pmatrix}.$$

7.1.1. *Let*

$$f(x) = f_1(x)\, f_2(x) \ldots f_r(x)$$

be the factorization of $f(x)$ into irreducible factors $f_i(x)$ in $\mathsf{F}[x]$. Then the orbits of the Galois group of $f(x)$ over F are precisely the roots of the $f_i(x)$. In particular, $f(x)$ is irreducible in $\mathsf{F}[x]$ if and only if the Galois group is transitive.

Proof. Let $\alpha_1, \ldots, \alpha_m$ denote the roots of the irreducible factor $f_1(x)$. For any element σ of the Galois group, $f_1(\alpha_1) = 0$ implies that $f_1(\alpha_1^\sigma) = 0$ so that α_1^σ is one of the roots $\alpha_1, \ldots, \alpha_m$. The same applies to the other irreducible factors. On the other hand, let $\alpha_1, \ldots, \alpha_m$ be an orbit of the Galois group P_G. The coefficients of the product

$$(x - \alpha_1)\,(x - \alpha_2) \ldots (x - \alpha_m)$$

are fixed under every element of the Galois group and are therefore contained in F. This proves our proposition.

The Galois group of a polynomial can also be characterized in another way. Let $H(x_1, \ldots, x_n)$ be a polynomial in the indeterminates $x_1, \ldots, x_n$ with coefficients in F. If $\alpha_1, \ldots, \alpha_n$ are elements of an extension field of F such that $H(\alpha_1, \ldots, \alpha_n) = 0$, then this equation is called a relation between $\alpha_1, \ldots, \alpha_n$ with coefficients in F.

7.1.2. Theorem. *The Galois group P_G of $f(x)$ over F consists precisely of all permutations of the roots $\alpha_1, \ldots, \alpha_n$ of $f(x)$ such that any relation $H(\alpha_1, \ldots, \alpha_n) = 0$ with coefficients in F remains valid.*

Proof. First, we show that any relation $H(\alpha_1, \ldots, \alpha_n) = 0$ with coefficients in F remains true if $\alpha_1, \ldots, \alpha_n$ undergo a permutation of the Galois group. Let

$$P_\sigma = \begin{pmatrix} \alpha_1 \ldots \alpha_n \\ \alpha_1^\sigma \ldots \alpha_n^\sigma \end{pmatrix}$$

be a permutation of the Galois group. Since σ is an automorphism of $\mathbf{E}/\mathbf{F}$ we obtain

$$H(\alpha_1, \ldots, \alpha_n)^\sigma = H(\alpha_1^\sigma, \ldots, \alpha_n^\sigma) = 0.$$

Conversely, let

$$P = \begin{pmatrix} \alpha_1 & \ldots & \alpha_n \\ \alpha_{1'} & \ldots & \alpha_{n'} \end{pmatrix}$$

be a permutation under which every relation between $\alpha_1, \ldots, \alpha_n$ with coefficients in $\mathbf{F}$ remains valid. We have to show that there is an element τ in G such that $P = P_\tau$.

We use the fact that the splitting field of $f(x)$ is a simple extension:

$$\mathbf{F}(\alpha_1, \ldots, \alpha_n) = \mathbf{F}(\vartheta).$$

The primitive element ϑ is a root of the Galois resolvent $R(x)$. There is a representation

$$\vartheta = h(\alpha_1, \ldots, \alpha_n)$$

as a polynomial in $\alpha_1, \ldots, \alpha_n$ with coefficients in $\mathbf{F}$. The equation $R(\vartheta) = 0$ gives

$$R\big(h(\alpha_1, \ldots, \alpha_n)\big) = 0$$

which is a relation between $\alpha_1, \ldots, \alpha_n$ with coefficients in $\mathbf{F}$. Since this relation remains valid under P we obtain

$$R\big(h(\alpha_{1'}, \ldots, \alpha_{n'})\big) = 0.$$

This shows that $h(\alpha_{1'}, \ldots, \alpha_{n'})$ is a root of $R(x)$. As we observed in § 6.6, the roots of $R(x)$ are precisely the g elements ϑ^σ where σ ranges over all elements of G. Hence, there is an element τ in G such that

$$\vartheta^\tau = h(\alpha_{1'}, \ldots, \alpha_{n'}).$$

Being an element of $\mathbf{F}(\vartheta)$, each α_i has a representation $\alpha_i = \varphi_i(\vartheta)$ as a polynomial in ϑ with coefficients in $\mathbf{F}$. So we have

$$\alpha_i = \varphi_i(\vartheta) = \varphi_i\big(h(\alpha_1, \ldots, \alpha_n)\big),$$

and applying P we obtain

$$\alpha_{i'} = \varphi_i\big(h(\alpha_{1'}, \ldots, \alpha_{n'})\big) = \varphi_i(\vartheta^\tau) = \alpha_i^\tau.$$

This shows that $P = P_\tau$ as was to be proved.

Let F be any field and $x, u_1, \ldots, u_n$ indeterminates. The polynomial

$$f(x) = x^n - u_1x^{n-1} + u_2x^{n-2} - \ldots + (-1)^n u_n$$

is called the *general polynomial* of degree n because every monic polynomial of degree n in $\mathsf{F}[x]$ can be obtained by substituting elements of F for the u_i. Our aim is to determine the Galois group of $f(x)$ over $\mathsf{F}(u_1, \ldots, u_n)$. Let $v_1, \ldots, v_n$ denote the roots of $f(x)$ in a suitable extension field of $\mathsf{F}(u_1, \ldots, u_n)$. Then we have the following factorization

$$f(x) = (x - v_1)(x - v_2) \ldots (x - v_n).$$

The indeterminates $u_1, \ldots, u_n$ are the elementary symmetric polynomials in $v_1, \ldots, v_n$, i.e.

$$u_1 = \sum_i v_i, \quad u_2 = \sum_{i<k} v_iv_k, \quad \ldots, \quad u_n = v_1 \ldots v_n.$$

Beside the polynomial $f(x)$ whose *coefficients* are indeterminates we consider a polynomial $g(x)$ whose *roots* are n indeterminates $y_1, \ldots, y_n$, i.e.

$$\begin{aligned} g(x) &= (x - y_1) \ldots (x - y_n) \\ &= x^n - z_1x^{n-1} + z_2x^{n-2} - \ldots + (-1)^n z_n \end{aligned}$$

where

$$z_1 = \sum_i y_i, \ z_2 = \sum_{i<k} y_iy_k, \ \ldots, \ z_n = y_1 \ldots y_n.$$

It is evident that $g(x)$ is separable and that its Galois group over $\mathsf{F}(z_1, \ldots, z_n)$ is the symmetric group on $y_1, \ldots, y_n$.

Now to any polynomial $p(u_1, \ldots, u_n)$ of the integral domain $\mathsf{F}[u_1, \ldots, u_n]$ we assign the polynomial $p(z_1, \ldots, z_n)$. This yields a homomorphism of $\mathsf{F}[u_1, \ldots, u_n]$ onto $\mathsf{F}[z_1, \ldots, z_n]$. It turns out that this homomorphism is actually an isomorphism. Indeed, suppose that $p(z_1, \ldots, z_n) = 0$, i.e.

$$p\left(\sum_i y_i, \sum_{i<k} y_iy_k, \ldots, y_1 \ldots y_n\right) = 0.$$

Since the y_i are indeterminates we may substitute the roots v_i of

$f(x)$ for the y_i to obtain

$$p\Big(\sum_i v_i, \sum_{i<k} v_i v_k, \ldots, v_1 \ldots v_n\Big) = p(u_1, \ldots, u_n) = 0.$$

Since the u_i are indeterminates it follows that p is the zero polynomial. Thus the mapping

$$p(u_1, \ldots, u_n) \to p(z_1, \ldots, z_n)$$

is an isomorphism of $\mathsf{F}[u_1, \ldots, u_n]$ onto $\mathsf{F}[z_1, \ldots, z_n]$. It is clear that this isomorphism can be extended to an isomorphism of the corresponding fields of quotients, i.e. to an isomorphism of $\mathsf{F}(u_1, \ldots, u_n)$ onto $\mathsf{F}(z_1, \ldots, z_n)$. Under this isomorphism $f(x)$ is mapped onto $g(x)$. As in the proof of Theorem 6.3.4, we conclude that our isomorphism can be extended to an isomorphism of the splitting fields of $f(x)$ and $g(x)$, respectively, i.e. to an isomorphism of $\mathsf{F}(v_1, \ldots, v_n)$ onto $\mathsf{F}(y_1, \ldots, y_n)$. As we observed above, the Galois group of $g(x)$ over $\mathsf{F}(z_1, \ldots, z_n)$ is the symmetric group. Due to our isomorphism it follows that the Galois group of $f(x)$ over $\mathsf{F}(u_1, \ldots, u_n)$ is also the symmetric group. This gives the following result:

7.1.3. Theorem. *The general polynomial*

$$f(x) = x^n - u_1 x^{n-1} + u_2 x^{n-2} - \ldots + (-1)^n u_n$$

is separable and irreducible over $\mathsf{F}(u_1, \ldots, u_n)$. *The Galois group of* $f(x)$ *over* $\mathsf{F}(u_1, \ldots, u_n)$ *is the symmetric group of degree n on the roots of* $f(x)$.

In § 5.7, we defined the discriminant of $f(x)$ as follows:

$$D(f) = \prod_{i<k} (v_i - v_k)^2.$$

It is obvious that $D(f)$ is fixed under the symmetric group on $v_1, \ldots, v_n$. From the definition of the alternating group A_n in § 3.10 it follows that

$$\sqrt{D(f)} = \prod_{i<k} (v_i - v_k)$$

is fixed under A_n but not under S_n provided that the characteristic of F is distinct from 2.

This gives the following corollary of the last theorem:

7.1.4. Corollary. *If* F *is a field whose characteristic is distinct from* 2, *then the Galois group of the general polynomial* $f(x)$ *over* $\mathsf{F}\left(u_1, \ldots, u_n, \sqrt{D(f)}\right)$ *is the alternating group.*

As we remarked above, to solve the equation $f(x) = 0$ means, in our context, to construct the splitting field $\mathsf{F}(\alpha_1, \ldots, \alpha_n)$ of $f(x)$. We shall build up the splitting field in several steps where each step requires the solution of an equation which is, in a certain sense, simpler than $f(x) = 0$. To do this one can proceed as follows:

Let $\beta \in \mathsf{F}(\alpha_1, \ldots, \alpha_n)$ and let $g(x)$ denote the minimal polynomial of β over F. The field $\mathsf{F}(\beta)$ belongs to a subgroup H of the Galois group G of $\mathsf{F}(\alpha_1, \ldots, \alpha_n)$ over F. The group H is the Galois group of $\mathsf{F}(\alpha_1, \ldots, \alpha_n)$ over $\mathsf{F}(\beta)$. We say that the adjunction of β to F leads to a reduction of the Galois group from G to H. If H is not the identity, i.e. if $\mathsf{F}(\beta)$ is a proper subfield of $\mathsf{F}(\alpha_1, \ldots, \alpha_n)$, we can choose an element γ of $\mathsf{F}(\alpha_1, \ldots, \alpha_n)$ that is not contained in $\mathsf{F}(\beta)$. Let $\varphi(x)$ denote the minimal polynomial of γ over $\mathsf{F}(\beta)$. The field $\mathsf{F}(\beta, \gamma)$ is a proper extension of $\mathsf{F}(\beta)$ and hence the Galois group of $\mathsf{F}(\alpha_1, \ldots, \alpha_n)$ over $\mathsf{F}(\beta, \gamma)$ is a proper subgroup K of H. Unless K is the identity we can again adjoin an element δ to $\mathsf{F}(\beta, \gamma)$ to obtain a further reduction of the Galois group. After a finite number of adjunctions the Galois group is finally reduced to the identity which means that $\mathsf{F}(\beta, \gamma, \delta, \ldots) = \mathsf{F}(\alpha_1, \ldots, \alpha_n)$. Therefore the solution of the equation $f(x) = 0$ can be achieved by solving the equations $g(x) = 0$, $\varphi(x) = 0, \ldots$ We shall refer to $g(x) = 0$, $\varphi(x) = 0, \ldots$ as auxiliary equations.

As an example we consider the 13th cyclotomic equation

$$x^{12} + x^{11} + \ldots + x + 1 = 0.$$

We use the notations and results of § 6.7. There we found that the Galois group $\langle\sigma\rangle$ of $\mathbf{Q}(\zeta)$ over $\mathbf{Q}$ is cyclic of order 12. The three roots η_0, η_1, η_2 of

$$x^3 + x^2 - 4x + 1 = 0$$

are fixed under the subgroup $\langle\sigma^3\rangle$. Therefore the Galois group of $\mathbf{Q}(\zeta)$ over $\mathbf{Q}(\eta_0)$ (or over $\mathbf{Q}(\eta_1)$ or $\mathbf{Q}(\eta_2)$ since these three fields coincide) is $\langle\sigma^3\rangle$. The periods of two terms ξ_0, ξ_3 are fixed under $\langle\sigma^6\rangle$.

They are roots of the equation

$$x^2 - \eta_0 x + \eta_0 = 0 .$$

The Galois group of $\boldsymbol{Q}(\zeta)$ over (ξ_0) is $\langle\sigma^6\rangle$. Finally, ζ is a root of

$$x^2 - \xi_0 x + 1 = 0 .$$

Thus the solution of the 13th cyclotomic equation can be achieved by solving the three auxiliary equations

$$x^3 + x^2 - 4x + 1 = 0 ,$$

$$x^2 - \eta_0 x + \eta_2 = 0 ,$$

$$x^2 - \xi_0 x + 1 = 0 .$$

The following diagram shows the Galois correspondence between the subfields of $\boldsymbol{Q}(\zeta)$ and the subgroups of $\langle\sigma\rangle$. Between any two subfields on the left-hand side there stands the minimal polynomial of a primitive element of the larger field over the smaller one.

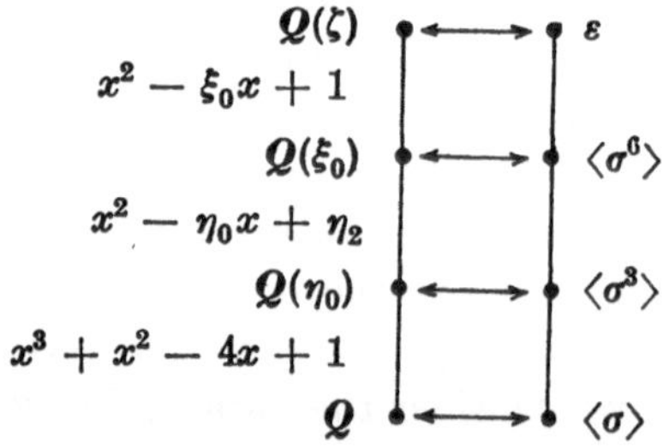

To reduce the solution of $f(x) = 0$ to the solution of a sequence of equations of a handy form it is occasionally expedient to make a detour by adjoining elements that are not contained in $\mathsf{F}(\alpha_1, \ldots, \alpha_n)$. If λ is such an element, then Theorem 6.6.10 tells us how much the adjunction of λ contributes to the reduction of the Galois group: The Galois group of $\mathsf{F}(\alpha_1, \ldots, \alpha_n)$ over $\mathsf{F}(\lambda)$ is isomorphic to the subgroup that belongs to the intersection $\mathsf{F}(\alpha_1, \ldots, \alpha_n) \cap \mathsf{F}(\lambda)$. Roughly speaking, only that part of $\mathsf{F}(\lambda)$ which is contained in $\mathsf{F}(\alpha_1, \ldots, \alpha_n)$ contributes to the reduction.

7.2. Solubility of equations by radicals

In the last section we observed that the splitting field of a polynomial $f(x)$ can be constructed in several steps or, in other words, that the solution of the equation $f(x) = 0$ can be achieved by solving a sequence of auxiliary equations $g(x) = 0$, $\varphi(x), = 0, \ldots$ In this section we characterize those equations $f(x) = 0$ for which these auxiliary equations can be chosen as so-called *pure equations*, i.e. equations of the form $x^n - a = 0$. Solutions of pure equations are called *radicals*, and equations that can be solved by a sequence of pure equations are said to be soluble by radicals.

For the sake of simplicity we shall restrict our attention to fields of characteristic 0. This enables us to use the results of § 6.7 on roots of unity. Moreover, we shall only consider equations with simple roots. In view of results at the end of § 5.4 this means no loss of generality.

Let $\mathbf{F}$ be a field of characteristic 0 and let $f(x)$ be a polynomial of positive degree in $\mathbf{F}[x]$. Suppose that $f(x)$ has no multiple roots in any extension field of $\mathbf{F}$. The equation $f(x) = 0$ is said to be *soluble by radicals* if there is a chain

$$\mathbf{F} = \mathbf{K}_0 \subset \mathbf{K}_1 \subset \mathbf{K}_2 \subset \ldots \subset \mathbf{K}_t$$

of extension fields $\mathbf{K}_i$ such that

$$\mathbf{K}_i = \mathbf{K}_{i-1}(\vartheta_i) \text{ where } \vartheta_i^{n_i} = c_i \in \mathbf{K}_{i-1} \quad (i = 1, \ldots, t)$$

and $\mathbf{K}_t$ contains the splitting field of $f(x)$.

To derive a criterion for solubility by radicals we first consider extension fields that are obtained by adjoining a single radical.

7.2.1. *Let n be a natural number and $\mathbf{K}$ a field of characteristic 0 that contains the n-th roots of unity. If ϑ is a root of a pure equation $x^n - a = 0$ with $a \in \mathbf{K}$, then the Galois group of $\mathbf{K}(\vartheta)$ over $\mathbf{K}$ is cyclic.*

Proof. All the roots of $x^n - a$ are

$$\vartheta, \zeta\vartheta, \ldots, \zeta^{n-1}\vartheta$$

where ζ denotes any primitive n-th root of unity. Since ζ is contained in $\mathbf{K}$ it follows that $\mathbf{K}(\vartheta)$ is a Galois extension of $\mathbf{K}$ because it is the splitting field of $x^n - a$ over $\mathbf{K}$. Let G denote the Galois group of $\mathbf{K}(\vartheta)$ over $\mathbf{K}$. Any element σ of G carries ϑ into some root of $x^n - a$, say $\vartheta^\sigma = \zeta^s\vartheta$. For $\tau \in G$ we have $\vartheta^\tau = \zeta^t\vartheta$ and $\vartheta^{\sigma\tau} = \zeta^{s+t}\vartheta$. This shows that the mapping $\sigma \to \zeta^s$ is an isomorphism of G into the multiplicative group of the n-th roots of unity. Since the latter group is cyclic it follows that G is cyclic as was to be proved.

We now prove the converse of the last proposition.

7.2.2. *Let n be a natural number and $\mathbf{K}$ a field of characteristic 0 that contains the n-th roots of unity. If $\mathbf{K}(\vartheta)$ is a Galois extension of $\mathbf{K}$ whose Galois group is cyclic of order n, then $\mathbf{K}(\vartheta)$ can be obtained by adjoining radicals to $\mathbf{K}$.*

Proof. Let $G = \langle\sigma\rangle$ be the Galois group of $\mathbf{K}(\vartheta)$ over $\mathbf{K}$ so that $\sigma^n = \varepsilon$ where ε denotes the identity of G. With an arbitrary n-th root of unity η we form the so-called *Lagrange resolvent*

$$(\eta, \vartheta) = \vartheta^\varepsilon + \eta\vartheta^\sigma + \eta^2\vartheta^{\sigma^2} + \ldots + \eta^{n-1}\vartheta^{\sigma^{n-1}}.$$

Applying σ we obtain

$$\begin{aligned}(\eta, \vartheta)^\sigma &= \vartheta^\sigma + \eta\vartheta^{\sigma^2} + \eta^2\vartheta^{\sigma^3} + \ldots + \eta^{n-2}\vartheta^{\sigma^{n-1}} + \eta^{n-1}\vartheta^\varepsilon \\ &= \eta^{-1}(\eta, \vartheta).\end{aligned}$$

This shows that the n-th power $(\eta, \vartheta)^n$ is fixed under σ and hence is contained in $\mathbf{K}$. In other words, the Lagrange resolvents are roots of pure equations in $\mathbf{K}[x]$. To prove our proposition we show that ϑ can be expressed as a linear combination of Lagrange resolvents with coefficients in $\mathbf{K}$. It is easy to see that

$$\sum_\eta \eta^i = 0 \quad (i = 1, \ldots, n-1) \tag{1}$$

where η ranges over all n-th roots of unity. Indeed, if ζ is a primitive n-th root of unity we find

$$\sum_\eta \eta^i = \sum_{k=0}^{n-1} (\zeta^k)^i = \sum_{k=0}^{n-1} (\zeta^i)^k = \frac{(\zeta^i)^n - 1}{\zeta^i - 1} = 0.$$

Due to (1) we obtain

$$\sum_\eta (\eta, \vartheta) = n\vartheta^\varepsilon$$

or

$$\vartheta = \frac{1}{n} \sum_{\eta} (\eta, \vartheta).$$

This completes the proof.

In 7.2.1 and 7.2.2 we assumed that the field $\mathbf{K}$ contains certain roots of unity. We will now show that, with respect to solubility by radicals, this is no loss of generality since the cyclotomic equations themselves are soluble by radicals. For our purpose it is sufficient to prove this only for the p-th roots of unity where p is a prime number. The proof in the general case is left as an exercise for the reader (cf. Exercise 10).

7.2.3. *Let p be a prime number. Then the p-th cyclotomic equation*

$$x^{p-1} + x^{p-2} + \dots + x + 1 = 0$$

over any field $\mathbf{F}$ of characteristic 0 is soluble by radicals.

Proof. For the prime number 2 there is nothing to prove because the second roots of unity ± 1 belong to $\mathbf{F}$. We assume that our proposition holds for all prime numbers less than p. Let ζ be a primitive p-th root of unity. As we saw in § 6.7, the Galois group of $\mathbf{F}(\zeta)$ over $\mathbf{F}$ is cyclic, and its order is equal to $p - 1$ or to a proper divisor of $p - 1$. Let

$$p - 1 = p_1^{a_1} p_2^{a_2} \dots p_r^{a_r}$$

be the factorization into powers of prime numbers $p_1, p_2, \dots, p_r$. We adjoin the p_1-th, $\dots$, p_r-th roots of unity to $\mathbf{F}$ to obtain the field $\mathbf{K}$. By our induction assumption this means the adjunction of radicals. From the results of § 6.7 it follows that the field $\mathbf{K}(\zeta)$ can be built up in several steps,

$$\mathbf{K} = \mathbf{K}_0 \subset \mathbf{K}_1 \subset \mathbf{K}_2 \subset \dots \subset \mathbf{K}_s = \mathbf{K}(\zeta),$$

where each $\mathbf{K}_i$ is a Galois extension of $\mathbf{K}_{i-1}$ and the Galois group of $\mathbf{K}_i$ over $\mathbf{K}_{i-1}$ has one of the prime numbers $p_1, \dots, p_r$ as its order. Since the necessary roots of unity are contained in $\mathbf{K}$ it follows from 7.2.2 that $\mathbf{K}_i$ can be obtained out of $\mathbf{K}_{i-1}$ by the adjunction of radicals. This proves our proposition.

We now come to the main theorem of this section.

7.2.4. Theorem. *Let* F *be a field of characteristic* 0 *and* $f(x)$ *a polynomial of positive degree in* $\mathsf{F}[x]$ *that has no multiple roots. Then, the equation* $f(x) = 0$ *is soluble by radicals if and only if the Galois group of* $f(x)$ *over* F *is soluble.*

Proof. First, suppose that the equation $f(x) = 0$ is soluble by radicals. Then, there is a chain of fields

$$\mathsf{F} = \mathsf{F}_0 \subset \mathsf{F}_1 \subset \mathsf{F}_2 \subset \ldots \subset \mathsf{F}_r$$

such that F_r contains the splitting field of $f(x)$ and

$$\mathsf{F}_i = \mathsf{F}_{i-1}(\vartheta_i), \ \vartheta_i^{n_i} = c_i \in \mathsf{F}_{i-1} \quad (i = 1, \ldots, r).$$

We may assume that the exponents n_i are prime numbers. For, if $n_i = kl$, $k > 1$, $l > 1$, we can refine the chain by inserting $\mathsf{F}_{i-1}\left(\sqrt[k]{c_i}\right)$ between F_{i-1} and F_i. By a finite number of such refinements we arrive at a chain in which the degree of each member over the preceding one is a prime number. We now adjoin the n_1-th, ..., n_r-th roots of unity to F. By the results of § 6.7, this means a sequence of Galois extensions with cyclic Galois groups. Next, we adjoin the n_i-th roots of the c_i. As far as these adjunctions yield proper extensions it follows from 7.2.1 that they are Galois extensions with cyclic Galois groups. The same applies to a sequence of further adjunction which form the last stage of our construction. Namely, we adjoin the n_i-th roots of all the conjugates of each c_i over F. It is easy to see that, after these adjunctions, we arrive at a field K which is a Galois extension of F. All these adjunctions yield a chain of extension fields

$$\mathsf{F} = \mathsf{K}_0 \subset \mathsf{K}_1 \subset \mathsf{K}_2 \subset \ldots \subset \mathsf{K}_t = \mathsf{K} \tag{2}$$

such that each K_i is a Galois extension of K_{i-1} whose Galois group over K_{i-1} is cyclic of prime order. Moreover, K contains the splitting field of $f(x)$. If G is the Galois group of K over F, then to the chain (2) there corresponds a chain of subgroups

$$G = G_0 \supset G_1 \supset G_2 \supset \ldots \supset G_t = \varepsilon$$

such that G_i belongs to K_i in the sense of Theorem 6.6.9. By the same theorem, G_i is a normal subgroup of G_{i-1}, and the factor group G_{i-1}/G_i is cyclic of prime order. By Theorem 3.8.3, the group

G is soluble. To the splitting field of $f(x)$ there belongs a normal subgroup H of G, and the Galois group of $f(x)$ over $\mathbf{F}$ is isomorphic to the factor group G/H. Thus, by Theorem 3.8.6, the Galois group of $f(x)$ is soluble.

Conversely, suppose that the Galois group L of $f(x)$ over $\mathbf{F}$ is soluble. Let $p_1, \ldots, p_s$ be the distinct prime divisors of the order of L. We first adjoin the p_1-th, ..., p_s-th roots of unity to $\mathbf{F}$ to obtain the field $\mathbf{K}$. Due to 7.2.3, this can be achieved by the adjunction of radicals. The Galois group of the splitting field $\mathbf{E}$ of $f(x)$ over $\mathbf{K}$ is a subgroup L_0 of L. By Theorem 3.8.6, L_0 is soluble so that there exists a composition series

$$L_0 \supset L_1 \supset \ldots \supset L_m = \varepsilon.$$

The orders of the factor groups L_{i-1}/L_i are prime numbers of the set $\{p_1, \ldots, p_s\}$. To this chain of subgroups there corresponds a chain of extension fields

$$\mathbf{K} = \mathbf{K}_0 \subset \mathbf{K}_1 \subset \ldots \subset \mathbf{K}_m = \mathbf{E}$$

such that $\mathbf{K}_i$ belongs to L_i in the sense of Theorem 6.6.9. Each $\mathbf{K}_i$ is a Galois extension of $\mathbf{K}_{i-1}$, and the Galois group of $\mathbf{K}_i$ over $\mathbf{K}_{i-1}$ is isomorphic to L_{i-1}/L_i and hence is cyclic of some order p_j. Since $\mathbf{K}$ contains the p_j-th roots of unity it follows from 7.2.2 that $\mathbf{K}_i$ can be obtained out of $\mathbf{K}_{i-1}$ by the adjunction of radicals. Therefore, $\mathbf{E} = \mathbf{K}_m$ itself can be built up by adjoining radicals to $\mathbf{K}$. Thus $f(x) = 0$ is soluble by radicals. This completes the proof.

By the *general equation* of degree n we understand the equation $f(x) = 0$ where $f(x)$ denotes the general polynomial. In § 3.10 we observed that the symmetric groups S_n are not soluble for $n > 4$. In view of Theorem 7.1.3 this gives the following corollary to the previous theorem:

7.2.5. Corollary. *The general equation of degree greater than* 4 *is not soluble by radicals.*

7.3. Quadratic, cubic, and quartic equations

As one learns at school, the quadratic equation

$$x^2 + px + q = 0 \tag{1}$$

where p and q are arbitrary real or complex numbers has the roots

$$\frac{1}{2}\left(-p \pm \sqrt{p^2 - 4q}\right). \tag{2}$$

If we confine ourselves to fields whose characteristic is distinct from 2 and regard (1) as the general quadratic equation, then the roots are also given by (2). For, if we substitute (2) for x in (1), the left-hand side becomes the zero polynomial in the indeterminates p, q. Conversely, if we can express the roots of a general polynomial as functions of its coefficients, then such a formula yields the roots of any particular polynomial by substitution.

In § 3.10 we observed that the symmetric groups S_2, S_3, S_4 are soluble. Therefore it follows from the Theorems 7.1.3 and 7.2.4 that the general equations of degrees 2, 3, and 4 are soluble by radicals. In this section we shall derive the formulae for the roots of these equations in the framework of the general theory. By ingenious computations these formulae can be deduced in a shorter way and they have actually been found before the Galois theory had been developed. But it might be more satisfying and instructive to follow the systematic way.

In order to apply our previous results we will, for the time being, confine ourselves to fields of characteristic 0. Later it will turn out that our formulae actually hold under less restrictive conditions.

Let $\mathbf{F}$ be any field of characteristic 0.

The Galois group of the general polynomial

$$f(x) = x^2 + px + q$$

over $\mathbf{F}(p, q)$ is of order 2. If x_1 and x_2 are the roots of $f(x)$, then we have

$$f(x) = (x - x_1)(x - x_2), \quad -p = x_1 + x_2\,, \quad q = x_1 x_2\,.$$

A simple calculation yields the expression

$$D = (x_1 - x_2)^2 = p^2 - 4q$$

for the discriminant D. By Corollary 7.1.4, the Galois group of $f(x)$ over the field $\mathbf{F}\left(p, q, \sqrt{D}\right)$ is the identity so that the roots x_1 and x_2 belong to this field. From

$$x_1 - x_2 = \sqrt{p^2 - 4q}, \quad x_1 + x_2 = -p$$

we find the familiar formula

$$x_1 = \frac{1}{2}\left(-p + \sqrt{p^2 - 4q}\right), \quad x_2 = \frac{1}{2}\left(-p - \sqrt{p^2 - 4q}\right).$$

It is easy to see that this formula also applies to all fields of characteristic $p > 2$.

Before we turn to cubic and quartic equations we state a general lemma which can be verified by a straightforward computation.

7.3.1. LEMMA. *If* $\mathbf{F}$ *is a field whose characteristic does not divide the natural number* n *and if*

$$f(x) = x^n + a_1x^{n-1} + a_2x^{n-2} + \dots + a_{n-1}x + a_n$$

is a polynomial in $\mathbf{F}[x]$, *then the substitution* $x = y - a_1/n$ *gives*

$$f(x) = y^n + b_2y^{n-2} + \dots + b_{n-1}y + b_n.$$

Due to this lemma we may restrict our attention to cubic polynomials of the form

$$x^3 + px + q \tag{3}$$

where p and q are indeterminates. Let x_1, x_2, x_3 denote the roots in an extension field of $\mathbf{K} = \mathbf{F}(p, q)$. It is evident that the Galois group of the polynomial (3) over $\mathbf{K}$ is the same as the Galois group of the general cubic polynomial, namely the symmetric group S_3. The composition series of S_3 is

$$S_3 \supset A_3 \supset \varepsilon.$$

To reduce the Galois group to A_3 we adjoin an element of $\mathbf{F}(x_1, x_2, x_3)$ that is fixed under A_3 but not under S_3. As we observed in § 7.1, the square root of the discriminant D, i.e.

$$\sqrt{D} = (x_1 - x_2)(x_1 - x_3)(x_2 - x_3)$$

is such an element. From (4) in § 5.6 we obtain

$$D = -4p^3 - 27q^2. \tag{4}$$

The Galois group of $\mathbf{F}(x_1, x_2, x_3)$ over $\mathbf{K}(\sqrt{D})$ is the alternating group A_3. Since A_3 is cyclic of order 3 we proceed as in the proof of 7.2.2. So we have first to adjoin the cubic roots of unity, i.e.

$$\varrho = \frac{1}{2}\left(-1+\sqrt{-3}\right), \quad \bar{\varrho} = \varrho^2 = \frac{1}{2}\left(-1-\sqrt{-3}\right).$$

Then we form the Lagrange resolvents

$$\begin{aligned}(1, x_1) &= x_1 + x_2 + x_3 = 0,\\ (\varrho, x_1) &= x_1 + \varrho x_2 + \bar{\varrho} x_3,\\ (\bar{\varrho}, x_1) &= x_1 + \bar{\varrho} x_2 + \varrho x_3.\end{aligned}$$

The third power of each resolvent is contained in $\mathbf{K}\left(\sqrt{D}, \varrho\right)$. We find

$$\begin{aligned}(\varrho, x_1)^3 &= x_1^3 + x_2^3 + x_3^3 + 3\varrho(x_1^2x_2 + x_2^2x_3 + x_3^2x_1)\\ &\quad + 3\bar{\varrho}(x_1x_2^2 + x_2x_3^2 + x_3x_1^2) + 6x_1x_2x_3\\ &= (x_1^3 + x_2^3 + x_3^3) - \frac{3}{2}(x_1^2x_2 + x_2^2x_3 + x_3^2x_1 + x_1x_2^2 + x_2x_3^2\\ &\quad + x_3x_1^2) + 6x_1x_2x_3 + \frac{3}{2}\sqrt{-3D}.\end{aligned}$$

The symmetric polynomials can be expressed by the elementary symmetric polynomials

$$\begin{aligned}x_1 + x_2 + x_3 &= 0,\\ x_1x_2 + x_2x_3 + x_3x_1 &= p,\\ x_1x_2x_3 &= -q.\end{aligned}$$

This gives

$$(\varrho, x_1)^3 = -\frac{27}{2}q + \frac{3}{2}\sqrt{-3D}.$$

Similarly, we find

$$(\bar{\varrho}, x_1)^3 = -\frac{27}{2}q - \frac{3}{2}\sqrt{-3D}.$$

So, we obtain

$$\begin{aligned}(\varrho, x_1) &= \sqrt[3]{-\frac{27}{2}q + \frac{3}{2}\sqrt{-3D}},\\ (\bar{\varrho}, x_1) &= \sqrt[3]{-\frac{27}{2}q - \frac{3}{2}\sqrt{-3D}}.\end{aligned} \tag{5}$$

Note that these cube roots are not independent of each other because a simple calculation shows that

$$(\varrho, x_1)\,(\bar{\varrho}, x_1) = -3p.$$

Therefore one of the cube roots can be chosen arbitrarily, and the other is then uniquely determined by the last equation. Due to $1 + \varrho + \bar{\varrho} = 0$ we obtain

$$x_1 = \frac{1}{3}\left((\varrho, x_1) + (\bar{\varrho}, x_1)\right)$$

$$x_2 = \frac{1}{3}\left(\bar{\varrho}(\varrho, x_1) + \varrho(\bar{\varrho}, x_1)\right)$$

$$x_3 = \frac{1}{3}\left(\varrho(\varrho, x_1) + \bar{\varrho}(\bar{\varrho}, x_1)\right)$$

and hence, by (4) and (5),

$$x_1 = w + w', \quad x_2 = \bar{\varrho} w + \varrho w', \quad x_3 = \varrho w + \bar{\varrho} w'$$

where

$$w = \sqrt[3]{-\frac{q}{2} + \sqrt{\left(\frac{q}{2}\right)^2 + \left(\frac{p}{3}\right)^3}}, \quad w' = \sqrt[3]{-\frac{q}{2} - \sqrt{\left(\frac{q}{2}\right)^2 + \left(\frac{p}{3}\right)^3}}.$$

These expressions for the roots are called *Cardano's formula.* It is easy to verify that Cardano's formula is also valid in all fields whose characteristic is greater than 3.

Due to Lemma 7.3.1, it is sufficient for the study of quartic equations to consider polynomials of the form

$$x^4 + px^2 + qx + r, \tag{6}$$

where p, q, r are indeterminates. Let x_1, x_2, x_3, x_4 denote the roots. The Galois group of (6) over $\mathbf{K} = \mathbf{F}(q, p, r)$ is the symmetric group S_4. In the notation of § 3.10,

$$S_4 \supset A_4 \supset V \supset C_1 \supset \varepsilon$$

is a composition series. Let

$$\mathbf{K} \subset \mathbf{K}(\sqrt{D}) \subset \mathbf{E} \subset \mathbf{L} \subset \mathbf{F}(x_1, x_2, x_3, x_4)$$

be the corresponding chain of subfields. Here D denotes the discriminant of x_1, x_2, x_3, x_4. The field $\mathbf{E}$ could be obtained from $\mathbf{K}(\sqrt{D})$ by the adjunction of an element that is fixed under V but not under A_4. However, it is expedient to proceed in a somewhat different way. We consider the element

$$\vartheta_1 = (x_1 + x_2)(x_3 + x_4).$$

It is evident that ϑ_1 is fixed under V. But it is easy to see that ϑ_1 is even fixed under the group U_1 of order 8 in the notation of § 3.10. If we apply S_4 to ϑ_1 we obtain

$$\vartheta_2 = (x_1 + x_3)(x_2 + x_4), \quad \vartheta_3 = (x_1 + x_4)(x_2 + x_3).$$

We find that ϑ_2 and ϑ_3 are fixed under the groups U_2 and U_3, respectively. The three elements $\vartheta_1, \vartheta_2, \vartheta_3$ are simultaneously fixed under $U_1 \cap U_2 \cap U_3 = V$ which means that $\mathbf{E} = \mathbf{K}(\sqrt{D}, \vartheta_1, \vartheta_2, \vartheta_3)$. The coefficients of the polynomial $(z - \vartheta_1)(z - \vartheta_2)(z - \vartheta_3)$ are fixed under S_4 and are therefore contained in $\mathbf{K}$. A straightforward computation gives

$$(z - \vartheta_1)(z - \vartheta_2)(z - \vartheta_3) = z^3 - 2pz^2 + (p^2 - 4r)z + q^2.$$

This polynomial is called the cubic resolvent of (6). By Cardano's formula, $\vartheta_1, \vartheta_2, \vartheta_3$ can be expressed by radicals.

The field $\mathbf{L}$ is obtained from $\mathbf{E}$ by adjoining an element that is fixed under the permutation $(x_1, x_2)(x_3, x_4)$. Such an element is the sum $\eta_1 = x_1 + x_2$. The other elements of V carry η_1 into $\eta_2 = x_3 + x_4$. Therefore η_1 and η_2 are roots of a polynomial with coefficients in $\mathbf{E}$. We find

$$\eta_1 + \eta_2 = x_1 + x_2 + x_3 + x_4 = 0,$$

$$\eta_1\eta_2 = (x_1 + x_2)(x_3 + x_4) = \vartheta_1.$$

This gives

$$(y - \eta_1)(y - \eta_2) = y^2 + \vartheta_1$$

and hence

$$x_1 + x_2 = \sqrt{-\vartheta_1}, \quad x_3 + x_4 = -\sqrt{-\vartheta_1}.$$

Similarly, we obtain

$$x_1 + x_3 = \sqrt{-\vartheta_2}, \quad x_2 + x_4 = -\sqrt{-\vartheta_2},$$

$$x_1 + x_4 = \sqrt{-\vartheta_3}, \quad x_2 + x_3 = -\sqrt{-\vartheta_3}.$$

This gives

$$\begin{aligned}
2x_1 &= \sqrt{-\vartheta_1} + \sqrt{-\vartheta_2} + \sqrt{-\vartheta_3}, \\
2x_2 &= \sqrt{-\vartheta_1} - \sqrt{-\vartheta_2} - \sqrt{-\vartheta_3}, \\
2x_3 &= -\sqrt{-\vartheta_1} + \sqrt{-\vartheta_2} - \sqrt{-\vartheta_3}, \\
2x_4 &= -\sqrt{-\vartheta_1} - \sqrt{-\vartheta_2} + \sqrt{-\vartheta_3}.
\end{aligned} \tag{7}$$

One easily verifies that

$$\sqrt{-\vartheta_1}\,\sqrt{-\vartheta_2}\,\sqrt{-\vartheta_3} = -q$$

so that one of the square roots can be expressed by the two others which agrees with $[\mathbf{F}(x_1, x_2, x_3, x_4) : \mathbf{E}] = 4$.

The equations (7) together with Cardano's formula for the cubic resolvent yield expressions for the roots of a quartic polynomial in terms of radicals.

Finally, we observe that

$$\begin{aligned}
\vartheta_1 - \vartheta_2 &= -(x_1 - x_4)(x_2 - x_3), \\
\vartheta_1 - \vartheta_3 &= -(x_1 - x_3)(x_2 - x_4), \\
\vartheta_2 - \vartheta_3 &= -(x_1 - x_2)(x_3 - x_4).
\end{aligned}$$

Therefore the discriminant of the cubic resolvent coincides with the discriminant of (6). From (4) in § 5.6 we thus obtain the following expression for the discriminant of (6):

$$16p^4r - 4p^3q^2 - 128p^2r^2 + 144pq^2r - 27q^4 + 256r^3.$$

7.4. Constructions by ruler and compass

Let P be a given set of points in a plane. To exclude trivial cases we assume that P contains at least two points. Our aim is to characterize all points that can be constructed from P by ruler and compass in a finite number of steps. This will provide answers to some classical geometric problems.

To begin with, we define what is meant by a constructible point. All points of P are regarded as constructible. Suppose that P' is a set of constructible points containing P. The ruler may be used to draw a so-called constructible line, i.e. a line joining two points of P'. The use of the compass is to draw a circle whose centre is a point of P' and whose radius is equal to the distance of two points of P'. Such a circle is said to be constructible. New constructible points are then obtained as the intersections of two constructible lines, of a constructible line and a constructible circle, or of two constructible circles. A point is said to be constructible from P if and only if it can be obtained in this way in a finite number of steps.

To give a survey of all constructible points we introduce a cartesian coordinate system with perpendicular axes in our plane. Since P contains at least two points we can choose the coordinate system in such a way that the points (0, 0) and (1, 0) belong to P. Then we have a line segment of length 1 between two given points. As is well known, this enables us to construct all points $(a, 0)$ where a is an arbitrary rational number. It follows that all points with rational coordinates are constructible.

Now let P consist of the points $P_1, \ldots, P_n$ and let (a_i, b_i) denote the coordinates of P_i. It is easy to verify that all points are constructible whose coordinates belong to the field $\mathbf{K} = \mathbb{Q}(a_1, \ldots, a_n, b_1, \ldots, b_n)$. But these are not all the constructible points because the compass enables us to construct points whose coordinates involve square roots of the coordinates of constructible points. The following theorem gives a characterization of all constructible points.

7.4.1. THEOREM. *A point is constructible from P if and only if both its coordinates belong to a Galois extension field of* $\mathbf{K}$ *whose degree over* $\mathbf{K}$ *is a power of* 2.

Proof. Suppose that P' is a set of constructible points and that all the coordinates of the points in P' belong to some field $\mathbf{L}$. The equation of the line joining two points of P' has coefficients in $\mathbf{L}$. The same holds for the equation of the circle whose centre is a point of P' and whose radius is equal to the distance of two given points in P'. The coordinates of the intersection of two such lines are contained in $\mathbf{L}$. To compute the coordinates of a line and a circle or of two circles, however, involves the extraction of square roots.

Therefore the coordinates of such intersections belong to $\mathbf{L}$ or to a field that is obtained from $\mathbf{L}$ by the adjunction of square roots. This shows that the coordinates of a constructible point P belong to a field that is obtained from $\mathbf{K}$ by a finite number of extensions where each extension consists of the adjunction of a square root. By adjoining also the square roots of the conjugates (over $\mathbf{K}$) one arrives at a Galois extension of $\mathbf{K}$ whose degree over $\mathbf{K}$ is a power of 2 and which contains the coordinates of P.

Conversely, suppose that the coordinates of a point P belong to a Galois extension $\mathbf{E}$ of $\mathbf{K}$ such that $[\mathbf{E}:\mathbf{K}] = 2^m$. Then the Galois group G of $\mathbf{E}$ over $\mathbf{K}$ is of order 2^m. From Theorem 3.11.7 it follows that G has a composition series whose indices are equal to 2. To this composition series there corresponds a chain of subfields

$$\mathbf{K} = \mathbf{K}_0 \subset \mathbf{K}_1 \subset \mathbf{K}_2 \subset \ldots \subset \mathbf{K}_m = \mathbf{E}$$

such that $[\mathbf{K}_i:\mathbf{K}_{i-1}] = 2$. Therefore we have $\mathbf{K}_i = \mathbf{K}_{i-1}(\sqrt{c_i})$ where $c_i \in \mathbf{K}_{i-1}$. In case c_i is a positive number, it is well known that $\sqrt{c_i}$ can be constructed by ruler and compass. But if c_i is negative or complex, then $\sqrt{c_i}$ can also be constructed as a point of the complex plane by ruler and compass because this requires only the construction of $\sqrt{|c_i|}$ and the bisection of the argument of c_i. Therefore all the numbers in $\mathbf{E}$, regarded as points of the complex plane, can be constructed by ruler and compass out of numbers in $\mathbf{K}$. This shows that, in particular, every real point whose coordinates belong to $\mathbf{E}$ can be constructed from P. This proves our theorem.

The first problem to which we are going to apply the theorem requires to construct, by ruler and compass, the edge of a cube whose volume is twice that of a given cube. Introducing a suitable scale we may assume that the edge of the given cube is of length 1. The edge of the cube of volume 2 is $\sqrt[3]{2}$. In the above notation we are given the points (0, 0) and (1, 0) so that the field $\mathbf{K}$ coincides with the field $\mathbf{Q}$ of the rational numbers. Since $x^3 - 2$ is irreducible in $\mathbf{Q}[x]$ it follows from 6.2.1 that $\sqrt[3]{2}$ is not contained in any field whose degree over $\mathbf{Q}$ is a power of 2. Therfore $\sqrt[3]{2}$ cannot be constructed by ruler and compass.

The quadrature of a circle requires the construction of a square whose area is equal to that of a given circle. The rectification of a circle means to construct a line segment whose length is equal to the circumference of a given circle. We may assume that the radius of the given circle is equal to 1. Then both problems are equivalent to the construction of line segments of lengths $\sqrt{\pi}$ and π, respectively. It can be proved that π is a transcendental number so that it is not contained in any algebraic extension of $\boldsymbol{Q}$. Therefore both constructions cannot be performed by ruler and compass.

The trisection of an angle α is equivalent to the construction of a line segment of length $\cos \alpha/3$ provided that a line segment of length $\cos \alpha$ is given. It is easy to verify that

$$4(\cos \alpha/3)^3 - 3 \cos \alpha/3 - \cos \alpha = 0.$$

Therefore the trisection of α by ruler and compass is possible if and only if the roots of the polynomial

$$4x^3 - 3x - \cos \alpha \tag{1}$$

are contained in a Galois extension field of degree 2^m over $\boldsymbol{Q}(\cos \alpha)$. Therefore the trisection of α is impossible if and only if the polynomial (1) is irreducible over $\boldsymbol{Q}(\cos \alpha)$. Taking $\alpha = \pi/3$ we have $\cos \pi/3 = \frac{1}{2}$. It is easy to see that the polynomial $4x^3 - 3x - \frac{1}{2}$ or, what amounts to the same, $8x^3 - 6x - 1$ is irreducible in $\boldsymbol{Q}[x]$. Therefore it is impossible to trisect the angle $\pi/3$ by ruler and compass. Clearly, the trisection of $\pi/2$ by ruler and compass is possible. In this case, (1) becomes $4x^3 - 3x$ which is reducible over $\boldsymbol{Q}$.

From the results of § 6.7, in particular from Theorem 6.7.4, it follows that a regular polygon with n sides can be constructed by ruler and compass if and only if $\varphi(n)$ is a power of 2. For

$$n = 2^a p_1^{a_1} \dots p_r^{a_r}$$

where $p_1, \dots, p_r$ are distinct odd prime numbers we find

$$\varphi(n) = 2^{a-1} p_1^{a_1-1} \dots p_r^{a_r-1} (p_1 - 1) \dots (p_r - 1).$$

Therefore $\varphi(n)$ is a power of 2 if and only if $a_i = 1$ and $p_i = 2^{k_i} + 1$ for $i = 1, \dots, r$. An integer of the form $2^k + 1$ cannot be a prime number unless k is a power of 2. For, if $k = tu$ where u is odd and greater than 1, then $2^k + 1 = (2^t)^u + 1$ is divisible by $2^t + 1$.

For $l = 0, 1, 2, 3, 4$ the numbers $2^{2^l} + 1$ are prime numbers, namely

$$3, 5, 17, 257, 65537. \tag{2}$$

Prime numbers of the form $2^{2^l} + 1$ are called *Fermat prime numbers*. It is not known whether there exist Fermat prime numbers other than (2). Certainly not all the numbers $2^{2^l} + 1$ are prime numbers, e.g. $2^{2^5} + 1$ is divisible by 641. Our result is that a regular n-gon can be constructed by ruler and compass if and only if

$$n = 2^a p_1 \dots p_r$$

where $p_1, \dots, p_r$ are distinct Fermat prime numbers.

Exercises

1. Find the Galois resolvent of $x^3 - 2$ over $\mathbf{Q}$.
2. Let p be a prime number and let $x^p - a$ be irreducible in $\mathbf{Q}[x]$. Show that the Galois group of $x^p - a$ over $\mathbf{Q}$ is isomorphic to the group of all linear transformations $z \to \alpha z + \beta$ where z is an indeterminate and $\alpha, \beta \in \mathrm{GF}(p)$, $\alpha \neq 0$.
3. Find a criterion to decide whether the Galois group of an irreducible cubic polynomial in $\mathbf{Q}[x]$ is isomorphic to S_3 or A_3.
4. Let $f(x) = x^4 + ax^2 + b$ with $a, b \in \mathbf{Q}$ be irreducible in $\mathbf{Q}[x]$. What groups can occur as the Galois group of $f(x)$ over $\mathbf{Q}$? Give criteria in terms of a and b to determine the Galois group.
5. Determine the Galois groups of the following polynomials.
   (i) $x^4 - 2$ over $\mathbf{Q}$, $\mathbf{Q}(\sqrt{2})$, and $\mathbf{Q}(i)$.
   (ii) $x^4 + 2$ over $\mathbf{Q}$ and $\mathbf{Q}(i)$.
6. Let G be the Galois group of a quartic polynomial with coefficients in a field whose characteristic is 0 or greater than 3. Show that the Galois group of the cubic resolvent is isomorphic to $G/(G \cap V)$.
7. (i) Let G be a transitive permutation group of prime degree p that contains a transposition. Show that G is isomorphic to S_p.
   (ii) Let $f(x)$ be an irreducible polynomial in $\mathbf{Q}[x]$ of prime degree p. Show that the Galois group of f over $\mathbf{Q}$ is isomorphic to S_p if $f(x)$ has $p - 2$ real and 2 complex roots.

8. Find the roots of the following polynomials:

$$x^2 - 3x + (3 - i), \quad x^3 - 6x - 9, \quad x^3 - 19x + 30.$$

9. Show that the field $\mathbf{K}(\vartheta)$ in 7.2.2 can be obtained by adjoining a single radical to $\mathbf{K}$.
10. Prove that every cyclotomic equation is soluble by radicals.

8

Order and valuations

In the field of the real numbers -1 cannot be expressed as a sum of squares. This property of the real numbers has been used by E. Artin and O. Schreier to develop a theory of so-called formally real fields. It turns out that precisely in the formally real fields one can introduce a complete order with the familiar properties with respect to addition and multiplication. Moreover, we shall generalize the concept of the absolute value to obtain the so-called valuations. In several branches of algebra the theory of valuations is an indispensabel tool. We must confine ourselves to the very rudiments of this theory.

8.1. Ordered fields

A field F is said to be *ordered* if a subset P of F is given such that

(i) $0 \notin P$.
(ii) If $a \in \mathsf{F}$, then either $a \in P$, $a = 0$, or $-a \in P$.
(iii) If $a \in P$ and $b \in P$, then $a + b \in P$ and $ab \in P$.

The elements of P are called positive, whereas a is negative if $-a \in P$.

In F we define a binary relation

$$a < b$$

to mean $b - a \in P$. We write $a \leqq b$ to indicate that $a = b$ or $a < b$. It is easy to see that $\leqq$ is a complete order in the sense of § 1.2. It is obvious that $\leqq$ is reflexive. If $a \leqq b$ and $b \leqq c$, then $a \leqq c$

holds trivially if $a = b$ or $b = c$. If $a < b$ and $b < c$, then we have $b - a \in P$ and $c - b \in P$; by (iii), this implies that $(b - a) + (c - b) = c - a$ belongs to P, i.e. $a < c$. Hence the relation $\leqq$ is transitive. Moreover, $a \leqq b$ and $b \leqq a$ implies that $a = b$; for, otherwise, both $b - a$ and $a - b$ would belong to P which contradicts (ii). Therefore the relation $\leqq$ is antisymmetric, Finally, if $a \neq b$, then precisely one of the elements $a - b$ or $b - a$ belongs to P so that $\leqq$ is a complete order.

As usual, we also write $b > a$ for $a < b$.

It is easy to verify that the relation $<$ has the familiar properties with respect to addition and multiplication:

$a < b$ implies that $a + c < b + c$ for any $c \in \mathbf{F}$,
$a < b$ implies that $ac < bc$ for every positive c,
if $a > 0$, $b > 0$, then $a < b$ implies that $a^{-1} > b^{-1}$.

The first two properties are immediate consequences of (iii). The third property follows from

$$ab(a^{-1} - b^{-1}) = b - a.$$

The absolute value $|a|$ of an element a in $\mathbf{F}$ is defined in the usual way, namely $|0| = 0$, $|a| = a$ if $a \in P$, and $|a| = -a$ if $-a \in P$. We have

$$|ab| = |a|\,|b|,$$
$$|a + b| \leqq |a| + |b|.$$

The equality is readily verified by considering the three cases

$$a \geqq 0,\ b \geqq 0;$$
$$a \geqq 0,\ b < 0;$$
$$a < 0,\ b < 0$$

separately. For $a \geqq 0$, $b \geqq 0$ we have

$$a + b = |a + b| = |a| + |b|,$$

and for $a < 0$, $b < 0$ we find

$$|a + b| = -(a + b) = -a + (-b) = |a| + |b|.$$

It obviously remains to consider the case $a \geqq 0$, $b < 0$. In this case we obtain

$$a + b < a < a - b = |a| + |b|,$$

$$-(a + b) \leqq -b \leqq a - b = |a| + |b|$$

which gives $|a + b| \leqq |a| + |b|$.

For any $a \in \mathsf{F}$ we have

$$a^2 = (-a)^2 \geqq 0$$

where $a^2 = 0$ implies that $a = 0$. Therefore an equation

$$\sum_{k=1}^{n} a_k^2 = 0$$

implies that $a_1 = a_2 = \ldots = a_n = 0$.

In particular, $1 = 1^2$ is positive since $1 \neq 0$. Therefore we have

$$m \cdot 1 = 1 + 1 + \ldots + 1 > 0$$

for every natural number m. This gives:

8.1.1. *Every ordered field is of characteristic* 0.

8.2. Formally real fields

A field F is called *formally real* if any equation

$$\sum_{k=1}^{n} a_k^2 = 0, \quad a_k \in \mathsf{F}$$

implies that $a_1 = a_2 = \ldots = a_n = 0$.

If $\sum_{k=1}^{n} a_k^2 > 0$ and, say $a_n \neq 0$, then we can divide by a_n^2 to obtain

$$\sum_{k=1}^{n-1} \left(\frac{a_k}{a_n}\right)^2 = -1.$$

Conversely,

$$\sum_{k=1}^{n-1} b_k^2 = -1$$

gives

$$\sum_{k=1}^{n-1} b_k^2 + 1^2 = 0.$$

Therefore a field $\mathbf{F}$ is formally real if and only if -1 cannot be expressed as a sum of squares of elements of $\mathbf{F}$.

Since, in an ordered field, every non-zero square is positive we have

8.2.1. *Every ordered field is formally real.*

Later we shall see that, conversely, every formally real field can be ordered.

A field $\mathbf{K}$ is called *real closed* if $\mathbf{K}$ is formally real and no proper algebraic extension of $\mathbf{K}$ is formally real. Examples of real closed fields are the field of the real numbers and the field of all real algebraic numbers.

8.2.2. Theorem. *Every real closed field can be ordered in one and only one way.*

Proof. Let $\mathbf{K}$ be real closed. We prove first that, for any element $w \neq 0$ of $\mathbf{K}$, either w or $-w$ is a square. Suppose that $w \in \mathbf{K}$, $w \neq 0$, and $w \neq v^2$ with $v \in \mathbf{K}$. If $\sqrt{w}$ denotes a root of the polynomial $x^2 - w$ in the algebraic closure of $\mathbf{K}$, then $\mathbf{K}(\sqrt{w})$ is a proper extension field of $\mathbf{K}$. Therefore $\mathbf{K}(\sqrt{w})$ is not formally real so that there exists a representation of -1 as a sum of squares of elements in $\mathbf{K}(\sqrt{w})$. Thus, there are elements $a_k\sqrt{w} + b_k$ with $a_k, b_k \in \mathbf{K}$ such that

$$-1 = \sum_{k=1}^{n} (a_k\sqrt{w} + b_k)^2$$

or

$$-1 = w\sum_{k=1}^{n} a_k^2 + \sum_{k=1}^{n} b_k^2 + 2\sqrt{w}\sum_{k=1}^{n} a_k b_k. \tag{1}$$

It follows that $\sum_{k=1}^{n} a_k b_k = 0$ for, otherwise, $\sqrt{w}$ would be contained in $\mathbf{K}$. Moreover, $\sum_{k=1}^{n} a_k^2 \neq 0$ since, otherwise, $\mathbf{K}$ would not be formally real.

Suppose that w can be expressed as a sum of squares of elements in $\mathbf{K}$. Then $w \sum_{k=1}^{n} a_k^2$ is also a sum of squares so that (1) yields a representation of -1 as a sum of squares. Since this contradicts the hypothesis that $\mathbf{K}$ is formally real it follows that w cannot be represented as a sum of squares. So we have the following result: If an element of $\mathbf{K}$ is not a square in $\mathbf{K}$ then it is not a sum of squares of elements in $\mathbf{K}$. In other words, every sum of squares in $\mathbf{K}$ is a square in $\mathbf{K}$.

From (1) we obtain

$$-w = \left(1 + \sum_{k=1}^{n} b_k^2\right)\left(\sum_{k=1}^{n} a_k^2\right)^{-1}.$$

As we just observed, both sums of squares on the right-hand side are squares in $\mathbf{K}$, i.e.

$$1 + \sum_{k=1}^{n} b_k^2 = c^2, \qquad \sum_{k=1}^{n} a_k^2 = d^2$$

where $c, d \in \mathbf{K}$. This gives $-w = (cd^{-1})^2$ as was to be shown.

To prove our theorem we observe that we obtain an ordering of $\mathbf{K}$ if we define the set P of the positive elements to consist precisely of the non-zero squares. We have to verify that P satisfies the conditions (i), (ii), (iii) of § 8.1. Clearly, (i) holds. As we just proved, if $w \neq 0$ then either w or $-w$ is a square. Hence (ii) is satisfied. Finally, the sum and the product of sums of squares are again sums of squares so that (iii) holds.

It is clear that this is the only order in $\mathbf{K}$ since every non-zero sum of squares is necessarily positive. This completes the proof.

From the rudiments of the theory of continuous functions it follows that every polynomial of odd degree with real coefficients has at least one real root. The next theorem shows that this property of the real numbers is shared by all real closed fields.

8.2.3. Theorem. *If* $\mathbf{K}$ *is real closed, then every polynomial in* $\mathbf{K}[x]$ *of odd degree has at least one root in* $\mathbf{K}$.

Proof. The theorem holds trivially for polynomials of degree 1. We assume that our proposition is true for every polynomial in $\mathbf{K}[x]$ whose degree is odd and less than the odd degree n of $f(x) \in \mathbf{K}[x]$.

If $f(x)$ is reducible in $\mathbf{K}[x]$, then at least one irreducible factor is of odd degree and hence has a root in $\mathbf{K}$. So we may assume that $f(x)$ is irreducible in $\mathbf{K}[x]$. Suppose that $f(x)$ has no root in $\mathbf{K}$. Let α be a root of $f(x)$ in the algebraic closure of $\mathbf{K}$. Since $\mathbf{K}$ is real closed it follows that the proper extension field $\mathbf{K}(\alpha)$ is not formally real. Therefore there exist elements $\gamma_1, \ldots, \gamma_m$ of $\mathbf{K}(\alpha)$ such that

$$-1 = \gamma_1^2 + \ldots + \gamma_m^2.$$

Each γ_k has a unique representation $\gamma_k = g_k(\alpha)$ as a polynomial in α with coefficients in $\mathbf{K}$ whose degree does not exceed $n - 1$. This gives

$$-1 = \sum_{k=1}^{m} \left(g_k(\alpha)\right)^2. \tag{2}$$

The polynomial

$$g(x) = -1 - \sum_{k=1}^{m} \left(g_k(x)\right)^2$$

belongs to $\mathbf{K}[x]$ and has the root α. Since $f(x)$ is irreducible we conclude that $g(x)$ is divisible by $f(x)$. Therefore we have

$$-1 = \sum_{k=1}^{m} \left(g_k(x)\right)^2 + f(x)\, h(x) \tag{3}$$

where $h(x) \in \mathbf{K}[x]$. The sum of the $\left(g_k(x)\right)^2$ is of even degree since the leading coefficient of each summand is a square and a sum of non-zero squares is distinct from zero. Moreover, this degree is positive for, otherwise, (2) would yield a representation of -1 as a sum of squares in $\mathbf{K}$. Hence it follows that the degree of $h(x)$ is odd and does not exceed $n - 2$. By the induction assumption, $h(x)$ has a root c in $\mathbf{K}$. Substituting c for x in (3) we obtain

$$-1 = \sum_{k=1}^{m} \left(g_k(c)\right)^2,$$

a contradiction since the $g_k(c)$ belong to $\mathbf{K}$ and $\mathbf{K}$ is formally real. This proves our theorem.

The field of the real numbers and the field of the real algebraic numbers are examples of real closed fields. A formally real field cannot be algebraically closed because $x^2 + 1$ has no root in any formally real field. But the field of the real numbers becomes algebraically closed by the adjunction of $\sqrt{-1}$, and the same is true for the field of all real algebraic numbers. The next theorem and its corollary show that all real closed fields have this property.

8.2.4. THEOREM. *Let* $\mathbf{K}$ *be an ordered field such that*

(a) *the positive elements of* $\mathbf{K}$ *have square roots in* $\mathbf{K}$,

(b) *every polynomial in* $\mathbf{K}[x]$ *of odd degree has at least one root in* $\mathbf{K}$.

Then $\mathbf{K}(\sqrt{-1})$ *is algebraically closed.*

Proof. To simplify the notation we write i for $\sqrt{-1}$. First, we show that every element of $\mathbf{K}(i)$ has square roots in $\mathbf{K}(i)$. Any element of $\mathbf{K}(i)$ has a unique representation $a + bi$ with $a, b \in \mathbf{K}$. We put $z = x + yi$ and attempt to determine $x, y \in \mathbf{K}$ such that $z^2 = a + bi$. This gives

$$(x + yi)^2 = (x^2 - y^2) + 2xyi = a + bi,$$

i.e.

$$x^2 - y^2 = a, \quad 2xy = b.$$

From the last two equations we obtain $(x^2 + y^2)^2 = a^2 + b^2$ so that $x^2 + y^2 = \sqrt{a^2 + b^2}$. The last equation and $x^2 - y^2 = a$ yield

$$x^2 = \frac{1}{2}\left(a + \sqrt{a^2 + b^2}\right), \quad y^2 = \frac{1}{2}\left(-a + \sqrt{a^2 + b^2}\right). \tag{4}$$

Suppose that $a + \sqrt{a^2 + b^2} < 0$. Then we obtain

$$\sqrt{a^2 + b^2} < -a,$$
$$a^2 + b^2 < a^2,$$
$$b^2 < 0$$

which is impossible. Hence, $a + \sqrt{a^2 + b^2} \geqq 0$. In the same way we find that $-a + \sqrt{a^2 + b^2} \geqq 0$. Hence, by (a), there exist elements $x, y \in \mathbf{K}$ that satisfy the equations (4), where the signs of x and y can be chosen arbitrarily. Multiplying both equations (4) we obtain $4x^2 y^2 = b^2$. Therefore we can determine the signs of x and y such that $2xy = b$. This shows that every element of $\mathbf{K}(i)$ has square roots in $\mathbf{K}(i)$.

It follows that there is no extension field of degree 2 over $\mathbf{K}(i)$.

To prove that $\mathbf{K}(i)$ is algebraically closed it is sufficient to show that every polynomial in $\mathbf{K}[x]$ splits into linear factors in $\mathbf{K}(i)\,[x]$. Indeed, if α is algebraic over $\mathbf{K}(i)$, then α is also algebraic over $\mathbf{K}$ since $[\mathbf{K}(i) : \mathbf{K}] = 2$. Therefore α is a root of a polynomial $f(x) \in \mathbf{K}[x]$. If we know that $f(x)$ splits into linear factors in $\mathbf{K}(i)\,[x]$ it follows that $\alpha \in \mathbf{K}(i)$ which means that $\mathbf{K}(i)$ is algebraically closed.

Now let $f(x)$ be any polynomial of positive degree in $\mathbf{K}[x]$, and let $\mathbf{E}$ be a splitting field of $(x^2 + 1)\,f(x)$ over $\mathbf{K}$ which contains $\mathbf{K}(i)$. Since $\mathbf{K}$ is of characteristic 0, $\mathbf{E}$ is a Galois extension of $\mathbf{K}$. Let G be the Galois group of $\mathbf{E}$ over $\mathbf{K}$ and put $|G| = 2^m u$ where u is odd. By Theorem 3.11.2, G has a Sylow 2-subgroup S of order 2^m. If $\mathbf{L}$ is the subfield of $\mathbf{E}$ that belongs to S, then $[\mathbf{E}:\mathbf{L}] = 2^m$ and $[\mathbf{L}:\mathbf{K}] = u$. From condition (b) on $\mathbf{K}$ it follows that there exists no extension field of odd degree over $\mathbf{K}$. This gives $\mathbf{L} = \mathbf{K}$, $[\mathbf{E}:\mathbf{K}] = 2^m$, and $|G| = 2^m$. By Theorem 3.11.7, G is soluble. Let G_1 be the subgroup of G that belongs to $\mathbf{K}(i)$. Then we have $|G : G_1| = 2$. In case $m > 1$, there exists a subgroup G_2 of G such that

$$G \supset G_1 \supset G_2$$

and $|G_1 : G_2| = 2$. Therefore there is a subfield $\mathbf{E}_2$ of $\mathbf{E}$ such that $[\mathbf{E}_2 : \mathbf{K}(i)] = 2$. This contradicts our previous result. Hence we have $m = 1$ and $\mathbf{E} = \mathbf{K}(i)$. Therefore $f(x)$ splits into linear factors in $\mathbf{K}(i)\,[x]$. As we observed above, this implies that $\mathbf{K}(i)$ is algebraically closed as was to be proved.

From the proof of Theorem 8.2.2 and from Theorem 8.2.3 it follows that every real closed field satisfies the conditions (a) and (b) of the last theorem. This gives:

8.2.5. Corollary. *If $\mathbf{K}$ is real closed, then $\mathbf{K}\left(\sqrt{-1}\right)$ is algebraically closed.*

One can prove that the field of the real numbers $\boldsymbol{R}$ satisfies the conditions (a) and (b) of Theorem 8.2.4. Therefore this Theorem can be used to prove the so-called

Fundamental Theorem of Algebra. *The field $\boldsymbol{C} = \boldsymbol{R}(i)$ of the complex numbers is algebraically closed.*

For the proof that $\boldsymbol{R}$ satisfies the condition (a) and (b) of Theorem 8.2.4 one has to make use of the particular properties of the field $\boldsymbol{R}$, e.g. the existence of least upper bounds. To prove the existence of least upper bounds one must refer to the construction of the real numbers by Dedekind cuts or by an equivalent device. Since this is beyond te scope of this book we omit the proof. The name "Fundamental Theorem of Algebra" has historical reasons. It refers to a previous epoch when the study of polynomials and their roots was regarded as the most important task of algebra.

Next we prove the converse of Theorem 8.2.4.

8.2.6. *If* $\mathbf{K}$ *is formally real and* $\mathbf{K}(\sqrt{-1})$ *algebraically closed, then* $\mathbf{K}$ *is real closed.*

Proof. Due to $[\mathbf{K}(\sqrt{-1}) : \mathbf{K}] = 2$ there is no algebraic extension of $\mathbf{K}$ other than $\mathbf{K}$ and $\mathbf{K}(\sqrt{-1})$. The field $\mathbf{K}(\sqrt{-1})$ is not formally real since -1 is a square. Therefore, $\mathbf{K}$ is real closed.

We conclude this section by a theorem on the existence of real closed fields.

8.2.7. Theorem. *If* $\mathbf{F}$ *is any formally real field, then there exists an algebraic extension field* $\mathbf{K}$ *of* $\mathbf{F}$ *that is real closed.*

Proof. Let $\mathbf{C}$ be the algebraic closure of $\mathbf{F}$. In $\mathbf{C}$ we consider the set $\mathfrak{S}$ of all formally real subfields that contain $\mathbf{F}$. This set is not empty because $\mathbf{F}$ belongs to it. If

$$\mathbf{L}_0 \subset \mathbf{L}_1 \subset \mathbf{L}_2 \subset \cdots$$

is a chain in $\mathfrak{S}$, then one easily verifies that the union $\cup\, \mathbf{L}_i$ belongs to $\mathfrak{S}$. Hence, by Zorn's Lemma, $\mathfrak{S}$ contains a maximal element $\mathbf{K}$. Then $\mathbf{K}$ is formally real. Any proper algebraic extension $\mathbf{K}_1$ of $\mathbf{K}$ (or rather an isomorphic image of $\mathbf{K}_1$) is contained in $\mathbf{C}$. Since $\mathbf{K}$ is maximal in $\mathfrak{S}$ it follows that $\mathbf{K}_1$ does not belong to $\mathfrak{S}$. This means that $\mathbf{K}_1$ is not formally real. Therefore $\mathbf{K}$ is real closed.

In view of Theorem 8.2.2 we have the following

8.2.8. Corollary. *Every formally real field can be ordered.*

8.3. Valuations

Let $\mathbf{F}$ be any field. A *valuation* φ of $\mathbf{F}$ is a function that is defined for all elements of $\mathbf{F}$ and whose values are real numbers such that for any $a, b \in \mathbf{F}$.

(i) $\varphi(a) \geqq 0$, $\varphi(a) = 0$ if and only if $a = 0$,

(ii) $\varphi(ab) = \varphi(a)\,\varphi(b)$,

(iii) $\varphi(a + b) \leqq \varphi(a) + \varphi(b)$ (triangle inequality).

In case $\mathbf{F}$ is the field of the complex numbers or a subfield of it, the absolute value is a familiar example of a valuation.

If we set $\varphi(a) = 1$ for any non-zero element a of $\mathbf{F}$ and $\varphi(0) = 0$, then we obviously obtain another example of a valuation. This valuation is said to be the *trivial* one.

To exhibit some less familiar examples we begin with a general remark.

Let $\mathbf{R}$ be an integral domain and suppose that we are given a valuation of $\mathbf{R}$, i.e. a real-valued function φ with the properties (i), (ii), (iii). Then it is possible to extend the function φ to the field of quotients $\mathbf{F}$ of $\mathbf{R}$ by putting

$$\varphi\left(\frac{a}{b}\right) = \frac{\varphi(a)}{\varphi(b)}\,.$$

It is easy to verify that this yields a valuation of $\mathbf{F}$. It is evident that (i) and (ii) are satisfied. Moreover,

$$\varphi\left(\frac{a}{b} + \frac{c}{d}\right) = \varphi\left(\frac{ad + bc}{bd}\right) = \frac{\varphi(ad + bc)}{\varphi(bd)} \leqq \frac{\varphi(ad) + \varphi(bc)}{\varphi(bd)}$$
$$= \varphi\left(\frac{ad}{bd}\right) + \varphi\left(\frac{bc}{bd}\right) = \varphi\left(\frac{a}{b}\right) + \varphi\left(\frac{c}{d}\right)$$

which shows that (iii) holds.

Example 1. Let p be a prime number. Any non-zero integer a has a unique representation of the form $a = p^k a'$ where $k \geqq 0$ and $(a', p) = 1$. We set $\varphi_p(a) = p^{-k}$ and, of course, $\varphi_p(0) = 0$. We will show that φ_p is a valuation of the integral domain $\boldsymbol{Z}$ of the integers. Clearly, (i) is satisfied. If $b \in \boldsymbol{Z}$, $b \neq 0$, and $b = p^l b'$ with $(b', p) = 1$, then we have

$$ab = p^{k+l}a'b'$$

where $(a'b', p) = 1$. This gives

$$\varphi_p(ab) = p^{-(k+l)} = p^{-k}p^{-l} = \varphi_p(a)\,\varphi_p(b)$$

which shows that (ii) holds. To verify (iii) it is no loss of generality to assume that $k \leqq l$. Then we have

$$a + b = p^k a' + p^l b' = p^k(a' + p^{l-k} b') = p^m c' \tag{1}$$

where $m \geqq k$ and $(c', p) = 1$. This gives

$$\varphi_p(a + b) = p^{-m} \leqq p^{-k} = \max(p^{-k}, p^{-l}).$$

Thus φ_p satisfies the inequality

$$\varphi_p(a + b) \leqq \max(\varphi_p(a), \varphi_p(b))$$

which is obviously sharper than (iii). We observe that for $k \neq l$ the factor $(a' + p^{l-k} b')$ in (1) is not divisible by p which implies that $m = k$ and

$$\varphi_p(a + b) = \max(\varphi_p(a), \varphi_p(b)) \quad \text{if} \quad \varphi_p(a) \neq \varphi_p(b).$$

As we observed above, the valuation φ_p of $\boldsymbol{Z}$ can be extended to the field of quotients $\boldsymbol{Q}$ of $\boldsymbol{Z}$. This valuation of the field of the rational numbers is called the *p-adic valuation.*

Example 2. Let F be any field and let $\mathsf{F}[x]$ denote the integral domain of the polynomials in an indeterminate x with coefficients in F. In $\mathsf{F}[x]$ we can define a valuation analogous to the p-adic valuation in $\boldsymbol{Z}$. Let $p(x)$ be an irreducible polynomial in $\mathsf{F}[x]$ of positive degree. Any polynomial $f(x) \neq 0$ in $\mathsf{F}[x]$ can be written in the form

$$f(x) = p(x)^k f_1(x)$$

where $k \geqq 0$ and $(f_1(x), p(x)) = 1$. Similar to the case of the p-adic valuation we set

$$\varphi_p(f) = c^k$$

where c is a real number such that $0 < c < 1$. Moreover, we set $\varphi_p(f) = 0$ if f is the zero polynomial. It is easy to verify that φ_p is a valuation of $\mathsf{F}[x]$. This valuation can be extended to the field of quotients of $\mathsf{F}[x]$, the field $\mathsf{F}(x)$ of the rational functions in x. Note that the valuation φ_p is trivial on the subfield F of $\mathsf{F}[x]$.

Example 3. As in the previous example let $\mathsf{F}[x]$ denote the integral domain of the polynomials in an indeterminate x with coefficients in a field F. For a non-zero polynomial $f(x)$ of degree m we set

$$\varphi_\infty(f) = c^{-m}$$

where c is a real number and $0 < c < 1$. Moreover, of course, $\varphi_\infty(0) = 0$. We leave it as an exercise for the reader to verify that φ_∞ is a valuation of $\mathbf{F}[x]$. Again, φ_∞ can be extended to a valuation of the field of the rational functions. As before, φ_∞ is trivial on $\mathbf{F}$.

Let e denote the unit element of the field $\mathbf{F}$. It follows from (i) and (ii) that for any valuation φ of $\mathbf{F}$

$$\varphi(e) = \varphi(-e) = 1$$

and hence $\varphi(a) = \varphi(-a)$ for any $a \in \mathbf{F}$.

From the triangle inequality (iii) it follows that

$$\varphi(a) = \varphi(b + a - b) \leqq \varphi(b) + \varphi(a - b)$$

or

$$\varphi(a) - \varphi(b) \leqq \varphi(a - b).$$

Interchanging the roles of a and b we obtain

$$\varphi(b) - \varphi(a) \leqq \varphi(b - a).$$

In view of $\varphi(a - b) = \varphi(b - a)$ this implies that

$$|\varphi(a) - \varphi(b)| \leqq \varphi(a - b).$$

If u is any root of unity in $\mathbf{F}$, i.e. $u^m = e$ for some natural number m, then it follows from (i) and (ii) that $\varphi(u) = 1$. Since the multiplicative group of a Galois field is finite, every non-zero element is a root of unity. This gives the following result.

8.3.1. *The only valuation of a Galois field is the trivial one.*

Two valuations φ_1 and φ_2 of a field $\mathbf{F}$ are called *equivalent* if, for arbitrary $a, b \in \mathbf{F}$, $\varphi_1(a) > \varphi_1(b)$ holds if and only if $\varphi_2(a) > \varphi_2(b)$.

In view of (ii), φ_1 and φ_2 are equivalent if and only if $\varphi_1(a) < 1$ implies that $\varphi_2(a) < 1$ and vice versa.

It is obvious that the trivial valuation is not equivalent to any non-trivial one.

For a real exponent s, let φ^s denote the function whose values are $\varphi(a)^s$.

8.3.2. Theorem. *If φ_1 and φ_2 are equivalent valuations of a field $\mathbf{F}$, then $\varphi_2 = \varphi_1$ with a positive exponent s.*

Proof. We may assume that both φ_1 and φ_2 are not trivial. Let a_0 be an element of $\mathbf{F}$ such that

$$0 < \varphi_1(a_0) < 1.$$

Such an element exists; since φ_1 is not trivial there is an element $c \neq 0$ in $\mathbf{F}$ such that $\varphi(c) \neq 1$. By (ii), either $a_0 = c$ or $a_0 = c^{-1}$ has the desired property.

Now let a be an arbitrary non-zero element of $\mathbf{F}$. We consider the set M_1 of all pairs (m, n) of integers such that $n > 0$ and

$$\varphi_1(a_0^m) < \varphi_1(a^n).$$

This inequality means $\varphi_1(a_0)^m < \varphi_1(a)^n$ or

$$\frac{m}{n} > \frac{\log \varphi_1(a)}{\log \varphi_1(a_0)}.$$

The right-hand side of the last inequality is the greatest lower bound of all the quotients m/n with $(m, n) \in M_1$. Similarly, we form the set M_2 of all pairs (m, n) of integers such that $n > 0$ and

$$\varphi_2(a_0^m) < \varphi_2(a^n).$$

We find that the greatest lower bound of the quotients m/n with $(m, n) \in M_2$ is equal to

$$\frac{\log \varphi_2(a)}{\log \varphi_2(a_0)}.$$

If φ_1 and φ_2 are equivalent, then the sets M_1 and M_2 coincide. This gives

$$\frac{\log \varphi_1(a)}{\log \varphi_1(a_0)} = \frac{\log \varphi_2(a)}{\log \varphi_2(a_0)}$$

or

$$\frac{\log \varphi_2(a)}{\log \varphi_1(a)} = \frac{\log \varphi_2(a_0)}{\log \varphi_1(a_0)} = s.$$

Here s is a real number that does not depend on a. Since φ_1 and φ_2 are equivalent we have not only $0 < \varphi_1(a_0) < 1$ but also $0 < \varphi_2(a_0) < 1$ which implies that s is positive. Thus we obtain $\varphi_2(a) = \varphi_1(a)^s$ for any $a \neq 0$ and, of course, also for $a = 0$. This proves our theorem.

A valuation φ is called *archimedean* if $\varphi(ne) > 1$ for some natural number n. Here ne stands for the sum $e + e + \dots + e$, n summands. If, however, $\varphi(ne) \leqq 1$ for any natural number n, then φ is said to be non-archimedean.

In what follows we shall simplify the notation and write, without ambiguity, n instead of ne. So we have $\varphi(n) \leqq 1$ for any natural number n if φ is a non-archimedean valuation.

8.3.3. Theorem. *A valuation φ of a field* **F** *is non-archimedean if and only if*

$$\varphi(a + b) \leqq \max\left(\varphi(a), \varphi(b)\right) \tag{2}$$

for any two elements a, b of **F**.

Proof. Suppose that (2) is satisfied. By induction we find

$$\varphi(a_1 + a_2 + \dots + a_n) \leqq \max\left(\varphi(a_1), \varphi(a_2), \dots, \varphi(a_n)\right).$$

Taking $a_1 = a_2 = \dots = a_n = e$ we obtain

$$\varphi(n) \leqq \varphi(e) = 1$$

which means that φ is non-archimedean.

Conversely, suppose that φ is non-archimedean. Then we have

$$\begin{aligned}\varphi(a + b)^n &= \varphi\left((a + b)^n\right)\\ &= \varphi\left(a^n + \binom{n}{1} a^{n-1}b + \dots + \binom{n}{n-1} ab^{n-1} + b^n\right)\\ &\leqq \varphi(a^n) + \varphi\left(\binom{n}{1} a^{n-1}b\right) + \dots + \varphi\left(\binom{n}{n-1} ab^{n-1}\right) + \varphi(b^n).\end{aligned}$$

Since φ is non archimedean it follows that for any natural number m

$$\varphi(mc) = \varphi(m)\,\varphi(c) \leqq \varphi(c).$$

Thus we obtain

$$\begin{aligned}\varphi(a + b)^n &\leqq \varphi(a^n) + \varphi(a^{n-1}b) + \dots + \varphi(ab^{n-1}) + \varphi(b^n)\\ &\leqq (n + 1)\left[\max\left(\varphi(a), \varphi(b)\right)\right]^n\end{aligned}$$

or

$$\varphi(a + b) \leqq \sqrt[n]{n + 1}\max\left(\varphi(a), \varphi(b)\right).$$

Since this holds for any natural number n and

$$\lim_{n\to\infty} \sqrt[n]{n + 1} = 1$$

the last inequality implies (2).

8.3.4. Corollary. *If φ is non-archimedean and $\varphi(a) \neq \varphi(b)$, then*

$$\varphi(a + b) = \max\left(\varphi(a), \varphi(b)\right).$$

Proof. Without loss of generality we may assume that $\varphi(b) < \varphi(a)$. Then we have to show that

$$\varphi(a + b) = \varphi(a) \quad \text{if} \quad \varphi(b) < \varphi(a).$$

Suppose that, on the contrary, $\varphi(a + b) < \varphi(a)$. Applying (2) to $a = (a + b) - b$ we obtain

$$\varphi(a) \leqq \max\big(\varphi(a + b), \varphi(b)\big) < \varphi(a),$$

a contradiction. This proves the corollary.

We shall now determine all non-trivial valuations of the field of the rational numbers.

8.3.5. Theorem. *Every archimedean valuation of the field of the rational numbers is equivalent to the usual absolute value.*

Proof. Let n and n' be natural numbers, $n > 1$, $n' > 1$. We express n' in the scale of n:

$$n' = a_0 + a_1 n + \dots + a_k n^k \tag{3}$$

where $0 \leqq a_i < n$ for $i = 0, \dots, k$ and $a_k \neq 0$. For any archimedean valuation φ of $\mathbf{Q}$ we obtain

$$\varphi(n') \leqq \varphi(a_0) + \varphi(a_1)\,\varphi(n) + \dots + \varphi(a_k)\,\varphi(n)^k.$$

By the triangle inequality we have $\varphi(a_i) \leqq a_i < n$. This gives

$$\varphi(n') < n\big(1 + \varphi(n) + \dots + \varphi(n)^k\big) \leqq n(k + 1)\,[\max\big(1, \varphi(n)\big)]^k.$$

From (3) it follows that $n' \geqq n^k$ or $k \leqq (\log n')/(\log n)$ and hence

$$\varphi(n') < n\left(\frac{\log n'}{\log n} + 1\right)[\max\big(1, \varphi(n)\big)]^{\frac{\log n'}{\log n}}.$$

We now replace n' by n'^t for a natural number t. This gives

$$\varphi(n')^t < n\left(t\,\frac{\log n'}{\log n} + 1\right)[\max\big(1, \varphi(n)\big)]^{t\frac{\log n'}{\log n}}$$

or

$$\varphi(n') < \left[n\left(t\,\frac{\log n'}{\log n} + 1\right)\right]^{1/t}[\max\big(1, \varphi(n)\big)]^{\frac{\log n'}{\log n}}.$$

Since

$$\lim_{t\to\infty}(\alpha t + \beta)^{1/t} = 1$$

for arbitrary real numbers α, β and $\alpha \neq 0$ we obtain

$$\varphi(n') \leqq [\max(1, \varphi(n))]^{\frac{\log n'}{\log n}}. \tag{4}$$

We now use the hypothesis that φ is archimedean. Therefore there exists a natural number $n_0 > 1$ such that $\varphi(n_0) > 1$. Substituting n_0 for n' in (4) we obtain

$$1 < [\max(1, \varphi(n))]^{\frac{\log n_0}{\log n}}$$

or

$$\varphi(n)^{\frac{\log n_0}{\log n}} > 1.$$

This shows that $\varphi(n) > 1$ for any natural number $n > 1$. Therefore it follows from (4) that

$$\varphi(n') \leqq \varphi(n)^{\frac{\log n'}{\log n}}$$

or

$$\frac{\log \varphi(n')}{\log n'} \leqq \frac{\log \varphi(n)}{\log n}. \tag{5}$$

Since n and n' are arbitrary natural numbers subject to the only condition $n > 1$, $n' > 1$ we may interchange the roles of n and n' in (5) to obtain the converse inequality. Thus we obtain

$$\frac{\log \varphi(n')}{\log n'} = \frac{\log \varphi(n)}{\log n}.$$

Hence the quotient

$$s = \frac{\log \varphi(n)}{\log n}$$

does not depend on n so that $\varphi(n) = n^s$ for any natural number $n > 1$. Moreover, $1 < \varphi(n) \leqq n$ implies that $0 < s \leqq 1$. It follows that $\varphi(a) = |a|^s$ for any rational number a. Therefore φ is equivalent to the absolute value as was to be shown.

We now come to the counterpart of the last theorem for non-archimedean valuations.

8.3.6. Theorem. *Any non-trivial non-archimedean valuation of the field of the rational numbers is equivalent to the p-adic valuation for some prime number p.*

Proof. Let φ be a non-trivial non-archimedean valuation of $\mathbf{Q}$. Since φ is non-archimedean we have $\varphi(a) \leqq 1$ for any integer a.

Since φ is not trivial there exists an integer $x_0 \neq 0$ such that $\varphi(x_0) < 1$. Let P denote the set of all integers x for which $\varphi(x) < 1$. We will show that P is an ideal in $\boldsymbol{Z}$. Let $x_1, x_2 \in P$. Then

$$\varphi(x_1 - x_2) \leqq \max\left(\varphi(x_1), \varphi(x_2)\right) < 1$$

so that $x_1 - x_2$ belongs to P. Moreover, for any integer a,

$$\varphi(ax_1) = \varphi(a)\,\varphi(x_1) \leqq \varphi(x_1) < 1$$

which shows that ax_1 is contained in P. Hence P is an ideal in $\boldsymbol{Z}$. Further, P is a prime ideal. Indeed, $xy \in P$, i.e. $\varphi(xy) = \varphi(x)\,\varphi(y) < 1$, implies that $\varphi(x) < 1$ or $\varphi(y) < 1$, i.e. $x \in P$ or $y \in P$. Since P contains the non-zero integer x_0 we have $P \neq (0)$. Hence $P = (p)$ for a prime number p. From $0 < \varphi(p) < 1$ it follows that $\varphi(p) = p^{-s}$ for some positive real number s. Let a be any non-zero integer and write $a = p^k a'$ where $k \geqq 0$ and $(a', p) = 1$. Then a' is not contained in $P = (p)$ which implies $\varphi(a') = 1$. Thus we obtain

$$\varphi(a) = \varphi(p^k)\,\varphi(a') = \varphi(p^k) = p^{-ks}$$

which shows that $\varphi = \varphi_p^s$ where φ_p denotes the p-adic valuation. This completes the proof.

Valuations are the main tool for the study of the arithmetic in algebraic number fields and for the theory of algebraic functions. In both cases one has to deal with the prolongations of valuations from a field to algebraic extensions. The more elaborate theory of valuations is beyond the scope of this book. We refer the reader to [11], [14], and [15].

Exercises

1. Show that the field of the complex numbers cannot be ordered.
2. Show that the only ordering of the field of the rational numbers is the usual one.
3. Show that the field $\boldsymbol{Q}(\sqrt{2})$ has precisely two distinct orderings.
4. Give an example of a formally real subfield of $\boldsymbol{C}$ that does not consist only of real numbers.
5. Let $\mathbf{K}$ be a real closed field. Prove that the only irreducible polynomials of positive degree in $\mathbf{K}[x]$ are of degree 1 or 2.

6. Let $f(x) \in \mathbf{K}[x]$ where $\mathbf{K}$ is real closed. Suppose that there are elements $\alpha, \beta \in \mathbf{K}$ such that $f(\alpha) < 0$, $f(\beta) > 0$. Show that there exists at least one $\gamma \in \mathbf{K}$ such that $f(\gamma) = 0$.
7. Prove that every valuation of a field of prime characteristic is non-archimedean.
8. Let φ be a non-archimedean valuation of a field $\mathbf{F}$. Show that $a_1 + a_2 + \ldots + a_n = 0$, $a_i \in \mathbf{F}$, implies that $\varphi(a_j) = \varphi(a_k)$ for some subscripts $j \neq k$.
9. Let $\mathbf{F}$ be any field and let $\mathbf{F}(x)$ denote the field of the rational functions in an indeterminate x with coefficients in $\mathbf{F}$.
   (i) Show that every valuation of $\mathbf{F}(x)$ that is trivial on $\mathbf{F}$ is non-archimedean.
   (ii) Prove that the valuations φ_p and φ_∞ considered in the Examples 2 and 3, respectively, are the only non-trivial valuations of $\mathbf{F}(x)$ that are trivial on $\mathbf{F}$. (Hint: If φ is a non-archimedean valuation of $\mathbf{F}(x)$ consider the cases $\varphi(x) \leqq 1$ and $\varphi(x) > 1$ separately.)
10. Let $C(z)$ denote the field of all rational functions of a complex variable z with coefficients in C.
    (i) Let $p(z) = z - a$ and let φ_p denote the valuation in Example 2. Exhibit the connection between the value $\varphi_p(f)$ of $f(z) \in C(z)$ and the behaviour of the function $f(z)$ at the point a of the complex plane.
    (ii) Find a similar connection between $\varphi_\infty(f)$ and the behaviour of $f(z)$ at the point ∞ of the complex plane. (Cf. Example 3.)
11. Let $\mathbf{F}$ be a field and let φ denote a non-trivial non-archimedean valuation of $\mathbf{F}$.
    (i) Show that the set R of all $a \in \mathbf{F}$ such that $\varphi(a) \leqq 1$ is a subring of $\mathbf{F}$.
    (ii) Prove that the subset P of all $x \in R$ such that $\varphi(x) < 1$ is a prime ideal in R.
    (iii) Prove that every element that is not contained in P is a unit.
    (iv) Show that P is a maximal ideal in R and that R contains no maximal ideal other than P.

9

Modules

Modules over a ring with unit element have been defined in § 3.13. In the present chapter we study modules over principal ideal rings. The results yield a classification of finitely generated abelian groups and of linear mappings of vector spaces. Another application concerns the arithmetic in algebraic number fields.

9.1. Elementary divisors

In this section we derive a theorem on matrices with elements in a principal ideal ring. This theorem is basic for the study of finitely generated modules.

Let $\mathbf{R}$ be a principal ideal ring. An $n \times n$ matrix P with elements in $\mathbf{R}$ is called *unimodular* if its determinant is a unit of $\mathbf{R}$. Due to the multiplication theorem for determinants the product of two unimodular matrices is again unimodular. By Cramer's Rule the inverse of a unimodular matrix is also unimodular. Therefore the unimodular $n \times n$ matrices with elements in $\mathbf{R}$ form a group under matrix multiplication.

Two $m \times n$ matrices A and A' with elements in $\mathbf{R}$ are said to be *equivalent* if there exist a unimodular $m \times m$ matrix U and a unimodular $n \times n$ matrix V, both with elements in $\mathbf{R}$, such that $UAV = A'$. It is easy to see that equivalence of matrices is an equivalence relation.

9.1.1. Theorem. *Let A be an $m \times n$ matrix with elements in a principal ideal ring* $\mathbf{R}$. *Then there exists a matrix equivalent to A of* *the form*

$$UAV = \begin{bmatrix} \varepsilon_1 & 0 & \dots & 0 & 0 & \dots & 0 \\ 0 & \varepsilon_2 & \dots & 0 & 0 & \dots & 0 \\ \dots & & & & & & \dots \\ 0 & 0 & \dots & \varepsilon_r & 0 & \dots & 0 \\ 0 & 0 & \dots & 0 & 0 & \dots & 0 \\ \dots & & & & & & \dots \\ 0 & 0 & \dots & 0 & 0 & \dots & 0 \end{bmatrix} \tag{1}$$

where the ε_i *are distinct from zero and*

$$\varepsilon_{i+1} \equiv 0 \,(\mathrm{mod}\ \varepsilon_i) \quad (i = 1, \dots, r-1). \tag{2}$$

The ideals (ε_i) *are uniquely determined, i.e. if*

$$\begin{bmatrix} \eta_1 & 0 & \dots & 0 & 0 & \dots & 0 \\ 0 & \eta_2 & \dots & 0 & 0 & \dots & 0 \\ \dots & & & & & & \dots \\ 0 & 0 & \dots & \eta_s & 0 & \dots & 0 \\ 0 & 0 & \dots & 0 & 0 & \dots & 0 \\ \dots & & & & & & \dots \\ 0 & 0 & \dots & 0 & 0 & \dots & 0 \end{bmatrix}$$

with

$$\eta_{i+1} \equiv 0 \,(\mathrm{mod}\ \eta_i) \quad (i = 1, \dots, s-1)$$

is also equivalent to A, *then* $r = s$ *and* $(\varepsilon_i) = (\eta_i)$ *for* $i = 1, \dots, r$.

Remark. The notation (1) is not meant to exclude the cases $r = m$ or $r = n$. In case $r = m$ the last $m - r$ rows of zeros do not occur and, similarly, for $r = n$ there are no zero columns.

Proof. By a permutation matrix we mean a square matrix which has precisely one entry 1 in each row and each column and zeros elsewhere. It is easy to see that the determinant of any permutation matrix is equal to 1 or -1 so that it is unimodular. Multiplication of A on the left by an $m \times m$ permutation matrix amounts to a permutation of the rows of A. Similarly, if we multiply A on the right by an $n \times n$ permutation matrix, then we obtain a permutation of the columns of A. Conversely, every permutation of the rows and columns of A can be performed by multiplying A on the left and on the right, respectively, by suitable permutation matrices.

By I_m we denote the $m \times m$ identity matrix. Let E_{ik} denote the $m \times m$ matrix whose entry in the (i, k) position is 1 whereas all other elements are equal to 0. It is evident that for any $\alpha \in \mathbf{R}$ and $i \neq k$ the sum $T_{ik}(\alpha) = I_m + \alpha E_{ik}$ is unimodular, namely of determinant 1. Left multiplication of A by $T_{ik}(\alpha)$ yields a matrix whose i-th row is obtained by multiplying the k-th row of A by α and adding it to the i-th row and whose other rows are the same as the corresponding rows of A.

In a similar way we define $n \times n$ matrices $T_{ik}(\alpha)$. Right multiplication of A by $T_{ik}(\alpha)$ gives a matrix whose k-th column is the sum of the k-th column of A and the i-th column of A multiplied by α and whose other columns coincide with those of A.

We need a third type of unimodular matrices. To simplify the notation we first consider a special case. Let α, β be the first two elements in the first row of A and suppose that $\alpha \neq 0, \beta \neq 0$. Since $\mathbf{R}$ is a principal ideal ring the greatest common divisor δ of α and β is unique up to a unit factor and can be represented in the form $\delta = \alpha\varkappa + \beta\lambda$ with $\varkappa, \lambda \in \mathbf{R}$. We put $\alpha' = \alpha/\delta$ and $\beta' = \beta/\delta$ so that $\alpha'\varkappa + \beta'\lambda = 1$. Therefore the matrix

$$D = \begin{bmatrix} \varkappa & -\beta' \\ \lambda & \alpha' \end{bmatrix}$$

is unimodular and hence so is the $n \times n$ matrix

$$D_{12} = \begin{bmatrix} \varkappa & -\beta' & 0 \ldots 0 \\ \lambda & \alpha' & 0 \ldots 0 \\ 0 & 0 & \\ \vdots & \vdots & I_{n-2} \\ 0 & 0 & \end{bmatrix}.$$

If we multiply A on the right by D_{12}, then the first two elements in the first row of the product are δ, 0.

If α, β are the first two elements in the first column of A then we can proceed in a similar way to find a unimodular $m \times m$ matrix D'_{12} such that left multiplication by D'_{12} gives a matrix whose first two elements in the first column are δ, 0.

Finally, if α and β are in the same row or in the same column of A, then we can combine the process just described with suitable permutations of the rows and columns of A to obtain a matrix equivalent to A in which δ, 0 are in the positions of α and β, respectively.

Let α be any non-zero element of $\mathbf{R}$. If α is not a unit, then α can be represented as a product of prime elements, $\alpha = \pi_1\pi_2 \dots \pi_l$. Due to the unique factorization in $\mathbf{R}$ the number l of prime factors of α is uniquely determined. For the purpose of our proof we define the *length* $L(\alpha)$ to be equal to l. In case α is a unit we put $L(\alpha) = 0$.

To prove our theorem we proceed by induction on n. For $n = 1$ there are no divisibility conditions on the ε_i so that our theorem is obviously true. Let us assume that the theorem holds for all matrices with less than n columns.

Now let A be a given $m \times n$ matrix with elements in $\mathbf{R}$. In the equivalence class of A we choose a matrix A_1 such that the minimal length $M(A_1)$ of the non-zero elements of A_1 is as small as possible. By permuting the rows and columns of A_1 in a suitable way, if necessary, we may assume that an element ε_1 of length $M(A_1)$ is in the (1,1) position of A_1. We maintain that all the elements in the first row and the first column of A_1 are divisible by ε_1. For, suppose that, on the contrary, α is a non-zero element in the first row or the first column of A_1 which is not divisible by ε_1. Then the length of the greatest common divisor δ of ε_1 and α is less than the length of ε_1. As we saw above, there is a matrix A_1' equivalent to A_1 which has the elements δ, 0 in the positions of ε_1 and α, respectively. This gives $M(A_1') < M(A_1) = L(\varepsilon_1)$ which contradicts the choice of A_1.

We can now subtract suitable multiples of the first row and the first column of A_1 from the other rows and columns, respectively, to arrive at a matrix of the form

$$A_2 = \begin{bmatrix} \varepsilon_1 & 0 \dots 0 \\ 0 & \\ \vdots & A_3 \\ 0 & \end{bmatrix}$$

where A_3 denotes an $(m - 1) \times (n - 1)$ matrix. Since these operations amount to multiplying A_1 by suitable unimodular matrices on the left and on the right it follows that A_2 is equivalent to A.

Next we show that all the elements of A_3 are divisible by ε_1. Suppose that, on the contrary, some element α of A_3 is not divisible by ε_1. If we add the row containing α to the first row of A_2 then we obtain a matrix with ε_1 in the (1,1) position such that not all the elements of the first row are divisible by ε_1. As we saw above, this leads to a contradiction.

By the induction assumption there exist unimodular matrices U_1 and V_1 such that

$$U_1 A_3 V_1 = \begin{bmatrix} \varepsilon_2 \, 0 \dots 0 & 0 \dots 0 \\ \dots\dots\dots\dots\dots\dots \\ 0\, 0 \dots \varepsilon_r & 0 \dots 0 \\ 0\, 0 \dots 0 & 0 \dots 0 \\ \dots\dots\dots\dots\dots\dots \\ 0\, 0 \dots 0 & 0 \dots 0 \end{bmatrix}$$

where $\varepsilon_2, \dots, \varepsilon_r$ are distinct from zero and

$$\varepsilon_{i+1} \equiv 0 \pmod{\varepsilon_i} \quad (i = 2, \dots, r-1).$$

Since ε_2 is a linear combination of elements of A_3 and the elements of A_3 are divisible by ε_1 it follows that

$$\varepsilon_2 \equiv 0 \pmod{\varepsilon_1}.$$

The matrices

$$U_2 = \begin{bmatrix} 1 & 0 \dots 0 \\ 0 & \\ \vdots & U_1 \\ 0 & \end{bmatrix}, \quad V_2 = \begin{bmatrix} 1 & 0 \dots 0 \\ 0 & \\ \vdots & V_1 \\ 0 & \end{bmatrix}$$

are obviously unimodular. Therefore the matrix

$$U_2 A_2 V_2 = \begin{bmatrix} \varepsilon_1 & 0 \dots 0 \\ 0 & \\ \vdots & U_1 A_3 V_1 \\ 0 & \end{bmatrix}$$

is equivalent to A and has the required properties.

It remains to show that the number r and the principal ideals $(\varepsilon_1), \dots, (\varepsilon_r)$ are uniquely determined by the equivalence class of A.

The uniqueness of r follows from the fact that r is equal to the rank of A and that the rank of a matrix is invariant under multiplication by non-singular (in particular unimodular) matrices on the right and on the left.

Let $\delta_k(A)$ denote a greatest common divisor of the k-rowed minors of A. If k exceeds the rank of A we put $\delta_k(A) = 0$, and by $\delta_0(A)$ we understand the unit element 1 of $\mathbf{R}$.

Since the k-rowed minors can be represented as linear combinations of $(k-1)$-rowed minors with coefficients in $\mathbf{R}$ it follows that

$$\delta_k(A) \equiv 0 \left(\operatorname{mod} \delta_{k-1}(A)\right).$$

Let U be any $m \times m$ matrix with elements in $\mathbf{R}$. The rows of UA are linear combinations of the rows of A with coefficients of $\mathbf{R}$. Therefore the k-rowed minors of UA are linear combinations of the k-rowed minors of A with coefficients in $\mathbf{R}$. This implies that

$$\delta_k(UA) \equiv 0 \left(\operatorname{mod} \delta_k(A)\right). \tag{3}$$

In case U is unimodular, U^{-1} has also elements in $\mathbf{R}$. Hence the same argument as before gives

$$\delta_k(A) = \delta_k(U^{-1}UA) \equiv 0 \left(\operatorname{mod} \delta_k(UA)\right). \tag{4}$$

Due to the Lemmas 4.3.4 and 4.3.5, (3) and (4) imply that

$$\left(\delta_k(A)\right) = \left(\delta_k(UA)\right)$$

for any unimodular $m \times m$ matrix U. In the same way we obtain

$$\left(\delta_k(A)\right) = \left(\delta_k(AV)\right)$$

for any unimodular $n \times n$ matrix V with elements in $\mathbf{R}$.

If A and A' are equivalent we thus obtain that

$$\left(\delta_k(A)\right) = \left(\delta_k(A')\right).$$

Due to the divisibility conditions (2) we find

$$(\delta_k) = \left(\delta_k(A)\right) = \left(\delta_k(UAV)\right) = (\varepsilon_1 \ldots \varepsilon_k) \quad (k = 1, \ldots, r)$$

and hence

$$(\varepsilon_1) = (\delta_1), (\varepsilon_2) = (\delta_2/\delta_1), \ldots, (\varepsilon_r) = (\delta_r/\delta_{r-1}).$$

Therefore $(\varepsilon_1), \ldots, (\varepsilon_r)$ are uniquely determined by the equivalence class of A.

This completes the proof.

The ideals $(\varepsilon_1), (\varepsilon_2), \ldots, (\varepsilon_r)$ and also the elements $\varepsilon_1, \varepsilon_2, \ldots, \varepsilon_r$ themselves are called the *elementary divisors* of A.

If ε_i' is any associate of ε_i, i.e. $\varepsilon_i' = \gamma\varepsilon_i$ with a unit γ of $\mathbf{R}$, then one can always find a matrix of the form (1) with ε_i' in the place of ε_i. Indeed, if we replace the 1 in the (i, i) position of I_m by γ, we obtain a unimodular matrix P_i, and the product P_iUAV coincides with UAV except for the element in the (i, i) position which is ε_i'.

It is not difficult to turn our proof into a constructive one. In case $\mathbf{R}$ is a euclidean ring one can arrive at the form (2) by permutation matrices and matrices of the form $T_{ik}(\alpha)$ only. In this case one minimizes the values of the function H with respect to which $\mathbf{R}$ is euclidean. For the ring of the integers, for example, this yields an effective method for computing the ε_i.

9.2. Modules over principal ideal rings

We recall the definition of a module in § 3.13. Let $\mathbf{R}$ be a ring with unit element and let M be an additively written abelian group with the operator domain $\mathbf{R}$. We write the operators on the left so that to any $\alpha \in \mathbf{R}$ and any $x \in M$ there corresponds a unique element αx of M and for arbitrary $x, y \in M$

$$\alpha(x + y) = \alpha x + \alpha y .$$

M is called a *left* $\mathbf{R}$*-module* if for arbitrary $\alpha, \beta \in \mathbf{R}$ and $x \in M$

$$(\alpha + \beta)\, x = \alpha x + \beta x ,$$

$$(\alpha\beta)\, x = \alpha(\beta x) ,$$

$$1x = x$$

where 1 denotes the unit element of $\mathbf{R}$.

In a similar way we can define right $\mathbf{R}$-modules. Since we shall consider only left $\mathbf{R}$-modules in this section we may simply speak of $\mathbf{R}$-modules or modules over $\mathbf{R}$. If there is no ambiguity we may also omit the reference to $\mathbf{R}$.

By a submodule of M we understand a subgroup of the additive group M that is admissible with respect to the operator domain $\mathbf{R}$. Since M is an abelian group, every subgroup is obviously normal. For any submodule S of M we can therefore form the factor group M/S. From the results of § 3.12 it follows that M/S becomes an

R-module if we define the action of **R** on the elements of the factor group M/S as follows. The elements of M/S are the cosets $x + S$. For $\alpha \in \mathbf{R}$ we define $\alpha(x + S)$ to be the coset $\alpha x + S$. This corresponds to (1) of § 3.12 in additive notation. The **R**-module thus obtained is called the *factor module* of M with respect to S and is also denoted by M/S.

If $S_1, \ldots, S_r$ are submodules of the **R**-module M, then the *sum*

$$S = S_1 + \ldots + S_r$$

consists of all elements

$$g_1 + \ldots + g_r$$

where $g_i \in S_i$, $i = 1, \ldots, r$. It is clear that S is a submodule of M. If every element of S has a unique representation in the form $g_1 + \ldots + g_r$ with $g_i \in S_i$, $i = 1, \ldots, r$, then we write

$$S = S_1 \oplus \ldots \oplus S_r$$

and call S the *direct sum* of $S_1, \ldots, S_r$.

If $g_1, \ldots, g_m$ are elements of M, then every expression of the form

$$\alpha_1 g_1 + \ldots + \alpha_m g_m \quad (\alpha \in \mathbf{R})$$

is called a *linear combination* of $g_1, \ldots, g_m$. The elements α_i are called the *coefficients* of the linear combination. Note that distinct linear combinations may represent one and the same element of M.

The set S of all linear combinations of $g_1, \ldots, g_m$ is obviously a submodule of M. We say that $g_1, \ldots, g_m$ *generate* S or are *generators* of S. An arbitrary **R**-module is said to be *finitely generated* if it is generated by a finite set of elements.

A *cyclic* module is one which is generated by a single element. For $g \in M$, the cyclic module generated by g is denoted by $\langle g \rangle$.

We consider the set $A(g)$ of all $\alpha \in \mathbf{R}$ such that $\alpha g = 0$. For $\alpha, \beta \in A(g)$, i.e. $\alpha g = \beta g = 0$ we have $(\alpha - \beta) g = \alpha g - \beta g = 0$ so that $\alpha - \beta \in A(g)$. This shows that $A(g)$ is a subgroup of the additive group $\mathbf{R}^+$ (cf. § 4.1). Moreover, if $\alpha \in A(g)$ and if ϱ is an arbitrary element of **R**, then

$$(\varrho\alpha) g = \varrho(\alpha g) = 0$$

which means that $\varrho\alpha$ belongs to $A(g)$. Therefore $A(g)$ is a subgroup of $\mathbf{R}^+$ which is admissible with respect to **R** as an operator domain where the action of $\varrho \in \mathbf{R}$ on $\mathbf{R}^+$ is defined as multiplication by ϱ on the left. Such a subgroup of $\mathbf{R}^+$ is called a *left ideal* of $\mathbf{R}$. In case $\mathbf{R}$ is commutative, $A(g)$ is an ideal of $\mathbf{R}$. We call $A(g)$ the *order ideal* or, briefly, the *order* of the element g or of the cyclic submodule $\langle g \rangle$.

It is evident that every abelian group G in additive notation can be regarded as a module over the ring $\mathbf{Z}$ of the integers. For an element g of G of finite order m the order ideal is the principal ideal (m). An element of G of infinite order has (0) as its order ideal.

A set of elements $x_1, \ldots, x_n$ of the $\mathbf{R}$-module M is called *linearly independent* if

$$\alpha_1 x_1 + \ldots + \alpha_n x_n = 0 \quad (\alpha_i \in \mathbf{R})$$

implies that $\alpha_1 = \ldots = \alpha_n = 0$. Otherwise $x_1, \ldots, x_n$ are said to be linearly dependent.

If an element of M can be represented as a linear combination of the linearly independent elements $x_1, \ldots, x_n$, then this representation is unique. Indeed,

$$\alpha_1 x_1 + \ldots + \alpha_n x_n = \beta_1 x_1 + \ldots + \beta_n x_n$$

gives

$$(\alpha_1 - \beta_1)\, x_1 + \ldots + (\alpha_n - \beta_n)\, x_n = 0$$

and hence due to the linear independence of $x_1, \ldots, x_n$

$$\alpha_i - \beta_i = 0 \quad (i = 1, \ldots, n).$$

A finitely generated $\mathbf{R}$-module M is said to be *free* if it is generated by finitely many linearly independent elements. Every set of linearly independent generators of M is called a *basis* of M. If $x_1, \ldots, x_n$ form a basis of M, then the cyclic submodules $\langle x_i \rangle$ have the order ideal (0), and M is the direct sum of these n cyclic submodules.

From now on we assume that $\mathbf{R}$ is a principal ideal ring.

9.2.1. *Let M be a finitely generated free module over a principal ideal ring $\mathbf{R}$. Then any two bases of M consist of the same number of elements.*

Proof. Let π be a prime element of $\mathbf{R}$. By πM we understand the subset of M consisting of all elements of the form πx with $x \in M$.

It is evident that πM is a submodule. If $x_1, \ldots, x_n$ are a basis of M, then $\alpha_1 x_1 + \ldots + \alpha_n x_n$ belongs to πM if and only if all the coefficients α_i are divisible by π. Moreover, $\alpha_1 x_1 + \ldots + \alpha_n x_n$ and $\beta_1 x_1 + \ldots + \beta_n x_n$ belong to the same coset of πM in the group theoretical sense if and only if $\alpha_i \equiv \beta_i \pmod{\pi}$ for $i = 1, \ldots, n$. From the fact that, by Theorem 4.4.2 and Corollary 4.4.3, the residue class ring $\mathbf{R}/(\pi)$ is a field it follows that the factor module $M/\pi M$ is isomorphic to a vector space of dimension n over $\mathbf{R}/(\pi)$. Since πM is defined without reference to a basis it follows that every basis of M must yield the same dimension of this vector space, i.e. every basis of M consists of n elements as was to be proved.

The number of elements in every basis of a free module M is called the *rank* of M. It is expedient to regard the zero module as a free module of rank 0.

9.2.2. Theorem. *Let M be a free module of rank n over a principal ideal ring $\mathbf{R}$. Then every submodule of M is a free module whose rank does not exceed n.*

Proof. We proceed by induction on the rank of the module. For the zero module there is nothing to prove. We assume that the theorem holds for the submodules of every free module of rank $n - 1$.

Let $x_1, \ldots, x_n$ be a basis of M and let S denote any submodule of M. We consider the set J of all coefficients of x_n that occur for elements of S. If $\alpha_n, \beta_n \in J$, then there are elements

$$a = \alpha_1 x_1 + \ldots + \alpha_n x_n, \quad b = \beta_1 x_1 + \ldots + \beta_n x_n$$

of S. Since S is a submodule it follows that $a - b \in S$ and $\lambda a \in S$ for any $\lambda \in \mathbf{R}$. Hence we have $\alpha_n - \beta_n \in J$ and $\lambda \alpha_n \in J$. This means that J is an ideal in $\mathbf{R}$. If $J = (0)$, then S is a submodule of the free $\mathbf{R}$-module M' generated by $x_1, \ldots, x_{n-1}$. By the induction assumption S is therfore a free module whose rank is less than or equal to $n - 1$.

If $J \neq (0)$ we use the fact that $\mathbf{R}$ is a principal ideal ring. Thus we have $J = (\gamma)$. Then S contains an element

$$c = \gamma_1 x_1 + \ldots + \gamma_{n-1} x_{n-1} + \gamma x_n$$

and for any element

$$s = \sigma_1 x_1 + \dots + \sigma_{n-1} x_{n-1} + \sigma_n x_n$$

of S we have $\sigma_n = \lambda\gamma$, $\lambda \in \mathbf{R}$. Therefore $s - \lambda c$ belongs to the intersection $S \cap M'$. By the induction assumption the submodule $S \cap M'$ of M' is a free module whose rank r does not exceed $n - 1$. Let $t_1, \dots, t_r$ be a basis of $S \cap M'$. We maintain that $t_1, \dots, t_r, c$ form a basis of S. First, $s - \lambda c$ belongs to $S \cap M'$ so that $s - \lambda c$ is a linear combination of $t_1, \dots, t_r$. Hence, any element s of S can be represented as a linear combination of $t_1, \dots, t_r, c$. It remains to show that $t_1, \dots, t_r, c$ are linearly independent. Suppose that

$$\varrho_1 t_1 + \dots + \varrho_r t_r + \varrho c = 0 \quad (\varrho_i, \varrho \in \mathbf{R}). \tag{1}$$

If we represent all elements on the left-hand side as linear combinations of $x_1, \dots, x_n$, then the coefficient of x_n in the representation of each term $\varrho_i t_i$ is equal to zero whereas in the representation of ϱc the element x_n occurs with the coefficient $\varrho\gamma$. So we obtain $\varrho\gamma = 0$, i.e. $\varrho = 0$ since $\gamma \neq 0$. Due to the linear independence of $t_1, \dots, t_r$ it now follows from (1) that $\varrho_1 = \dots = \varrho_r = 0$. Therefore $t_1, \dots, t_r$, c form a basis of S. Due to $r \leqq n - 1$, the rank $r + 1$ of S does not exceed n. This completes the proof.

We observe that, in contrast to the second proposition in 3.13.9 for vector spaces, a free module of rank n in general contains proper submodules of the same rank. For instance, in the free module of rank n over the integers the elements with even coefficients form a proper submodule of rank n.

The following theorem gives an indication of the importance of free modules.

9.2.3. Theorem. *Every* **R**-*module with n generators is isomorphic to a factor module of a free* **R**-*module of rank n.*

Proof. Let G be any **R**-module generated by $g_1, \dots, g_n$. We consider the free **R**-module M with the basis $x_1, \dots, x_n$. To any element

$$\alpha_1 x_1 + \dots + \alpha_n x_n$$

of M we assign the element

$$\alpha_1 g_1 + \dots + \alpha_n g_n$$

of G. It is clear that this mapping is an $\mathbf{R}$-homomorphism of M onto G in the sense of § 3.12. By Theorem 3.12.1, the kernel is a submodule S of M, and G is $\mathbf{R}$-isomorphic to the factor module M/S. This proves our theorem.

9.2.4. COROLLARY. *Let $\mathbf{R}$ be a principal ideal ring. Then every submodule of an $\mathbf{R}$-module with n generators can be generated by at most n elements.*

Proof. Let G be an $\mathbf{R}$-module with n generators and let G_0 be a submodule of G. By the previous theorem, G is isomorphic to a factor module M/S of a free $\mathbf{R}$-module M of rank n. From Theorem 3.7.5, or rather from its generalization to groups with operators, it follows that G_0 is isomorphic to a factor module M_0/S where M_0 is a submodule of M. By Theorem 9.2.2, the rank of M_0 does not exceed n. Therefore G_0 can be generated by at most n elements as was to be proved.

The utility of Theorem 9.2.3 is mainly due to the fact that one can choose bases of a free module M and of its submodule S in a suitable way to obtain information about the structure of the factor module M/S. Before persuing this subject we have to consider the connection between distinct bases of a free module.

9.2.5. *Let $x_1, \ldots, x_n$ be a basis of a free $\mathbf{R}$-module M over a principal ideal ring $\mathbf{R}$. The elements*

$$y_i = \sum_{k=1}^{n} \gamma_{ik} x_k \quad (i = 1, \ldots, n) \tag{2}$$

form a basis of M if and only if the $n \times n$ matrix $C = [\gamma_{ik}]$ is unimodular.

Proof. Introducing the columns

$$\boldsymbol{x} = \begin{bmatrix} x_1 \\ \vdots \\ x_n \end{bmatrix}, \quad \boldsymbol{y} = \begin{bmatrix} y_1 \\ \vdots \\ y_n \end{bmatrix}$$

we can write the equations (2) as follows:

$$\boldsymbol{y} = C\boldsymbol{x}. \tag{3}$$

Suppose first that $y_1, \ldots, y_n$ form a basis of M. Then $x_1, \ldots, x_n$ can be expressed as linear combinations of $y_1, \ldots, y_n$, i.e.

$$\boldsymbol{x} = D\boldsymbol{y} \tag{4}$$

where D is an $n \times n$ matrix with elements in **R**. Substituting (3) in (4) we obtain $\boldsymbol{x} = DC\boldsymbol{x}$. Since the x_i are linearly independent the last equation gives $DC = I_n$. Taking the determinants we find $|D|\,|C| = 1$. This means that $|C|$ is a unit of **R**.

Conversely, suppose that C is unimodular. Then by Cramer's Rule, the inverse matrix C^{-1} has elements in **R**. From (3) we obtain $\boldsymbol{x} = C^{-1}\boldsymbol{y}$ so that each x_i is a linear combination of $y_1, \ldots, y_n$ with coefficients in **R**. This shows that $y_1, \ldots, y_n$ generate M. Moreover, if $y_1, \ldots, y_n$ were linearly dependent, then $\boldsymbol{x} = C^{-1}\boldsymbol{y}$ would imply that $x_1, \ldots, x_n$ are linearly dependent which contradicts the hypothesis that the x_i form a basis of M. This proves our proposition.

9.2.6. Corollary. *Let M be a finitely generated module over a principal ideal ring* **R**. *If $g_1, \ldots, g_n$ are generators of M and*

$$h_i = \sum_{k=1}^{n} \gamma_{ik} g_k \quad (i = 1, \ldots, n)$$

where the $n \times n$ matrix $C = [\gamma_{ik}]$ is unimodular, then $h_1, \ldots, h_n$ also generate M.

Proof. As in the second part of the proof of 9.2.5 we conclude that each g_k can be represented as a linear combination of $h_1, \ldots, h_n$ with coefficients in **R**. Therefore every linear combination of the g_k can also be expressed as a linear combination of the h_i as was to be shown.

Now let M be a free module and S a submodule of M. We shall show that one can choose particular bases of M and S which enable us to describe the structure of the factor module M/S.

9.2.7. Theorem. *Let* **R** *be a principal ideal ring. If M is a free* **R**-*module of rank n and S a submodule of rank r, then one can choose a basis $y_1, \ldots, y_n$ of M and a basis $t_1, \ldots, t_r$ of S such that*

$$t_i = \varepsilon_i y_i \quad (i = 1, \ldots, r)$$

and

$$\varepsilon_{i+1} \equiv 0 \pmod{\varepsilon_i} \quad (i = 1, \ldots, r-1).$$

The ideals $(\varepsilon_1), \ldots, (\varepsilon_r)$ *are uniquely determined.*

Proof. Let $x_1, \ldots, x_n$ and $s_1, \ldots, s_r$ be arbitrary bases of M and S, respectively. Each s_i has a unique representation as a linear combination of $x_1, \ldots, x_n$, i.e.

$$s_i = \sum_{k=1}^{n} \alpha_{ik} x_k \quad (i = 1, \ldots, r). \tag{5}$$

If we denote the columns with the elements $x_1, \ldots, x_n$ and $s_1, \ldots, s_r$ by $\boldsymbol{x}$ and $\boldsymbol{s}$, respectively, and introduce the $r \times n$ matrix $A = [\alpha_{ik}]$, the equations (5) can be written as follows:

$$\boldsymbol{s} = A\boldsymbol{x}. \tag{6}$$

Now let $y_1, \ldots, y_n$ and $t_1, \ldots, t_r$ be other bases of M and S, respectively, and let $\boldsymbol{y}$ and $\boldsymbol{t}$ denote the columns of the respective elements. Due to 9.2.5 we have

$$\boldsymbol{x} = V\boldsymbol{y}, \quad \boldsymbol{s} = W\boldsymbol{t}$$

where V and W are unimodular matrices. Introducing the new bases in (6) we obtain $W\boldsymbol{t} = AV\boldsymbol{y}$ or

$$\boldsymbol{t} = UAV\boldsymbol{y}$$

where $U = W^{-1}$ is also unimodular. This shows that to the pair M, S there corresponds a unique equivalence class of $r \times n$ matrices in the sense of § 9.1. By Theorem 9.1.1 we can choose the unimodular matrices U and V such that

$$UAV = \begin{bmatrix} \varepsilon_1 & 0 & \ldots & 0 & 0 & \ldots & 0 \\ 0 & \varepsilon_2 & \ldots & 0 & 0 & \ldots & 0 \\ \ldots & \ldots & \ldots & \ldots & \ldots & \ldots & \ldots \\ 0 & 0 & \ldots & \varepsilon_r & 0 & \ldots & 0 \end{bmatrix}.$$

(Since the s_i are linearly independent, the matrix A is of rank r. This means $r = m$ in the notation of Theorem 9.1.1. Therefore the zero rows in (1) of § 9.1 do not occur.) For this choice of the matrices U and V the corresponding bases $y_1, \ldots, y_n$ and $t_1, \ldots, t_r$ of M and S, respectively, have the required property. The uniqueness of $(\varepsilon_1), \ldots, (\varepsilon_r)$ follows from Theorem 9.1.1.

This completes the proof.

We now use the previous theorem to derive a general theorem about the structure of finitely generated modules over a principal ideal ring.

Let $g_1, \ldots, g_n$ be generators of an **R**-module G. As in the proof of Theorem 9.2.3 we consider the free **R**-module M with the basis $x_1, \ldots, x_n$. The mapping

$$\alpha_1 x_1 + \ldots + \alpha_n x_n \to \alpha_1 g_1 + \ldots + \alpha_n g_n \quad (\alpha_i \in \mathbf{R})$$

is an **R**-homomorphism of M onto G. The kernel S consists precisely of the elements

$$\varrho_1 x_1 + \ldots + \varrho_n x_n$$

of M such that

$$\varrho_1 g_1 + \ldots + \varrho_n g_n = 0.$$

We refer to such an equation as a relation between the generators $g_1, \ldots, g_n$ of G.

By Theorem 9.2.7 we can choose a basis $y_1, \ldots, y_n$ of M and a basis $t_1, \ldots, t_r$ of S such that

$$t_i = \varepsilon_i y_i \quad (i = 1, \ldots, r)$$

and

$$\varepsilon_{i+1} \equiv 0 \ (\mathrm{mod}\ \varepsilon_i) \quad (i = 1, \ldots, r-1).$$

The transformation which connects the two bases of M is of the form

$$y_i = \sum_{k=1}^{n} \gamma_{ik} x_k \quad (i = 1, \ldots, n)$$

where the $n \times n$ matrix $[\gamma_{ik}]$ is unimodular. Hence, by Corollary 9.2.6, the elements

$$h_i = \sum_{k=1}^{n} \gamma_{ik} g_k \quad (i = 1, \ldots, n)$$

also generate G.

An element

$$\varrho_1 y_1 + \ldots + \varrho_r y_r + \varrho_{r+1} y_{r+1} + \ldots + \varrho_n y_n$$

of M belongs to S if it can be represented as a linear combination of $t_1, \ldots, t_r$, i.e. if and only if

$$\varrho_i \equiv 0 \ (\mathrm{mod}\ \varepsilon_i) \quad (i = 1, \ldots, r) \tag{7}$$

and

$$\varrho_{r+1} = \dots = \varrho_n = 0. \tag{8}$$

Due to $M/S \cong G$, it follows that

$$\varrho_1 h_1 + \dots + \varrho_r h_r + \varrho_{r+1} h_{r+1} + \dots + \varrho_n h_n = 0$$

if and only if the conditions (7) and (8) are satisfied. By Theorem 9.1.1., the ideals $(\varepsilon_1), \dots, (\varepsilon_r)$ are uniquely determined.

Some of the ε_i, possibly all of them, may be units of **R**. Suppose that precisely $\varepsilon_1, \dots, \varepsilon_k$ are units. Then the relations $\varepsilon_i h_i = 0$, $i = 1, \dots, k$, mean that $h_1 = \dots = h_k = 0$. For $j = k + 1, \dots, r$, the cyclic module $\langle h_j \rangle$ has (ε_j) as its order ideal. The order ideals of $\langle h_{r+1} \rangle, \dots, \langle h_n \rangle$ are equal to (0).

Any element g of G has a representation of the form

$$g = \alpha_{k+1} h_{k+1} + \dots + \alpha_r h_r + \alpha_{r+1} h_{r+1} + \dots + \alpha_n h_n. \tag{9}$$

If

$$g = \alpha'_{k+1} h_{k+1} + \dots + \alpha'_r h_r + \alpha'_{r+1} h_{r+1} + \dots + \alpha'_n h_n$$

is a representation of the same element, then we obtain

$$(\alpha_{k+1} - \alpha'_{k+1})\, h_{k+1} + \dots + (\alpha_r - \alpha'_r)\, h_r$$
$$+ (\alpha_{r+1} - \alpha'_{r+1})\, h_{r+1} + \dots + (\alpha_n - \alpha'_n)\, h_n = 0.$$

As we observed above this gives

$$\alpha_j - \alpha'_j \equiv 0 \pmod{\varepsilon_j} \quad \text{for } j = k + 1, \dots, r,$$
$$\alpha_j - \alpha'_j = 0 \qquad\qquad \text{for } j = r + 1, \dots, n$$

which implies

$$\alpha_j h_j = \alpha'_j h_j \quad (j = k + 1, \dots, n).$$

Hence the representation (9) of g is unique. Therefore G is the direct sum of the cyclic submodules $\langle h_j \rangle$:

$$G = \langle h_{k+1} \rangle \oplus \dots \oplus \langle h_r \rangle \oplus \langle h_{r+1} \rangle \oplus \dots \oplus \langle h_n \rangle. \tag{10}$$

An element a of an **R**-module H is called a *torsion element* if there exists a non-zero element γ of **R** such that $\gamma a = 0$. If a and b are torsion elements, then so are $a + b$ and ϱa for any $\varrho \in$ **R**. Indeed, if $\delta b = 0$, $\delta \neq 0$, then

$$\gamma\delta(a + b) = \gamma\delta a + \gamma\delta b = 0$$

and $\gamma\varrho a = 0$. Therefore all the torsion elements of H form a submodule H_t of H, the *torsion submodule.* In case $H_t = H$, H is called a *torsion module.*

From the direct decomposition (10) it follows that

$$G_t = \langle h_{k+1}\rangle \oplus \ldots \oplus \langle h_r\rangle.$$

Moreover, the submodule

$$\langle h_{r+1}\rangle \oplus \ldots \oplus \langle h_n\rangle$$

is free.

Summarizing our results we obtain the following.

9.2.8. THEOREM. *Let* **R** *be a principal ideal ring.*

(i) *Any finitely generated* **R**-*module* G *has a direct decomposition*

$$G = T \oplus F$$

where T *is the torsion submodule of* G *and* F *a free module. The rank of* F *is uniquely determined.*

(ii) *Any torsion* **R**-*module* T *has a direct decomposition*

$$T = \langle a_1\rangle \oplus \ldots \oplus \langle a_s\rangle$$

i nto cyclic submodules $\langle a_i\rangle$ *where the order ideals* (η_i) *of the* $\langle a_i\rangle$ *satisfy the conditions*

$$\eta_{i+1} \equiv 0 \pmod{\eta_i} \quad (i = 1, \ldots, s - 1).$$

In case G itself is free, then, of course, the torsion module T in the direct decomposition (i) does not occur. Similarly, if G is a torsion module, then the direct summand F does not occur.

An abelian group in additive notation can always be regarded as a module over the ring $\mathbf{Z}$ of the integers. Therefore the previous theorem applies, in particular, to finitely generated abelian groups. In this case, F is a direct sum of cyclic groups of infinite order. The η_i in (ii) can be choosen as natural numbers so that $\langle a_i\rangle$ is a cyclic group of order η_i in the usual sense.

Now let $\langle a\rangle$ be a cyclic **R**-module whose order ideal (η) is distinct from (0) and (1), Suppose that η admits a factorization $\eta = \eta'\eta''$ where η' and η'' are no units and $(\eta', \eta'') = 1$. Then there exist elements λ' and λ'' of **R** such that

$$\lambda'\eta' + \lambda''\eta'' = 1.$$

We consider the elements

$$a' = \eta''a, \quad a'' = \eta'a.$$

It is evident that the order ideals of the cyclic modules $\langle a'\rangle$ and $\langle a''\rangle$ are (η') and (η'') respectively. We have $\langle a'\rangle \cap \langle a''\rangle = 0$; for if d belongs to this intersection, then we have $\eta' d = \eta'' d = 0$ and hence $d = (\lambda'\eta' + \lambda''\eta'')\, d = \lambda'\eta'\, d + \lambda''\eta''\, d = 0$. Moreover, a' and a'' generate $\langle a\rangle$ since

$$a = (\lambda'\eta' + \lambda''\eta'')\, a = \lambda'\eta'a + \lambda''\eta''a = \lambda'a'' + \lambda''a'.$$

By 3.9.2, $\langle a\rangle$ is therefore the direct sum of $\langle a'\rangle$ and $\langle a''\rangle$.

In case one of the factors η', η'' or both of them admit a further factorization into relatively prime elements other than units we can proceed in a similar way. If

$$\eta = \pi_1^{e_1} \ldots \pi_t^{e_t}$$

is the factorization of η into powers of distinct prime elements π_i we finally arrive at a direct decomposition

$$\langle a\rangle = \langle a_1^*\rangle \oplus \ldots \oplus \langle a_t^*\rangle$$

where the order ideal of $\langle a_i^*\rangle$ is $(\pi_i^{e_i})$. This direct decomposition is unique because $\langle a_i^*\rangle$ consists precisely of all elements of $\langle a\rangle$ whose order ideals are of the form (π_i^f). A module whose order ideal is generated by a power of a prime element is called a *primary module*.

9.2.9. Theorem. *Every cyclic torsion module over a principal ideal ring admits a unique decomposition into a direct sum of primary cyclic submodules.*

Combining the two previous theorems we obtain:

9.2.10. Corollary. *Every finitely generated torsion module over a principal ideal ring can be represented as a direct sum of primary cyclic submodules whose order ideals are uniquely determined.*

In particular, every finite abelian group is the direct sum of its Sylow subgroups; and every finite abelian p-group can be decomposed into a direct sum of cyclic p-groups.

9.3. Endomorphisms of vector spaces

Let V be an n-dimensional vector space over a field $\mathbf{K}$. As we defined in § 3.13, a *linear mapping* or a $\mathbf{K}$-*endomorphism* of V is a mapping:

$$a : v \dashrightarrow va \quad (v \in V)$$

such that for arbitrary elements v, w of V and γ of $\mathbf{K}$

$$(v + w)\, a = va + wa, \quad (\gamma v)\, a = \gamma(va). \tag{1}$$

If $u_1, \ldots, u_n$ is a basis of V and

$$x = \xi_1 u_1 + \ldots . + \xi_n u_n$$

is any element of V then (1) gives

$$\begin{aligned} xa &= (\xi_1 u_1 + \ldots + \xi_n u_n)\, a = (\xi_1 u_1)\, a + \ldots + (\xi_n u_n)\, a \\ &= \xi_1(u_1 a) + \ldots + \xi_n(u_n a) \end{aligned}$$

so that the action of a is uniquely determined by the images

$$u_i a = \sum_{k=1}^{n} \alpha_{ik} u_k \tag{2}$$

of a basis. Introducing the notation

$$\boldsymbol{u} = \begin{bmatrix} u_1 \\ \vdots \\ u_n \end{bmatrix}, \quad \boldsymbol{u}a = \begin{bmatrix} u_1 a \\ \vdots \\ u_n a \end{bmatrix}, \quad A = \begin{bmatrix} \alpha_{11} \ldots \alpha_{1n} \\ \ldots\ldots\ldots\ldots \\ \alpha_{n1} \ldots \alpha_{nn} \end{bmatrix}$$

we can write the equations (2) as follows

$$\boldsymbol{u}a = A\boldsymbol{u}. \tag{3}$$

Conversely, any $n \times n$ matrix A with elements in $\mathbf{K}$ defines a $\mathbf{K}$-endomorphism of V.

If $v_1, \ldots, v_n$ is another basis of V we have

$$v_i = \sum_{k=1}^{n} \sigma_{ik} u_k \quad (i = 1, \ldots, n)$$

or

$$\boldsymbol{v} = S\boldsymbol{u} \tag{4}$$

where $\boldsymbol{v}$ denotes the column of the elements $v_1, \ldots, v_n$ and $S = [\sigma_{ik}]$ is an $n \times n$ matrix with entries in $\mathbf{K}$. Since $u_1, \ldots, u_n$ can also be expressed as linear combinations of $v_1, \ldots, v_n$, the matrix S is non-singular and $\boldsymbol{u} = S^{-1}\boldsymbol{v}$. Due to (1) we obtain from (3)

$$\boldsymbol{v}a = (S\boldsymbol{u})\, a = S(\boldsymbol{u}a) = SA\boldsymbol{u} = SAS^{-1}\boldsymbol{v}.$$

Thus, SAS^{-1} is the matrix that corresponds to the **K**-endomorphism a with respect to the basis $v_1, \ldots, v_n$.

Conversely, two matrices A and SAS^{-1} describe the same **K**-endomorphism a in terms of the bases $\boldsymbol{u}$ and $\boldsymbol{v} = S\boldsymbol{u}$ respectively.

Two $n \times n$ matrices A and A' with elements in **K** are said to be *similar* if there exists a non-singular $n \times n$ matrix S with elements in **K** such that $A' = SAS^{-1}$. It is easy to see that similarity is an equivalence relation.

Our results about $n \times n$ matrices and **K**-endomorphisms of V can be summarized as follows:

9.3.1. THEOREM. *There is a one-to-one correspondence between the* **K**-*endomorphisms of V and the similarity classes of $n \times n$ matrices with elements in* **K**.

The condition that S has elements in **K** is redundant in the following sense: If there is a non-singular $n \times n$ matrix T with elements in any extension field of **K** such that $A' = TAT^{-1}$, then there also exists a non-singular $n \times n$ matrix S with entries in **K** for which $A = SAS^{-1}$. Indeed, the existence of such a matrix T amounts to the solubility of the system $A'T = TA$ of n^2 linear equations for the elements of T under the condition that T is non-singular. The coefficients of these n^2 equations are elements of A and A', respectively, and are hence contained in **K**. As is well known, every system of linear equations with coefficients in a field **K** has a general solution in **K**.

Let **R** be a ring with unit element. We assume that the n-dimensional vector space V over **K** is at the same time a right **R**-module.

This means that to any element a of **R** and to any element v of V there corresponds a unique element va of V, and that for arbitrary $a, b \in \mathbf{R}$ and $v, w \in V$

$$(v + w)\, a = va + wa\,, \tag{5}$$

$$v(a + b) = va + vb\,, \tag{6}$$

$$v(ab) = (va)\, b\,, \tag{7}$$

$$v1 = v \tag{8}$$

where 1 denotes the unit element of **R**. Moreover, we assume that for arbitrary $\gamma \in \mathbf{K}$, $v \in V$, $a \in \mathbf{R}$

$$\gamma(va) = (\gamma v)\, a. \tag{9}$$

By (5) and (9), any element a of **R** gives rise to a **K**-endomorphism of V. Moreover, (6) and (7) show that the sum and the product of two elements of **R** induce the sum and the product, respectively, of the corresponding **K**-endomorphisms in the sense of § 3.12. Finally, (8) shows that the unit element of **R** induces the identity automorphism of V. Thus we obtain a homomorphism of **R** into the ring of the **K**-endomorphisms of V.

Using the previous notation we have

$$ua = Au, \quad ub = Bu, \ldots,$$

$$u(a + b) = ua + ub = Au + Bu = (A + B)\, u,$$

$$u(ab) = (ua)\, b = (Au)\, b = A(ub) = A(Bu) = ABu$$

where $A, B, \ldots$ are $n \times n$ matrices with elements in **K**. If we assign to a the matrix A, to b the matrix B, etc. we therefore obtain a homomorphism of **R** into the ring of all $n \times n$ matrices with elements in **K**. Such a homomorphism is called a *representation* of **R** of degree n. A vector space V which is a right **R**-module and satisfies the condition (9) is called a *representation module* for **R**. If we change the basis of V by (4), then we obtain a so-called *equivalent* representation by the matrices SAS^{-1}, SBS^{-1}, ...

Suppose that V, regarded as an additive group, admits a decomposition into a direct sum

$$V = V_1 \oplus V_2 \oplus \ldots \oplus V_r$$

where the V_i are admissible with respect to **K** as left operators and to **R** as right operators. Then the V_i are representation submodules. By Theorem 3.13.10, there exists a basis of V that consists of bases of the subspaces $V_1, V_2, \ldots, V_r$. If we use this basis then we obtain a representation of **R** equivalent to that by $A, B, \ldots$ whose matrices are of the form

$$\begin{bmatrix} A_1 & 0 & \ldots & 0 \\ 0 & A_2 & \ldots & 0 \\ \ldots & \ldots & \ldots & \ldots \\ 0 & 0 & \ldots & A_r \end{bmatrix}, \quad \begin{bmatrix} B_1 & 0 & \ldots & 0 \\ 0 & B_2 & \ldots & 0 \\ \ldots & \ldots & \ldots & \ldots \\ 0 & 0 & \ldots & B_r \end{bmatrix}, \ldots$$

where for each i, $1 \leqq i \leqq r$, the matrices A_i, B_i, ... form a representation corresponding to the representation module V_i.

In Chapter 10 we shall deal with representations of algebras and in particular, of group algebras. Here we shall study the representations of the ring $\mathbf{K}[x]$ of the polynomials in an indeterminate x with coefficients in $\mathbf{K}$. This will lead to a characterization of the similarity classes of $n \times n$ matrices.

Let A be any $n \times n$ matrix with elements in $\mathbf{K}$ and let V be an n-dimensional vector space over $\mathbf{K}$. Then V becomes a representation module for the integral domain $\mathbf{K}[x]$ if we define the action of a polynomial

$$f(x) = \gamma_0 + \gamma_1 x + \ldots + \gamma_m x^m \quad (\gamma_i \in \mathbf{K})$$

as a right operator as follows:

$$\boldsymbol{u} f(x) = f(A)\, \boldsymbol{u} \tag{10}$$

where

$$f(A) = \gamma_0 I_n + \gamma_1 A + \ldots + \gamma_m A^m$$

and I_n denotes the n-rowed identity matrix. It is easy to verify that due to this definition all the postulates for a representation module are satisfied. Note that, in particular, for $f(x)$, $g(x) \in \mathbf{K}[x]$ and $f(x)\, g(x) = h(x)$ we have $f(A)\, g(A) = h(A)$ since the powers of A commute. Condition (9) means that

$$\gamma\big(\boldsymbol{u} f(x)\big) = (\gamma \boldsymbol{u})\, f(x)$$

or

$$\gamma\big(f(A)\boldsymbol{u}\big) = f(A)\, (\gamma \boldsymbol{u})$$

which is satisfied since $\gamma f(A) = f(A)\, \gamma$. Using (10) for $f(x) = \gamma_0$ we obtain $\boldsymbol{u}\gamma_0 = \gamma_0 \boldsymbol{u}$.

Introducing a new basis by (4) amounts to replacing A by a similar matrix SAS^{-1} because $Sf(A)\, S^{-1} = f(SAS^{-1})$.

Since $\mathbf{K}[x]$ is a principal ideal ring we can make use of the results of § 9.2 which are, of course, also valid for right modules. By Theorem 9.2.8, V is a direct sum

$$V = \langle w_1 \rangle \oplus \ldots \oplus \langle w_r \rangle \tag{11}$$

of cyclic $\mathbf{K}[x]$-modules $\langle w_i \rangle$ whose order ideals are (0) or principal ideals $\big(\eta_i(x)\big)$ with $\eta_i(x) \in \mathbf{K}[x]$. In the present case the order ideal

(0) cannot occur since, for any $v \in V$, the $n+1$ elements $v, vx, \ldots, vx^n$ are linearly dependent over $\mathbf{K}$, i.e.

$$\beta_0 v + \beta_1 vx + \ldots + \beta_n vx^n = v(\beta_0 + \beta_1 x + \ldots + \beta_n x^n) = 0$$

which means that v is a torsion element. Therefore the order ideals $(\eta_i(x))$ of the $\langle w_i \rangle$ are generated by non-zero polynomials $\eta_i(x) \in \mathbf{K}[x]$. Again by Theorem 9.2.8 we have

$$\eta_{i+1}(x) \equiv 0 \pmod{\eta_i(x)} \quad (i = 1, \ldots, r-1). \tag{12}$$

Since only the principal ideals $(\eta_i(x))$ are relevant, each $\eta_i(x)$ may be replaced by an associated polynomial in $\mathbf{K}[x]$, i.e. $\eta_i(x)$ may be multiplied by any non-zero element of $\mathbf{K}$. Therefore we may assume that the $\eta_i(x)$ are monic polynomials of $\mathbf{K}[x]$.

Clearly, each cyclic $\mathbf{K}[x]$ module $\langle w_i \rangle$ is a vector space over $\mathbf{K}$. It is easy to see that the dimension of $\langle w_i \rangle$ as a vector space is equal to the degree n_i of $\eta_i(x)$. Indeed, $\eta_i(x)$ is the unique monic polynomial of minimal degree in $\mathbf{K}[x]$ such that $w_i \eta_i(x) = 0$. Therefore

$$w_i, w_i x, \ldots, w_i x^{n_i - 1} \tag{13}$$

are a maximal system of elements of $\langle w_i \rangle$ that are linearly independent over $\mathbf{K}$; hence they form a basis of $\langle w_i \rangle$ as a vector space over $\mathbf{K}$.

Comparing the dimensions on both sides of (11) we obtain $n = n_1 + \ldots + n_r$.

Let

$$\eta_i(x) = x^{n_i} + \alpha_1 x^{n_i - 1} + \ldots + \alpha_{n_i}.$$

Then we have

$$(w_i x^{n_i - 1})\, x = w_i x^{n_i} = -\alpha_{n_i} w_i - \ldots - \alpha_1 w_i x^{n_i - 1}.$$

If $\boldsymbol{w}_i$ denotes the column of the elements (13) we therefore obtain

$$\boldsymbol{w}_i x = A_i \boldsymbol{w}_i$$

where

$$A_i = \begin{vmatrix} 0 & 1 & 0 & \ldots & 0 \\ 0 & 0 & 1 & \ldots & 0 \\ \multicolumn{5}{c}{\dotfill} \\ 0 & 0 & 0 & \ldots & 1 \\ -\alpha_{n_i} & -\alpha_{n_i - 1} & -\alpha_{n_i - 2} & \ldots & -\alpha_1 \end{vmatrix}. \tag{14}$$

For each i, $1 \leqq i \leqq r$, we now take the n_i elements (13) to build up a new basis of V. Using this basis amounts to replacing A by a similar matrix

$$\begin{bmatrix} A_1 & 0 & \dots & 0 \\ 0 & A_2 & \dots & 0 \\ \dots & \dots & \dots & \dots \\ 0 & 0 & \dots & A_r \end{bmatrix} \tag{15}$$

where each A_i is of the form (14) and the 0 denote zero matrices of a suitable size. The matrices $A_1, \ldots, A_r$ are called the *companion matrices* of A.

To simplify the notation we shall write

$$A_1 \dotplus A_2 \dotplus \ldots \dotplus A_r$$

instead of (15).

Under the assumption that $\mathbf{K}$ is algebraically closed we shall now decompose the cyclic $\mathbf{K}[x]$-modules $\langle w_i \rangle$ into direct sums of primary cyclic submodules.

Since $\mathbf{K}$ is algebraically closed the prime elements of $\mathbf{K}[x]$ are precisely the polynomials of degree 1. Therefore the factorization of $\eta_i(x)$ is of the form

$$\eta_i(x) = (x - \beta_1)^{k_1} \ldots (x - \beta_t)^{k_t}$$

and correspondingly we obtain the direct decomposition

$$\langle w_i \rangle = \langle w_{i1} \rangle \oplus \ldots \oplus \langle w_{it} \rangle \tag{16}$$

where $\langle w_{ij} \rangle$ has $\left((x - \beta_j)^{k_j}\right)$ as its order ideal. As above we find that $\langle w_{ij} \rangle$ regarded as a vector space over $\mathbf{K}$ is of dimension k_j. In view of (16) we obviously have

$$\sum_{j=1}^{t} k_j = n_i .$$

For each direct summand in (16), regarded as a vector space over $\mathbf{K}$ we will now choose a particular basis. To simplify the notation, let $\langle w \rangle$ be any of the direct summands on the right-hand side of (16) and let $\left((x - \beta)^k\right)$ denote its order ideal.

Since there is no polynomial $f(x)$ of degree less than k such that $wf(x) = 0$ it follows that the k elements

$$z_1 = w(x - \beta)^{k-1},\ \ z_2 = w(x - \beta)^{k-2}, \ldots, z_{k-1} = w(x - \beta),\ z_k = w$$

are linearly independent over **K** and hence form a basis of the vector space $\langle w \rangle$. We have

$$z_1(x - \beta) = 0,\ \ z_2(x - \beta) = z_1, \ldots, z_k(x - \beta) = z_{k-1}$$

or

$$z_1 x = \beta z_1,\ \ z_2 x = z_1 + \beta z_2, \ldots, z_k x = z_{k-1} + \beta z_k.$$

If z denotes the column with the elements $z_1, \ldots, z_k$ the last equations can be written as follows

$$zx = Bz$$

where

$$B = \begin{bmatrix} \beta\,0\,0\ldots 0\,0 \\ 1\,\beta\,0\ldots 0\,0 \\ 0\,1\,\beta\ldots 0\,0 \\ \ldots\ldots\ldots\ldots \\ 0\,0\,0\ldots 1\,\beta \end{bmatrix} \tag{17}$$

for $k > 1$ and $B = \beta$ for $k = 1$.

Choosing such a basis for each direct summand in (16) we obtain a new basis of $\langle w_i \rangle$. The matrix in the similarity class of A_i that corresponds to this new basis is of the form

$$B_1 \dotplus \ldots \dotplus B_t$$

where each B_j is of the form (17) with $\beta = \beta_j$.

Applying this process to each direct summand in (11) we finally arrive at a matrix similar to A of the following form

$$(B_{11} \dotplus \ldots \dotplus B_{1t_1}) \dotplus \ldots \dotplus (B_{r1} \dotplus \ldots \dotplus B_{rt_r}). \tag{18}$$

Here all the B_{ij} are of the form (17). The diagonal elements in the blocks B_{ij} are the roots of $\eta_i(x)$. The matrix (18) is called the *Jordan normal form* of A. The B_{ij} are called the *elementary blocks* of A.

As we shall see later the elementary blocks of A are uniquely determined by A except for the order in which they occur in (18). A permutation of the B_{ij} in (18) leads to a matrix in the same similarity class for it corresponds to a permutation of the elements of a basis of V. Anticipating the uniqueness of the elementary blocks in this sense we can state the following:

9.3.2. **Theorem.** *Two $n \times n$ matrices with elements in an algebraically closed field are similar if and only if the elementary blocks in their Jordan normal forms coincide except for their order.*

By the *characteristic polynomial* of the $n \times n$ matrix A we understand the determinant

$$\varphi(x) = |xI_n - A|.$$

Clearly, $\varphi(x)$ is a monic polynomial of degree n. The roots of $\varphi(x)$ are called the *characteristic roots* of A. Due to

$$|xI_n - A| = |S|\,|xI_n - A|\,|S^{-1}| = |xI_n - SAS^{-1}|$$

the characteristic polynomials of similar matrices coincide.

It is evident that the characteristic roots of a triangular matrix are equal to the elements in its principal diagonal. Since the elementary blocks in the Jordan normal form of A are triangular matrices we find

$$\varphi(x) = \prod_{i=1}^{r} \eta_i(x).$$

As we observed above, a permutation of the elementary blocks in the Jordan normal form leads to a similar matrix. We shall rearrange the elementary blocks in such a way that blocks which correspond to the same characteristic root are adjacent. Moreover, we will arrange the blocks belonging to the same characteristic root such that the numbers of the rows do not increase. To indicate the characteristic root we shall now denote the matrix (17) by $B(\beta)$.

Let $\beta_1, \ldots, \beta_s$ be the distinct characteristic roots of A and let $m_1, \ldots, m_s$ denote their respective multiplicities as roots of $\varphi(x)$. To each β_i there correspond certain elementary blocks

$$B_1(\beta_i), \ldots, B_{q_i}(\beta_i).$$

If m_{ij} denotes the number of rows (and columns) of $B_j(\beta_i)$, we have

$$\sum_{j=1}^{q_i} m_{ij} = m_i$$

and

$$m_{i1} \geqq m_{i2} \geqq \dots \geqq m_{iq_i}.$$

There exists a non-singular $n \times n$ matrix R such that

$$RAR^{-1} = \big(B_1(\beta_1) \dotplus \dots \dotplus B_{q_1}(\beta_1)\big) \dotplus \dots \dotplus \big(B_1(\beta_s) \dotplus \dots \dotplus B_{q_s}(\beta_s)\big).$$

(We observe that this decomposition of A into elementary blocks coincides with (18) except for the order of the blocks.) We now collect the elementary blocks with the same characteristic root to obtain the $m_i \times m_i$ matrix

$$C_i = B_1(\beta_i) \dotplus \dots \dotplus B_{q_i}(\beta_i) \tag{19}$$

and

$$RAR^{-1} = C_1 \dotplus \dots \dotplus C_s. \tag{20}$$

In the ring of all $n \times n$ matrices with elements in $\mathbf{K}[x]$ the non-singular matrices with elements in $\mathbf{K}$ are unimodular since their determinants are non-zero elements of $\mathbf{K}$, i.e. units of $\mathbf{K}[x]$. Hence, by Theorem 9.1.1, the elementary divisors of $xI_n - A$ and

$$R(xI_n - A)\, R^{-1} = xI_n - RAR^{-1} \tag{21}$$

coincide. Due to (20) we have

$$xI_n - RAR^{-1} = (xI_{m_1} - C_1) \dotplus \dots \dotplus (xI_{m_s} - C_s). \tag{22}$$

We shall now express the numbers m_{ij} in terms of the powers of $x - \beta_i$ which divide the elementary divisors of $xI_n - A$. This implies that the m_{ij} are uniquely determined by A.

For $k \neq i$, $|xI_{m_k} - C_k|$ is not divisible by $x - \beta_i$. Therefore the positive powers of $x - \beta_i$ that divide the elementary divisors of $xI_n - A$ coincide with those that divide the elementary divisors of $xI_{m_i} - C_i$.

For a fixed characteristic root β_i let π_k denote the highest power of $x - \beta_i$ that divides all k-rowed minors of $xI_{m_i} - C_i$. As we saw at the end of § 9.1, the quotients π_k/π_{k-1} are the precise powers of $x - \beta_i$ that divide the elementary divisors of $xI_{m_i} - C_i$. To determine the π_k we use (19).

To begin with, it is clear that

$$\pi_{m_i} = (x - \beta_i)^{m_i}.$$

To obtain π_{m_i-1} we have to find an $(m_i - 1)$-rowed minor of $xI_{m_i} - C_i$ which is divisible by a minimal power of $x - \beta_i$. It is easy to see that we get such a minor by cancelling that row and that column of $xI_{m_i} - C_i$ which pass through the first row and the last column of $B_1(\beta_i)$, respectively. This gives

$$\pi_{m_i-1} = (x - \beta_i)^{m_i - m_{i1}}.$$

Similarly, we obtain an $(m_i - 2)$-rowed minor of $xI_{m_i} - C_i$ which is divisible by a minimal power of $x - \beta_i$ if we also cancel that row and that column which contain the first row and the last column of $B_2(\beta_i)$, respectively. So we find

$$\pi_{m_i-2} = (x - \beta_i)^{m_i - m_{i1} - m_{i2}}.$$

Continuing in this way we obtain $\pi_{m_i-3}, \pi_{m_i-4}, \ldots$ In case $m_i > q_i$ we have $\pi_k = 1$ for $k \leqq m_i - q_i$. Forming the quotients we find the exact powers of $x - \beta_i$ that divide the elementary divisors of $xI_n - A$, namely

$$1, \ldots, 1, \quad (x - \beta_i)^{m_{iq_i}}, \quad (x - \beta_i)^{m_{i,q_i-1}}, \ldots, (x - \beta_i)^{m_{i1}} \qquad (23)$$

where the 1's do not necessarily occur.

This shows that the m_{ij}, i.e. the sizes of the elementary blocks in the Jordan normal form of A are uniquely determined. Thus the proof of Theorem 9.3.2 is now complete.

Moreover, if we form the sequences (23) for all the characteristic roots $\beta_1, \beta_2, \ldots, \beta_s$ and multiply all the first terms of each sequence, all the second terms of each sequence, etc., then we obtain the sequence

$$1, \ldots, 1, \quad \eta_1(x), \quad \eta_2(x), \ldots, \eta_r(x).$$

So the $\eta_i(x)$ can be characterized as follows:

9.3.3. *The polynomials $\eta_1(x), \ldots, \eta_r(x)$ are the elementary divisors of $xI_n - A$ that are distinct from units.*

Combining this with our previous result about the relation between the sizes of the elementary blocks of A and the linear factors of the elementary divisors of $xI_n - A$ we obtain the following:

9.3.4. THEOREM. *The $n \times n$ matrices A and A' are similar if and only if the elementary divisors of $xI_n - A$ and $xI_n - A'$ coincide.*

The degrees m_{ij} of the elementary blocks in the Jordan normal form of A can also be expressed in terms of the ranks of certain polynomials in A.

For convenience of notation we set for any $m \times m$ matrix M with elements in the field $\mathbf{K}$

$$\text{nul. } M = m - \text{rank } M$$

and call nul. M the *nullity* of M. Since similar matrices have the same rank it follows that nul. M has the same value for all matrices in the similarity class of M. In case

$$M = M_1 \dotplus \ldots \dotplus M_s$$

it is evident that

$$\text{nul. } M = \text{nul. } M_1 + \ldots + \text{nul. } M_s .$$

Therefore it follows from (21) and (22) that for any $\xi \in \mathbf{K}$

$$\text{nul. } (\xi I_n - A) = \text{nul. } (\xi I_{m_1} - C_1) + \ldots + \text{nul. } (\xi I_{m_s} - C_s).$$

In particular, if ξ is equal to a characteristic root β_i of A, then we have

$$|\beta_i I_{m_k} - C_k| \neq 0 \text{ for } k \neq i,$$

i.e.

$$\text{nul. } (\beta_i I_{m_k} - C_k) = 0 \text{ for } k \neq i$$

and hence

$$\text{nul. } (\beta_i I_n - A) = \text{nul. } (\beta_i I_{m_i} - C_i). \tag{24}$$

A straightforward calculation shows that for an elementary $m \times m$ block $B(\beta_i) = B$

$$\text{rank } (\beta_i I_m - B)^k = m - k \text{ for } k = 1, \ldots, m - 1$$

and

$$(\beta_i I_m - B)^k = 0 \text{ for } k \geqq m.$$

This gives

$$\text{nul. } (\beta_i I_m - B)^k = k \text{ for } k = 1, \ldots, m - 1, \tag{25}$$

$$\text{nul. } (\beta_i I_m - B)^k = m \text{ for } k \geqq m. \tag{26}$$

We now consider the diagonal block C_i of RAR^{-1} which splits into elementary blocks according to (19). Let l_k denote the number of elementary $k \times k$ blocks in (19) where $l_k = 0$ if there occurs no elementary $k \times k$ block. Using (25) and (26) one easily verifies that

$$\begin{aligned}
\text{nul.}\,(\beta_i I_{m_i} - C_i) &= l_1 + l_2 + l_3 + \ldots,\\
\text{nul.}\,(\beta_i I_{m_i} - C_i)^2 &= l_1 + 2l_2 + 2l_3 + \ldots,\\
\text{nul.}\,(\beta_i I_{m_i} - C_i)^3 &= l_1 + 2l_2 + 3l_3 + \ldots,\\
&\ldots\ldots\ldots\ldots\ldots\ldots\ldots\ldots
\end{aligned}$$

By these equations the numbers $l_1, l_2, l_3, \ldots$ can be expressed in terms of the values nul. $(\beta_i I_{m_i} - C_i)^k$ and therefore in terms of rank $(\beta_i I_{m_i} - C_i)^k$. An easy calculation gives

$$\begin{aligned}
l_j &= 2\,\text{nul.}\,(\beta_i I_{m_i} - C_i)^j - \text{nul.}\,(\beta_i I_{m_i} - C_i)^{j+1} - \text{nul.}\,(\beta_i I_{m_i} - C_i)^{j-1}\\
&= \text{rank}\,(\beta_i I_{m_i} - C_i)^{j+1} + \text{rank}\,(\beta_i I_{m_i} - C_i)^{j-1} - 2\,\text{rank}\,(\beta_i I_{m_i} - C_i)^j.
\end{aligned}$$

Due to (24) the numbers l_k, i.e. the sizes of the elementary blocks in the Jordan normal form of A are uniquely determined by the ranks of the powers $(\beta_i I_n > A)^k$ for $i = 1, \ldots, s$. In view of Theorem 9.3.2 we thus obtain the following

9.3.5. Theorem. *The $n \times n$ matrices A and A' with elements in a field $\mathbf{K}$ are similar if and only if for any element β of the algebraic closure of $\mathbf{K}$:*

$$\text{rank}\,(\beta I_n - A)^k = \text{rank}\,(\beta I_n - A')^k, \quad k = 1, 2, \ldots$$

Our previous considerations show that it is sufficient to take for β the characteristic roots β_i of A and to restrict k to the interval $1 \leqq k \leqq m_i$ where m_i denotes the multiplicity of β_i as a characteristic root of A.

It also follows from the results obtained above that the least exponent k such that

$$(\beta_i I_{m_i} - C_i)^k = 0$$

is equal to the number m_{i1} of the rows of the largest elementary block of C_i. Hence

$$\psi_i(x) = (x - \beta_i)^{m_{i1}}$$

is the monic polynomial of least degree such that

$$\psi_i(C_i) = 0.$$

As we saw above,

$$\eta_r(x) = \prod_{i=1}^{s} (x - \beta_i)^{m_{i1}}.$$

Thus $\psi(x) = \eta_r(x)$ is the monic polynomial of least degree for which

$$\psi(A) = 0.$$

Making use of the definition of $\eta_r(x)$ as elementary divisor of $xI_n - A$ we can express this fact as follows:

9.3.6. THEOREM. *Let $\varphi(x) = |xI_n - A|$ be the characteristic polynomial of the $n \times n$ matrix A and let $\delta(x)$ denote the monic greatest common divisor of all $(n-1)$-rowed minors of $xI_n - A$. Then $\psi(x) = \varphi(x)/\delta(x)$ is the monic polynomial of least degree such that $\psi(A) = 0$.*

As a corollary we obtain

9.3.7. CAYLEY-HAMILTON'S THEOREM. *If $\varphi(x)$ is the characteristic polynomial of the $n \times n$ matrix A, then $\varphi(A) = 0$.*

9.4. Finiteness conditions

Let M be a right or left $\mathbf{R}$-module where $\mathbf{R}$ is an arbitrary ring with unit element.

We say that M satisfies the *ascending chain condition* if for any ascending sequence

$$M_1 \subseteq M_2 \subseteq M_3 \subseteq \ldots$$

of submodules there exists a natural number r such that $M_r = M_{r+1} = \ldots$

Analogously, M satisfies the *descending chain condition* if for any descending sequence

$$M_1 \supseteq M_2 \supseteq M_3 \supseteq \ldots$$

of submodules there exists a natural number s such that $M_s = M_{s+1} = \ldots$

It is easy to see that the *ascending chain condition* is equivalent to the so-called *maximum condition*:

Every non-empty collection of submodules of M contains a maximal member, i.e. a submodule that is not properly contained in any other submodule of this collection.

To establish this equivalence we first assume that M satisfies the ascending chain condition. Let Ω be any collection of submodules of M. We choose a member M_1 of Ω. If M_1 is not maximal, then there exists a submodule M_2 in Ω such that $M_1 \subset M_2$. In case M_2 is not maximal, we can choose a submodule M_3 in Ω such that $M_2 \subset M_3$. After a finite number of steps we arrive at a maximal member of Ω since, otherwise, we would obtain an infinite chain $M_1 \subset M_2 \subset M_3 \subset \ldots$ which contradicts the assumption that M satisfies the ascending chain condition.

Conversely, suppose that M satisfies the maximum condition, and let $M_1 \subseteq M_2 \subseteq \ldots$ be an ascending chain of submodules. Due to the maximum condition there exists a maximal member M_r in the set $\{M_i\}$. It is obvious that $M_r = M_{r+1} = \ldots$

In a similar way it follows that the descending chain condition is equivalent to the *minimum condition*:

Every non-empty collection of submodules of M contains a minimal member, i.e. a submodule that does not properly contain any other submodule of this collection.

The maximum condition implies the following principle of induction: Let P be a property of the module M such that P holds for a submodule A of M if P holds for every submodule $B \supset A$. Then P holds for all submodules of M. Indeed, suppose that P is not true for all submodules of M. Then there exists a maximal submodule N for which P does not hold. This leads to a contradiction since then P holds for every submodule $B \supset N$.

If M satisfies the maximum or the minimum condition, then so does any factor module M/N because, by Theorem 3.7.5 for groups with operators, there is a one-to-one correspondence between the submodules of M/N and the submodules H of M such that $N \subseteq H \subseteq M$.

If $\mathbf{R}$ is a commutative ring, then every ideal of $\mathbf{R}$ can be regarded as a left or right $\mathbf{R}$-module. To show that the descending chain condition is very restrictive we prove the following proposition:

9.4.1. *If* **R** *is an integral domain that satisfies the descending chain condition for ideals, then* **R** *is a field.*

Proof. For any $a \in \mathbf{R}$, $a \neq 0$, we consider the descending chain of the principal ideals

$$(a) \supseteq (a^2) \supseteq (a^3) \supseteq \dots$$

By the descending chain condition we have $(a^n) = (a^{n+1})$ for some natural number n. Therefore there exists an element b of **R** such that $a^n = ba^{n+1}$ or $a^n(1 - ba) = 0$. Since **R** contains no divisors of zero it follows that $ba = 1$. Hence, for any non-zero element of **R**, there is a multiplicative inverse so that **R** is a field.

The next theorem gives another characterization of modules that satisfy the ascending chain condition. In accordance with the notation for cyclic submodules and for subspaces of a vector space, let $\langle a_1, \dots, a_r \rangle$ denote the submodule of the module M that is generated by the elements $a_1, \dots, a_r$ of M.

9.4.2. Theorem. *A module M satisfies the ascending chain condition if and only if every submodule of N is finitely generated.*

Proof. Suppose that M satisfies the ascending chain condition and let S be any submodule. If $S = 0$ our proposition is obviously true. In case $S \neq 0$ let a_1 be any non-zero element of S. If $\langle a_1 \rangle = S$, then S is generated by a single element. Otherwise we can choose an element a_2 of S that is not contained in $\langle a_1 \rangle$. Then $\langle a_1, a_2 \rangle$ properly contains $\langle a_1 \rangle$. In case $\langle a_1, a_2 \rangle = S$ our proposition is proved. Otherwise we choose an element a_3 of S such that $a_3 \notin \langle a_1, a_2 \rangle$. Then we have $\langle a_1, a_2 \rangle \subset \langle a_1, a_2, a_3 \rangle$. After a finite number of steps we arrive at $\langle a_1, a_2, \dots, a_n \rangle = S$. For, otherwise, we would obtain an infinite strictly ascending chain

$$\langle a_1 \rangle \subset \langle a_1, a_2 \rangle \subset \langle a_1, a_2, a_3 \rangle \subset \dots$$

Conversely, suppose that every submodule of M is finitely generated. Let

$$M_1 \subseteq M_2 \subseteq M_3 \subseteq \dots$$

be an ascending chain of submodules. We form the set theoretical union

$$U = M_1 \cup M_2 \cup M_3 \cup \dots$$

It is easy to see that U is a submodule of M. Indeed, if $x, y \in U$, then both x and y belong to some M_k with a sufficiently large subscript k. Then $x + y \in M_k$ and $\alpha x \in M_k$ for any $\alpha \in \mathbf{R}$; hence $x + y$ and αx are also contained in U. Due to the assumption on M, U is finitely generated, say $U = \langle a_1, a_2, \ldots, a_n \rangle$. Each a_i is contained in some submodule M_{k_i}. For $m = \max(k_1, k_2, \ldots, k_n)$ we have $a_i \in M_m$ for $i = 1, 2, \ldots, n$ so that $U = M_m$, i.e. $M_m = M_{m+1} = \ldots$ This completes the proof.

Now let $\mathbf{R}$ be any ring with unit element. By a left ideal of $\mathbf{R}$ we understand a subgroup of the additive group $\mathbf{R}^+$ that is admissible with respect to $\mathbf{R}$ as left operator domain where the action of an element $a \in \mathbf{R}$ as an operator is defined as multiplication by a on the left. Thus every left ideal of $\mathbf{R}$ can be regarded as a left $\mathbf{R}$-module. In a similar way we can define right ideals of $\mathbf{R}$.

We will now show that the chain conditions for left ideals in $\mathbf{R}$ imply the same conditions for finitely generated left $\mathbf{R}$-modules.

9.4.3. THEOREM. *Let* $\mathbf{R}$ *be a ring that satisfies the ascending (descending) chain condition for left ideals. Then any finitely generated left* $\mathbf{R}$*-module satisfies the ascending (descending) chain condition for submodules.*

Proof. Let $g_1, \ldots, g_n$ be a set of generators for the left $\mathbf{R}$-module M. For any submodule S of M we consider the following subsets $J_1(S), J_2(S), \ldots, J_n(S)$ of $\mathbf{R}$:

$J_k(S)$ consists of all elements $\alpha \in \mathbf{R}$ such that there exists an element

$$\alpha g_k + \alpha_{k+1} g_{k+1} + \ldots + \alpha_n g_n$$

in S.

From this definition it is obvious that $J_k(S)$ is a left ideal of $\mathbf{R}$. Moreover, if S and T are submodules of M such that $S \subseteq T$, then $J_k(S) \subseteq J_k(T)$ for $k = 1, \ldots, n$.

We will now show that from $S \subseteq T$ and $J_k(S) = J_k(T)$ for $k = 1, \ldots, n$ it follows that $S = T$. Let

$$t = \tau_1 g_1 + \tau_2 g_2 + \ldots + \tau_n g_n$$

be any element of T. Then $J_1(S) = J_1(T)$ implies that $\tau_1 \in J_1(S)$;

hence S contains an element

$$s_1 = \tau_1 g_1 + \alpha_2 g_2 + \dots + \alpha_n g_n.$$

Forming the difference we obtain

$$t - s_1 = \sigma_2 g_2 + \dots + \sigma_n g_n$$

where $\sigma_i = \tau_i - \alpha_i$. Due to $J_2(S) = J_2(T)$ there exists an element s_2 in S such that

$$s_2 = \sigma_2 g_2 + \beta_3 g_3 + \dots + \beta_n g_n.$$

Thus we obtain

$$t - s_1 - s_2 = \varrho_3 g_3 + \dots + \varrho_n g_n.$$

From $J_3(S) = J_3(T)$ we now conclude that S contains an element

$$s_3 = \varrho_3 g_3 + \gamma_4 g_4 + \dots + \gamma_n g_n.$$

We then form $t - s_1 - s_2 - s_3$ and continue in the same way. After n steps we obtain elements $s_1, s_2, \dots, s_n$ of S such that $t - s_1 - s_2 - \dots - s_n = 0$, i.e. $t \in S$ and, hence, $S = T$.

Now let

$$M_1 \subseteq M_2 \subseteq M_3 \subseteq \dots$$

be an ascending chain of submodules of M. For each k, $1 \leqq k \leqq n$, we obtain an ascending chain

$$J_k(M_1) \subseteq J_k(M_2) \subseteq J_k(M_3) \subseteq \dots$$

of left ideals of **R**. Due to the ascending chain condition for **R** there exists, for each k, a natural number m_k such that

$$J_k(M_{m_k}) = J_k(M_{m_k+1}) = \dots$$

Taking $m = \max(m_1, m_2, \dots, m_n)$ we have

$$J_k(M_m) = J_k(M_{m+1}) = \dots \text{ for } k = 1, \dots, n.$$

As we observed above this implies that

$$M_m = M_{m+1} = \dots.$$

Therefore M satisfies the ascending chain condition.

In the same way one can prove that M satisfies the descending chain condition if the descending chain condition for left ideals holds for **R**.

This proves our theorem.

For any commutative ring **R** with unit element we can form the ring $\mathbf{R}[x]$ of the polynomials in an indeterminate x with coefficients in **R**.

By a *Noetherian ring* we understand a commutative ring with unit element that satisfies the ascending chain condition for ideals.

9.4.4. THEOREM. *If* **R** *is a Noetherian ring then so is* $\mathbf{R}[x]$.

Proof. By Theorem 9.4.2, it suffices to show that any ideal A of $\mathbf{R}[x]$ is finitely generated.

For $k = 0, 1, 2, \dots$ let $J_k(A)$ denote the set that consists of 0 and the leading coefficients if all polynomials of degree k in A. It is obvious that $J_k(A)$ is an ideal in **R**. Moreover, let $\gamma \in J_k(A)$, $\gamma \neq 0$, so that there is a polynomial

$$\gamma x^k + \gamma_{k-1} x^{k-1} + \dots + \gamma_0$$

in A. Since A is an ideal the polynomial

$$x(\gamma x^k + \dots + \gamma_0) = \gamma x^{k+1} + \dots + \gamma_0 x$$

also belongs to A. Therefore γ is also contained in $J_{k+1}(A)$. So we obtain

$$J_0(A) \subseteq J_1(A) \subseteq J_2(A) \subseteq \dots \tag{1}$$

Due to the ascending chain condition for **R** there is a natural number m such that

$$J_m(A) = J_{m+1}(A) = \dots$$

By Theorem 9.4.2, each $J_k(A)$ is finitely generated, say

$$J_k(A) = \langle \alpha_{k1}, \dots, \alpha_{kr_k} \rangle, \quad k = 0, 1, \dots, m.$$

For $k = 0, 1, \dots, m$ and $l = 1, \dots, r_k$ we choose polynomials

$$f_{kl}(x) = \alpha_{kl} x^k + \beta_{kl} x^{k-1} + \dots$$

in A. We contend that these polynomials generate A. Let

$$g(x) = \gamma_n x^n + \gamma_{n-1} x^{n-1} + \dots + \gamma_0$$

be any polynomial in A. We may assume that every polynomial in A of degree less than n can be represented as a linear combination of the $f_{kl}(x)$ with coefficients in $\mathbf{R}[x]$. In case $n \leqq m$ we have

$$\gamma_n = \varrho_1 \alpha_{n1} + \dots + \varrho_{r_n} \alpha_{nr_n}$$

for suitable $\varrho_i \in \mathbf{R}$. Therefore

$$g(x) - \sum_{i=1}^{r_n} \varrho_i f_{ni}(x)$$

is a polynomial in A whose degree is less than n so that, by the induction assumption, it can be represented as a linear combination of the $f_{kl}(x)$. If $n > m$, then

$$\gamma_n = \sigma_1 \alpha_{m1} + \dots + \sigma_{r_m} \alpha_{mr_m}$$

for suitable $\sigma_i \in \mathbf{R}$. Hence

$$g(x) - \sum_{i=1}^{r_m} \sigma_i x^{n-m} f_{mi}(x)$$

belongs to A and is of degree less than n. As above it follows from the induction assumption that $g(x)$ can be expressed in the desired way. This completes the proof.

By repeated application of the last theorem it follows that if $\mathbf{R}$ is a Noetherian ring then so is the ring $\mathbf{R}[x_1, \dots, x_n]$ of the polynomials in several indeterminates with coefficients in $\mathbf{R}$. Every principal ideal ring is obviously a Noetherian ring and, in particular, the ring of all polynomials in one indeterminate with coefficients in a field is Noetherian. As a corollary of the last theorem we therefore obtain.

9.4.5. HILBERT'S BASIS THEOREM. *If* $\mathbf{R}$ *is a principal ideal ring or a field, then every ideal of the ring* $\mathbf{R}[x_1, \dots, x_n]$ *of the polynomials in the indeterminates* $x_1, \dots, x_n$ *with coefficients in* $\mathbf{R}$ *is finitely generated.*

9.5. Algebraic integers

Let $\mathbf{F}$ be an algebraic number field, i.e. a extension field of the field $\mathbf{Q}$ of the rational numbers of finite degree $n = [\mathbf{F}: \mathbf{Q}]$.

An element α of $\mathbf{F}$ is called an *algebraic integer* if α satisfies an equation

$$\alpha^m + c_1 \alpha^{m-1} + \dots + c_m = 0 \tag{1}$$

whose coefficients c_i belong to the ring $\mathbf{Z}$ of the integers.

In this section we shall be much more concerned with algebraic integers than with integers in the usual sense, i.e. elements of $\mathbf{Z}$. Therefore we shall simplify the terminology and use the term integer to denote any algebraic integer of F. To the elements of $\mathbf{Z}$ we shall refer as rational integers.

9.5.1. *If α is an integer, then the minimal polynomial of α in $\mathbf{Q}[x]$ has coefficients in $\mathbf{Z}$.*

Proof. Let $f(x)$ be any polynomial in $\mathbf{Z}[x]$ such that $f(\alpha) = 0$. The minimal polynomial $g(x)$ of α in $\mathbf{Q}[x]$ divides $f(x)$, i.e.

$$f(x) = g(x)\,h(x) \tag{2}$$

where $h(x) \in \mathbf{Q}\,[x]$. There are natural numbers a and b such that

$$g(x) = a^{-1}g^*(x), \quad h(x) = b^{-1}h^*(x)$$

where $g^*(x)$ and $h^*(x)$ are primitive polynomials in $\mathbf{Z}[x]$ (cf. § 5.3). From (2) we obtain

$$abf(x) = g^*(x)\,h^*(x).$$

By Lemma 5.3.1, the product $g^*(x)\,h^*(x)$ is primitive. This gives $ab = 1$, i.e. $a = 1$ so that $g(x)$ has coefficients in $\mathbf{Z}$ as was to be shown.

9.5.2. Corollary. *If a rational number is an integer, then it belongs to $\mathbf{Z}$.*

We shall consider $\mathbf{Z}$-modules in F. The $\mathbf{Z}$-module generated by the elements $\alpha_1, \ldots, \alpha_r$ of F is denoted by $(\alpha_1, \ldots, \alpha_r)$.

We now give another characterization of the integers in F.

9.5.3. Theorem. *An algebraic number α of F is an integer if and only if all the powers α^k, $k = 1, 2, \ldots$, are contained in a finitely generated $\mathbf{Z}$-module.*

Proof. First suppose that α is an integer. Then α satisfies an equation of the form (1).

$$\alpha^m = -c_1\alpha^{m-1} - \ldots - c_m \quad (c_i \in \mathbf{Z}) \tag{3}$$

shows that α^m is contained in the $\mathbf{Z}$-module M generated by 1, $\alpha, \ldots, \alpha^{m-1}$. Let us assume that for some natural number $k > m$,

α^k is contained in M, i.e.

$$\alpha^k = d_1\alpha^{m-1} + \dots + d_{m-1}\alpha + d_m, \quad d_i \in \boldsymbol{Z}.$$

Multiplying the last equation by α we obtain a representation of α^{k+1} as a linear combination of $\alpha^m, \dots, \alpha^2, \alpha$ with coefficients in $\boldsymbol{Z}$. If we replace α^m by the right-hand side of (3) we get a representation of α^{k+1} as a linear combination of $\alpha^{m-1}, \dots, \alpha, 1$ with coefficients in $\boldsymbol{Z}$, i.e. $\alpha^{k+1} \in M$. Therefore all powers α^k, $k = 1, 2, \dots$ belong to M.

Conversely, suppose that all the powers α^k, $k = 1, 2, \dots$ are contained in the finitely generated $\boldsymbol{Z}$-module M. By Theorem 9.4.3, M satisfies the ascending chain condition. In the chain

$$(1) \subseteq (1, \alpha) \subseteq (1, \alpha, \alpha^2) \subseteq \dots$$

there exists therefore a natural number m such that $(1, \dots, \alpha^{m-1}) = (1, \dots, \alpha^{m-1}, \alpha^m)$. Hence α^m can be expressed as a linear combination of $1, \dots, \alpha^{m-1}$ with coefficients in $\boldsymbol{Z}$, i.e. α is an integer.

This completes the proof.

9.5.4. Theorem. *The integers of the field* F *form an integral domain.*

Proof. Let α and β be integers in F. It is obvioulsy sufficient to show that $\alpha \pm \beta$ and $\alpha\beta$ are integers.

By Theorem 9.5.3, the powers α^k and β^k, $k = 1, 2, \dots,$ are contained in finitely generated $\boldsymbol{Z}$-modules, say

$$\alpha^k \in (\alpha_1, \dots, \alpha_r), \beta^k \in (\beta_1, \dots, \beta_s).$$

Therefore all the products $\alpha^k\beta^l$ belong to the $\boldsymbol{Z}$-module

$$M = (\alpha_1\beta_1, \dots, \alpha_1\beta_s, \alpha_2\beta_1, \dots, \alpha_2\beta_s, \dots, \alpha_r\beta_s).$$

Hence all the powers $(\alpha \pm \beta)^k$ and $(\alpha\beta)^k$, $k = 1, 2, \dots,$ belong to M so that $\alpha \pm \beta$ and $\alpha\beta$ are integers as was to be proved.

9.5.5. Lemma. *For any* $\xi \in \mathsf{F}$ *there exists a natural number* d *such that* $d\xi$ *is an integer.*

Proof. Since F is of finite degree over $\boldsymbol{Q}$, ξ is algebraic over $\boldsymbol{Q}$. Therfore ξ satisfies an equation

$$\xi^r + a_1\xi^{r-1} + \dots + a_r = 0 \tag{4}$$

where the a_i are rational numbers, say $a_i = b_i/d_i$ with $b_i, d_i \in \mathbf{Z}$. Let d denote the least common multiple of the denominators d_i. Multiplication of (4) by d^r gives

$$(d\xi)^r + a_1 d(d\xi)^{r-1} + \dots + a_r d^r = 0.$$

The coefficients $a_1d, \dots, a_rd^r$ are rational integers which shows that $d\xi$ is an integer.

By Theorem 6.5.2, F is a simple extension of $\mathbf{Q}$, i.e. $\mathsf{F} = \mathbf{Q}(\vartheta_0)$. Due to Lemma 9.5.5 there exists a natural number d such that $d\vartheta_0$ is an integer. It is evident that $d\vartheta_0 = \vartheta$ is also a primitive element for F over $\mathbf{Q}$, i.e. $\mathsf{F} = \mathbf{Q}(\vartheta)$.

9.5.6. THEOREM. *The integral domain* I *of the integers in the algebraic number field* $\mathsf{F} = \mathbf{Q}(\vartheta)$ *is a free* $\mathbf{Z}$*-module of rank* n *where* $n = [\mathsf{F}\colon \mathbf{Q}]$.

Proof. As we just observed we may assume that ϑ is an integer. Any element γ of I has a unique representation of the form

$$\gamma = c_0 + c_1\vartheta + \dots + c_{n-1}\vartheta^{n-1} \tag{5}$$

with coefficients $c_i \in \mathbf{Q}$. The minimal polynomial of ϑ in $\mathbf{Q}[x]$ is of degree n. Therefore there are n distinct conjugates of ϑ, say $\vartheta_1 = \vartheta, \vartheta_2, \dots, \vartheta_n$. It is evident that the ϑ_i are algebraic integers of the field $\mathbf{Q}(\vartheta_1, \dots, \vartheta_n)$. Replacing ϑ by the ϑ_i in (5) we obtain the n equations

$$\gamma_i = \sum_{k=0}^{n-1} c_k\vartheta_i^k \quad (i = 1, \dots, n;\ \gamma_1 = \gamma, \vartheta_1 = \vartheta). \tag{6}$$

As is well known from the theory of determinants,

$$D = \begin{vmatrix} 1\ \vartheta_1 \dots \vartheta_1^{n-1} \\ 1\ \vartheta_2 \dots \vartheta_2^{n-1} \\ \dots\dots\dots\dots \\ 1\ \vartheta_n \dots \vartheta_n^{n-1} \end{vmatrix} = \prod_{i>j} (\vartheta_i - \vartheta_j)$$

so that $D \neq 0$ since $\vartheta_1, \vartheta_2, \dots, \vartheta_n$ are distinct from each other. Moreover, D^2 is fixed under all permutations of the ϑ_i which means that D^2 is fixed under the Galois group of $\mathbf{Q}(\vartheta_1, \dots, \vartheta_n)$ over $\mathbf{Q}$. Therefore D^2 is a rational number; D^2 is even a rational integer since

D^2 is obviously an integer of the field $Q(\vartheta_1, \ldots, \vartheta_n)$. By Cramer's Rule, the equations (6) yield

$$c_k = A_k/D \quad (k = 0, 1, \ldots, n - 1) \tag{7}$$

where A_k is a certain determinant with the elements γ_i and ϑ_i^j. Since the γ_i and the ϑ_i^j are integers of $Q(\vartheta_1, \ldots, \vartheta_n)$, the same holds for A_k. From (7) we obtain

$$D^2 c_k = A_k D.$$

Since $D^2 c_k$ is a rational number and $A_k D$ is an integer of $Q(\vartheta_1, \ldots, \vartheta_n)$ the last equation shows that $D^2 c_k = q_k$ is a rational integer. Hence any element γ of $\mathbf{I}$ has a unique representation

$$\gamma = g_0 \frac{1}{D^2} + g_1 \frac{\vartheta}{D^2} + \cdots + g_{n-1} \frac{\vartheta^{n-1}}{D^2}$$

with $g_i \in \mathbf{Z}$. This means that $\mathbf{I}$ is a submodule of the free $\mathbf{Z}$-module with the basis $1/D^2, \vartheta/D^2, \ldots, \vartheta^{n-1}/D^2$.

By Theorem 9.2.2, $\mathbf{I}$ is a free $\mathbf{Z}$-module. Moreover, the rank of $\mathbf{I}$ as a free $\mathbf{Z}$-module is equal to n since $1, \vartheta, \ldots, \vartheta^{n-1}$ are linearly independent. This completes the proof.

From the Theorems 9.2.2 and 9.4.2 we conclude that $\mathbf{I}$ satisfies the ascending chain condition.

Any ideal of $\mathbf{I}$ is obviously a $\mathbf{Z}$-module so that, by Theorem 9.2.2, it is a free $\mathbf{Z}$-module. Moreover, if A is an ideal of $\mathbf{I}$ other than (0), then A is a free $\mathbf{Z}$-module of rank n since, for any $\alpha \in A$, $\alpha \neq 0$, the n elements $\alpha, \alpha\vartheta, \ldots, \alpha\vartheta^{n-1}$ belong to A and are linearly independent.

9.5.7. Theorem. *Every prime ideal of* $\mathbf{I}$ *other than* (0) *and* (1) *is maximal.*

Proof. Let P be a prime ideal in $\mathbf{I}$, $P \neq (0)$, $P \neq \mathbf{I}$, and let α be a non-zero element of P. Then we have $(\alpha) \subseteq P$. The minimal polynomial

$$x^l + b_1 x^{l-1} + \ldots + b_l$$

of α in $Q[x]$ has coefficients in $\mathbf{Z}$, and $b_l \neq 0$ since the minimal polynomial is irreducible. From

$$\alpha^l + b_1 \alpha^{l-1} + \ldots + b_l = 0$$

it follows that $b_l \in (\alpha)$ and hence $b_l \in P$. This gives $b_l \in P \cap \boldsymbol{Z}$.

It is clear that $P \cap \boldsymbol{Z}$ is an ideal in $\boldsymbol{Z}$. Moreover, $P \cap \boldsymbol{Z}$ is a prime ideal in $\boldsymbol{Z}$. Indeed, let a and b be rational integers such that $ab \in P \cap \boldsymbol{Z}$; since P is a prime ideal it follows that $a \in P$ or $b \in P$ and therefore $a \in P \cap \boldsymbol{Z}$ or $b \in P \cap \boldsymbol{Z}$. Also $P \cap \boldsymbol{Z} \neq \boldsymbol{Z}$ because $1 \notin P$ since $P \neq \mathbf{I}$.

Now let S be an ideal in $\mathbf{I}$ such that $P \subset S \subset \mathbf{I}$. Since P is properly contained in S we can choose an element β of S that does not belong to P. Being an integer, β satisfies an equation

$$\beta^t + c_1\beta^{t-1} + \ldots + c_t = 0$$

with coefficients $c_i \in \boldsymbol{Z}$. Therefore β also satisfies a congruence

$$\beta^u + d_1\beta^{u-1} + \ldots + d_u \equiv 0 \pmod{P} \tag{8}$$

whose coefficients belong to $\boldsymbol{Z}$. We may assume that this is the congruence of least degree. Then we have

$$d_u \not\equiv 0 \pmod{P}$$

because, otherwise, β would satisfy a congruence mod P of degree less than n. From (8) we conclude that $d_u \in (\beta)$ and hence, $d_u \in S$. This gives $d_u \in S \cap \boldsymbol{Z}$ but $d_u \notin P \cap \boldsymbol{Z}$. Therefore $P \cap \boldsymbol{Z}$ is properly contained in $S \cap \boldsymbol{Z}$. As we observerd above, $P \cap \boldsymbol{Z}$ is a prime ideal in $\boldsymbol{Z}$. By Corollary 4.4.3, $P \cap \boldsymbol{Z}$ is a maximal ideal in $\boldsymbol{Z}$. Therefore the ideal $S \cap \boldsymbol{Z}$ of $\boldsymbol{Z}$ coincides with $\boldsymbol{Z}$, i.e. $1 \in S \cap \boldsymbol{Z}$ so that $S = \mathbf{I}$.

This proves our theorem.

Let A and B be $\boldsymbol{Z}$-modules in $\mathbf{F}$. By $A + B$ we understand the set of all sums $\alpha + \beta$ with $\alpha \in A$, $\beta \in B$. It is clear that $A + B$ is a $\boldsymbol{Z}$-module. If A and B are ideals in $\mathbf{I}$, then so is $A + B$. We also define the product of A and B:

AB consists of all finite sums $\Sigma\alpha_i\beta_i$, $\alpha_i \in A$, $\beta_i \in B$. It is obvious that AB is a $\boldsymbol{Z}$-module. If A and B are ideals in $\mathbf{I}$, then so is AB.

In case A and B are finitely generated, say

$$A = (\alpha_1, \ldots, \alpha_r), \ B = (\beta_1, \ldots, \beta_s),$$

then $A + B$ and AB are also finitely generated, namely

$$A + B = (\alpha_1, \ldots, \alpha_r, \beta_1, \ldots, \beta_s),$$

$$AB = (\alpha_1\beta_1, \ldots, \alpha_r\beta_1, \alpha_1\beta_2, \ldots, \alpha_r\beta_2, \ldots, \alpha_r\beta_s).$$

If A is a principal ideal in $\mathbf{I}$, say $A = (\alpha)$, then $(\alpha)B$ consists of all products $\alpha\beta$ with $\beta \in B$. Instead of $(\alpha)B$ we write αB.

It is easy to establish the distributive law

$$A(B + C) = AB + AC.$$

$A(B + C)$ consists of all finite sums $\Sigma\alpha_i(\beta_i + \gamma_i)$ with $\alpha_i \in A$, $\beta_i \in B$, $\gamma_i \in C$. We have

$$\sum \alpha_i(\beta_i + \gamma_i) = \sum \alpha_i\beta_i + \sum \alpha_i\gamma_i$$

which shows that $A(B + C) \subseteq AB + AC$. On the other hand, $AB \subseteq A(B + C)$ and $AC \subseteq A(B + C)$ so that $AB + AC \subseteq A(B+C)$.

9.5.8. Lemma. *If the $\mathbf{Z}$-modules A and B are contained in $\mathbf{I}$ and if P is a prime ideal of $\mathbf{I}$, then $AB \subseteq P$ implies that $A \subseteq P$ or $B \subseteq P$ (or both).*

Proof. Suppose that the lemma is false. Then there exist elements $\alpha \in A$ and $\beta \in B$ such that $\alpha \notin P$, $\beta \notin P$. Since P is a prime ideal this contradicts $\alpha\beta \in P$.

9.5.9. Lemma. *Let A be an ideal of $\mathbf{I}$, $A \neq (0)$. Then there exist prime ideals $P_1, \ldots, P_r$ of $\mathbf{I}$ such that $A \subseteq P_i$ and*

$$P_1P_2 \ldots P_r \subseteq A.$$

Proof. In case $A = \mathbf{I}$ there is nothing to prove. Since $\mathbf{I}$ satisfies the maximal condition for ideals we can apply the induction principle mentioned in the beginning of § 9.4. Accordingly we assume that the lemma is true for all ideals H such that $A \subset H \subseteq \mathbf{I}$.

If A itself is a prime ideal, then there is nothing to prove. So we may assume that A is not a prime ideal. Then there exist elements $\sigma, \tau \in \mathbf{I}$ such that $\sigma\tau \in A$ but $\sigma \notin A$, $\tau \notin A$. The ideals

$$S = A + (\sigma), \quad T = A + (\tau)$$

contain A properly. From the induction assumption we conclude that there exist prime ideals $P_1, \ldots, P_s, P_{s+1}, \ldots, P_r$ such that

$$P_1 \ldots P_s \subseteq S, \quad P_{s+1} \ldots P_r \subseteq T.$$

We have

$$ST = A + \sigma A + A\tau + (\sigma\tau) \subseteq A.$$

Therefore we obtain

$$P_1 \dots P_s P_{s+1} \dots P_r \subseteq ST \subseteq A$$

which proves the lemma.

Let P be a prime ideal of $\mathbf{I}$, $P \neq (0)$, $P \neq \mathbf{I}$. By P^{-1} we understand the set of all $\xi \in \mathbf{F}$ such that $\xi\pi \in \mathbf{I}$ for any $\pi \in P$. It is obvious that P^{-1} is a $\mathbf{Z}$-module.

9.5.10. Lemma. *P^{-1} is not contained in $\mathbf{I}$.*

Proof. Let $\gamma \in P$, $\gamma \neq 0$. By Lemma 9.5.9 there are prime ideals $P_1, P_2, \dots, P_r$ such that

$$P_1 P_2 \dots P_r \subseteq (\gamma).$$

We may assume that the product on the left-hand side is minimal in the following sense: If one of the factors is cancelled, then the product of the remaining factors is not contained in (γ).

Due to $\gamma \in P$ we have $P_1 P_2 \dots P_r \subseteq P$. Since P is a prime ideal it follows that at least one of the factors $P_1, P_2, \dots, P_r$ is contained in, and hence equal to, P, say $P_1 = P$. So we have

$$PP_2 \dots P_r \subseteq (\gamma), \; P_2 \dots P_r \nsubseteq (\gamma).$$

Therefore there exists an integer β such that $\beta \in P_2 \dots P_r$ but $\beta \notin (\gamma)$. This gives

$$P\beta \subseteq PP_2 \dots P_r \subseteq (\gamma).$$

Therefore $\pi(\beta/\gamma)$ is contained in $\mathbf{I}$ for any $\pi \in P$ so that $\beta/\gamma \in P^{-1}$. But $\beta \notin (\gamma)$ means that β/γ does not belong to $\mathbf{I}$ as was to be shown.

9.5.11. Theorem. *If P is a prime ideal in $\mathbf{I}$, $P \neq (0)$, $P \neq \mathbf{I}$, then $P^{-1}P = \mathbf{I}$.*

Proof. From the definition of P^{-1} it is clear that $\mathbf{I} \subseteq P^{-1}$. This gives $P = \mathbf{I}P \subseteq P^{-1}P$. Since P is a maximal ideal and $P^{-1}P \subseteq \mathbf{I}$ it follows that $P^{-1}P = P$ or $P^{-1}P = \mathbf{I}$. We have to exclude the first possibility.

Suppose that, on the contrary, $P^{-1}P = P$. Then we obtain

$$(P^{-1})^2 P = P^{-1}P^{-1}P = P^{-1}P = P$$

and, by induction,

$$(P^{-1})^k P = P \tag{9}$$

for all natural numbers k. Let $\pi \in P$, $\pi \neq 0$ and let α be an arbitrary element of P^{-1}. Then (9) shows that, for any natural number k, $\alpha^k \pi$ belongs to P and hence to I. Therefore, all the powers α^k, $k = 1, 2, \ldots$ are contained in the finitely generated $\boldsymbol{Z}$-module $\pi^{-1}\mathsf{I}$. By Theorem 9.5.3, this implies that α is an integer. Since α is an arbitrary element of P^{-1} this contradicts Lemma 9.5.10, and the theorem is proved.

9.5.12. Theorem. *Let A be an ideal of I, $A \neq (0)$, $A \neq \mathsf{I}$. Then A can be represented as a product of prime ideals. This representation is unique up to the order of the factors.*

Proof. It is evident that the theorem also holds for I itself if we define the empty product of prime ideals to be equal to I. Let us assume that any ideal B such that $A \subset B \subseteq \mathsf{I}$ can be represented as a product of prime ideals. By Lemma 9.5.9, there are prime ideals $P_1, P_2, \ldots, P_m$ such that $A \subseteq P_i$ and

$$P_1 P_2 \ldots P_m \subseteq A. \tag{10}$$

We may assume that the product on the left-hand side is minimal, i.e. by cancelling any one of the factors we obtain a product which is not contained in A. Now let P be a prime ideal such that $A \subseteq P$. Then we have $P_1 P_2 \ldots P_m \subseteq P$. Hence one of the factors P_i coincides with P, say $P_1 = P$. Due to Theorem 9.5.11 multiplication of (10) by P^{-1} gives

$$P_2 \ldots P_m \subseteq P^{-1}A = B.$$

Clearly, B is an ideal in I. Since the product on the left-hand side of (10) is minimal it follows that A is properly contained in B. By the induction assumption, B has a representation

$$B = Q_1 \ldots Q_t$$

as a product of prime ideals Q_i. Thus we obtain $P^{-1}A = Q_1 \ldots Q^t$ which gives $A = PQ_1 \ldots Q_t$.

To prove the uniqueness, let us suppose that

$$A = P_1 \ldots P_r = Q_1 \ldots Q_s \tag{11}$$

are two representations of A as a product of prime ideals. For $r = 1$ the uniqueness is obvious. We may assume that in any product of fewer than r prime ideals the factors are unique except for their order. From Lemma 9.5.8 it follows that at least one of the Q_i coincides with P_1, say $Q_1 = P_1$. Multiplying (11) by P_1^{-1} we obtain

$$P_2 \dots P_r = Q_2 \dots Q_s.$$

Due to the induction assumption this implies that $r - 1 = s - 1$ and that $P_2, \dots, P_r$ coincide with $Q_2, \dots, Q_r$ except for the order.

This completes the proof.

If a is an element of a principal ideal ring that is neither a unit nor equal to zero then, by Theorem 4.4.6, a has a factorization

$$a = p_1 p_2 \cdots p_r$$

into prime elements p_i where the p_i are uniquely determined except for their order and arbitrary units as factors. This statement can equivalently be expressed by saying that the principal ideal (a) has a unique factorization

$$(a) = (p_1)(p_2) \dots (p_r)$$

into prime ideals (p_i). Thus, Theorem 9.5.12 can be regarded as a generalization of the Unique Factorization Theorem 4.4.6 to the integral domain $\mathbf{I}$ of the algebraic integers in an algebraic number field.

It is easy to see (cf. Exercise 19 at the end of the present chapter) that the integral domain $\mathbf{I}$ of the algebraic integers of the field $\mathbf{Q}(\sqrt{-5})$ consists precisely of the numbers $a_1 + a_2\sqrt{-5}$ with $a_1, a_2 \in \mathbf{Z}$. The example at the end of § 4.3 shows that $\mathbf{I}$ is not a principal ideal ring.

One readily verifies (cf. Exercise 20) that the ideals

$$P = (2, 1 + \sqrt{-5}), \quad \bar{P} = (2, 1 - \sqrt{-5}),$$
$$Q = (3, 1 + \sqrt{-5}), \quad \bar{Q} = (3, 1 - \sqrt{-5})$$

are prime ideals because the residue class rings $\mathbf{I}/P$, $\mathbf{I}/\bar{P}$, $\mathbf{I}/Q$, $\mathbf{I}/\bar{Q}$ are isomorphic to GF(2) and GF(3), respectively. A simple computation shows that the products $P\bar{P}$, $Q\bar{Q}$, PQ, $\bar{P}\bar{Q}$ are principal ideals, namely

$$P\bar{P} = (2), \quad Q\bar{Q} = (3), \quad PQ = (1 + \sqrt{-5}), \quad \bar{P}\bar{Q} = (1 - \sqrt{-5}).$$

Thus (1) in § 4.3 gives

$$\begin{aligned}(6) = (2)\,(3) = (P\bar{P})\,(Q\bar{Q}) &= (1 + \sqrt{-5})\,(1 - \sqrt{-5}) \\ = (PQ)\,(\bar{P}\bar{Q}) &= P\bar{P}Q\bar{Q}\end{aligned}$$

which explains the two essentially distinct factorizations of 6 into prime elements. In particular, the principal ideal generated by a prime element is not necessarily a prime ideal so that Theorem 4.4.2 does not hold in arbitrary integral domains.

Exercises

1. Let S, T, U be submodules of an $\mathbf{R}$-module M. If $S \subseteq T$ prove that

$$T \cap (U + S) = (T \cap U) + S.$$

2. Regarding an $\mathbf{R}$-module M as an additive group with the operator domain $\mathbf{R}$ one can introduce the concept of a composition series.
   (i) Show that Theorem 3.8.2 (Jordan-Hölder's Theorem) also holds in this case. (Hint: Use Theorem 3.12.1.)
   (ii) Prove that M has a composition series if and only if M satisfies both chain conditions.
   (iii) Suppose that M has a composition series. Show that every submodule of M also has a composition series. For any submodule S of M let $l(S)$ denote the common length of all composition series of S. Show that

$$l(M) = l(S) + l(M/S).$$

   If S and T are submodules of M prove that

$$l(S) + l(T) = l(S + T) + l(S \cap T).$$

   (Hint: Use Theorem 3.7.4.)

3. The definition of a left $\mathbf{R}$-module can obviously be extended to

the case of an arbitrary ring **R** (possibly without unit element) by dropping the condition that the unit element acts as identity operator. If M is such a left **R**-module what are the elements of the submodule generated by a given subset S of M?

4. Let **R** be a ring with unit element 1. Drop the condition that 1 acts as the identity operator on a left **R**-module M. Show that M is a direct sum $M = M_0 \oplus M_1$ of left **R**-modules such that $1x_0 = 0$ for every $x_0 \in M_0$ and $1x_1 = x_1$ for every $x_1 \in M_1$.
5. An **R**-module M is called simple if it contains no **R**-submodule other than M and 0. Prove the following proposition (Schur's Lemma): If M is a non-zero simple **R**-module, then the ring of the **R**-endomorphisms of M is a division ring. (Show that every **R**-endomorphism of M is an automorphism of M or carries every element of M into 0.)
6. Show that every additively written abelian group can be regarded as a $\boldsymbol{Z}$-module.
   (i) Show that the additive group of the integers satisfies the ascending chain condition but not the descending chain condition.
   (ii) Show that a group of type p^∞ (cf. Example 6 in § 3.2) satisfies the descending chain condition but not the ascending chain condition.
7. Let an abelian group be defined by a system of relations

$$\sum_{k=1}^{m} \alpha_{ik} g_k = 0 \quad (i = 1, \ldots, l)$$

on the generating elements $g_1, \ldots, g_m$.
   (i) Find a necessary and sufficient condition on the $l \times m$ matrix $A = [\alpha_{ik}]$ for the group to be finite and express the order of the group in terms of A.
   (ii) Let

$$A = \begin{bmatrix} 5 & 4 & 1 & 5 \\ 7 & 6 & 5 & 11 \\ 2 & 2 & 10 & 12 \\ 10 & 8 & -4 & 4 \end{bmatrix}$$

   Represent the group as a direct sum of cyclic groups.

8. Let G be an abelian group of finite order g.
   (i) Show that for every positive divisor n of g there exists at least one subgroup of order n of G.
   (ii) If for every positive divisor n of g there exists precisely one subgroup of order n prove that G is cyclic.
9. Show that the direct product of two cyclic groups of finite relatively prime orders is cyclic.
10. Let A and B be finite dimensional subspaces of a vector space. Let dim A denote the dimension of A. Prove that

$$\dim (A + B) + \dim (A \cap B) = \dim A + \dim B.$$

11. Let A be an $m \times n$ matrix of rank r with elements in a field $\mathbf{F}$. Show that there exist a non-singular $m \times m$ matrix U and a non-singular $n \times n$ matrix V, both with elements in $\mathbf{F}$, such that

$$UAV = \begin{bmatrix} I_r & 0 \\ 0 & 0 \end{bmatrix}.$$

12. Find the elementary divisors of the following matrices:

$$\begin{bmatrix} 1 & 4 & 6 & 1 \\ 1 & 6 & 6 & 1 \\ 2 & 10 & 18 & 2 \end{bmatrix}, \quad \begin{bmatrix} 2 & 1 & 2 \\ 1 & 1 & 1 \\ 3 & 2 & 4 \end{bmatrix}, \quad \begin{bmatrix} 5 & 10 & 3 \\ 1 & 4 & 1 \\ 1 & 2 & 1 \end{bmatrix}.$$

13. Show that every matrix which commutes with a matrix B of the form (17) in § 9.3 can be expressed as a polynomial in B.
14. Find the Jordan normal form of the following matrices:

$$\begin{bmatrix} 0 & 0 & 1 \\ 1 & 0 & 0 \\ 0 & 0 & 1 \end{bmatrix}, \quad \begin{bmatrix} 1 & 0 & 2 \\ 1 & 1 & 0 \\ 0 & 0 & 2 \end{bmatrix}, \quad \begin{bmatrix} 0 & 0 & 1 \\ 0 & 1 & 0 \\ 1 & 0 & 0 \end{bmatrix}, \quad \begin{bmatrix} 3 & -3 & 1 \\ 1 & 0 & 0 \\ 0 & 1 & 0 \end{bmatrix},$$

an $n \times n$ matrix all of whose elements are equal to 1.
15. Let A be any $n \times n$ matrix with elements in a field $\mathbf{F}$.
   (i) Show that A is similar to its transpose A^T,
   (ii) Show that there is even a symmetric matrix S with elements in $\mathbf{F}$ such that $S^{-1}AS = A^T$.
16. Let A be any $n \times n$ matrix with elements in a field $\mathbf{F}$, and let $\alpha_1, \ldots, \alpha_n$ denote the characteristic roots of A. Show that,

for any polynomial $f(x) \in \mathbf{F}[x]$, the characteristic roots of $f(A)$ are equal to $f(\alpha_1), \ldots, f(\alpha_n)$.

17. Let A be an $n \times n$ matrix with elements in an algebraically closed field $\mathbf{K}$. Prove that A is similar to a diagonal matrix if and only if there is a polynomial $f(x) \in \mathbf{K}[x]$ with simple roots such that $f(A) = 0$.
18. If A and B are two $n \times n$ matrices with elements in C, show that AB and BA have the same characteristic polynomial.
19. Let m be a rational integer that is not divisible by the square of any prime number.
    (i) Show that $\alpha = \frac{1}{2}\left(a + b\sqrt{m}\right)$ is an algebraic integer of the field $Q\left(\sqrt{m}\right)$ if and only if a and b are rational integers such that

$$a \equiv b \pmod 2 \qquad \text{if } m \equiv 1 \pmod 4,$$

$$a \equiv b \equiv 0 \pmod 2 \text{ if } m \equiv 2 \text{ or } 3 \pmod 4.$$

    (ii) Let

$$\omega = \frac{1}{2}\left(1 + \sqrt{m}\right) \text{ if } m \equiv 1 \pmod 4,$$

$$\omega = \sqrt{m} \qquad \text{if } m \equiv 2 \text{ or } 3 \pmod 4.$$

    Show that $x + y\omega$ is an algebraic integer of $Q\left(\sqrt{m}\right)$ if and only if x and y are rational integers.
20. Let $\mathbf{I}$ be the integral domain of the algebraic integers in the field $Q\left(\sqrt{-5}\right)$. We consider the following ideals in $\mathbf{I}$:

$$P = \left(2, 1 + \sqrt{-5}\right), \ \bar{P} = \left(2, 1 - \sqrt{-5}\right),$$

$$Q = \left(3, 1 + \sqrt{-5}\right), \ \bar{Q} = \left(3, 1 - \sqrt{-5}\right).$$

    Prove that
    (i) $\mathbf{I}/P$ and $\mathbf{I}/\bar{P}$ are isomorphic to GF(2).
    (ii) $\mathbf{I}/Q$ and $\mathbf{I}/\bar{Q}$ are isomorphic to GF(3).
21. Let $\mathbf{I}$ be the integral domain of the algebraic integers of an algebraic number field. Two ideals A and B of $\mathbf{I}$ are said to be relatively prime if $A + B = \mathbf{I}$.

(i) Show that for any two ideals A, B of $\mathbf{I}$

$$AB \subseteq A \cap B.$$

(ii) If A and B are relatively prime prove that

$$AB = A \cap B.$$

(Hint: show that $(A \cap B)(A + B) \subseteq AB$.)

22. Show that an algebraic integer ε of an algebraic number field $\mathbf{F}$ is a unit of the integral domain of the algebraic integers in $\mathbf{F}$ if and only if $N_{\mathbf{F}/Q}(\varepsilon) = \pm 1$.

23. Let P be a prime ideal in the integral domain $\mathbf{I}$ of the algebraic integers in some algebraic number field, $P \neq \mathbf{I}$, $P \neq (0)$.
    (i) Show that there is one and only one prime number p such that $p \in P$.
    (ii) Show that $\mathbf{I}/P$ is some $\mathrm{GF}(p^n)$ where $p \in P$.

10

Algebras

Algebras over a commutative ring with unit element have been defined in Example 9 of § 4.1. This chapter is devoted to a more detailed study of algebras of finite rank over a field.

10.1. Basic definitions

Let **K** be any field. By an *algebra* of finite dimension n over **K** we understand a ring **A** with the following properties:

(i) **A** is a vector space of dimension n over **K**.

(ii) For any $\alpha \in \mathbf{K}$ and arbitrary elements a, b of **A**

$$\alpha(ab) = a(\alpha b) = (\alpha a)\, b\,.$$

If $v_1, \ldots, v_n$ is a basis of **A** as a vector space, then any element a of **A** has a unique representation

$$a = \alpha_1 v_1 + \ldots + \alpha_n v_n$$

with coefficients $\alpha_i \in \mathbf{K}$. Since **A** is a ring the products $v_i v_j$ are defined, namely

$$v_i v_j = \sum_{k=1}^{n} \gamma_{ijk} v_k \quad (\gamma_{ijk} \in \mathbf{K})\,. \tag{1}$$

Due to the distributive laws in **A** and (ii) the product of arbitrary elements of **A** is uniquely defined by (1). We have

$$\left(\sum_{i=1}^{n} \alpha_i v_i\right)\left(\sum_{j=1}^{n} \beta_j v_j\right) = \sum_{i=1}^{n}\sum_{j=1}^{n} \alpha_i \beta_j v_i v_j = \sum_{k=1}^{n}\left(\sum_{i=1}^{n}\sum_{j=1}^{n} \alpha_i \beta_j \gamma_{ijk}\right) v_k\,.$$

A subspace R of $\mathbf{A}$ is called a *right ideal* if $r \in R$ implies that $ra \in R$ for any $a \in \mathbf{A}$. Similarly, a subspace L is said to be a *left ideal* if for arbitrary elements l of L and a of $\mathbf{A}$ the product al belongs to L. A *two-sided ideal* is one which is simultaneously a right and a left ideal.

Since all ideals are subspaces it follows from 3.13.9 that $\mathbf{A}$ satisfies the maximum and the minimum condition for right ideals and for left ideals. It turns out that the most important theorems about the structure of algebras are also valid for a much wider class of rings, namely for rings that satisfy the *minimum condition for right ideals*. Therefore we shall prove these theorems under this weaker hypothesis following the monograph [23].

To exclude trivialities we shall only deal with rings that contain at least one non-zero element.

Let $\mathbf{A}$ be a ring. A subgroup R of the additive group $\mathbf{A}^+$ is called a *right ideal* if R is admissible with respect to $\mathbf{A}$ as a right operator domain, i.e. if $r \in R$ implies that $ra \in R$ for any $a \in \mathbf{A}$. Similarly we define left ideals and two-sided ideals of $\mathbf{A}$.

By the sum $R_1 + R_2$ of two right ideals we understand the subgroup of $\mathbf{A}^+$ that consists of all sums $r_1 + r_2$ where $r_1 \in R_1$, $r_2 \in R_2$. It is evident that $R_1 + R_2$ is a right ideal. Similarly, we can define the sum of any finite number of right ideals. If R_λ, where λ ranges over some index set Λ, is any collection of right ideals, then the sum of the R_λ is defined as the right ideal that consists of all finite sums

$$r_{\lambda_1} + r_{\lambda_2} + \dots + r_{\lambda_m}$$

where $r_{\lambda_i} \in R_{\lambda_i}$, $\lambda_i \in \Lambda$. If such sums are direct sums in the group-theoretical sense we write $\oplus$ instead of $+$. Sums of left and two-sided ideals are defined analogously.

If S and T are any non-empty subsets of $\mathbf{A}$, then we define the product ST to be the set of all finite sums $\Sigma s_i t_i$ where $s_i \in S$, $t_i \in T$. It is evident that ST is a right ideal if T is a right ideal; ST is a left ideal if S is a left ideal; and ST is a two-sided ideal if T is a right ideal and S is a left ideal.

In particular, for any element c of $\mathbf{A}$, the products $c\mathbf{A}$, $\mathbf{A}c$, and $\mathbf{A}c\mathbf{A}$ are right, left, and two-sided ideals, respectively. These ideals are said to be generated by c.

In case S, T, U are (one-sided or two-sided) ideals it is easy to verify the associative and the distributive laws

$$(ST)U = S(TU),$$

$$S(T + U) = ST + SU, \quad (T + U)\,S = TS + US.$$

By a ring with minimum condition we understand a ring that satisfies the minimum condition for right ideals.

10.2. The radical

An element a of any ring is called *nilpotent* if $a^m = 0$ for some natural number m. An element c is said to be *idempotent* if $c \neq 0$ and $c^2 = c$. We simply refer to such an element c as an idempotent.

A right, left, or two sided ideal J is called *nilpotent* if $J^m = (0)$ for some natural number m. This means that the product of m arbitrary elements of J, distinct or not, is equal to 0. In particular, the m-th power of every element of J is equal to 0 so that a nilpotent ideal cannot contain an idempotent element. It is of great importance that the converse is also true:

10.2.1. THEOREM. *If $\mathbf{A}$ is a ring with minimum condition, then every non-nilpotent right ideal contains an idempotent.*

Proof. Let J be a non-nilpotent right ideal of $\mathbf{A}$. We consider the set $\mathfrak{R}$ of all non-nilpotent right ideals of $\mathbf{A}$ contained in J. $\mathfrak{R}$ is not empty since $J \in \mathfrak{R}$. Let M be a minimal member of $\mathfrak{R}$. Then any right ideal of $\mathbf{A}$ properly contained in M is nilpotent. Since $M^2 \subseteq M$ and M^2 is not nilpotent we have $M^2 = M$.

Now let $\mathfrak{R}_1$ denote the collection of all right ideals R of $\mathbf{A}$ such that $R \subseteq M$ and $RM \neq (0)$. Clearly, $\mathfrak{R}_1$ is not empty since $M \subseteq \mathfrak{R}_1$. Let M_1 denote a minimal member of $\mathfrak{R}_1$. Due to $M_1M \neq (0)$ there is an element a in M_1 such that $aM \neq (0)$. We have $aM \subseteq M_1 \subseteq M$ and $aMM = aM \neq (0)$ so that aM belongs to $\mathfrak{R}_1$. Since M_1 is minimal we obtain $aM = M_1$. Hence there exists an element c in M such that $ac = a$. This gives $ac^m = a$ for any natural number m. Hence, c is not nilpotent. Moreover, $a(c^2 - c) = 0$.

We consider the set V of all $v \in M$ such that $av = 0$. It is clear that V is a right ideal. Due to $aM = M_1 \neq (0)$ we conclude that V

is properly contained in M. Since M is minimal this implies that V is nilpotent. The element $v_1 = c^2 - c$ belongs to V. In case $v_1 = 0$, c is an idempotent contained in M and hence in J, and this ends the proof.

If $v_1 \neq 0$, we proceed as follows. Let

$$c_1 = c + v_1 - 2cv_1.$$

It is clear that c, c_1, and v_1 commute with each other. Therefore c_1 is not nilpotent since, otherwise, $c = c_1 - v_1 + 2cv_1$ would also be nilpotent. A straightforward computation gives

$$c_1^2 - c_1 = 4v_1^3 - 3v_1^2.$$

We now consider the element $v_2 = c_1^2 - c_1$. In case $v_2 = 0$, c_1 is an idempotent contained in M, and we are finished. Otherwise we observe that v_2 is nilpotent, commutes with c_1 and contains v_1^2 as a factor. We put

$$c_2 = c_1 + v_2 - 2c_1v_2$$

and find that $c_2^2 - c_2$ contains $v_1^{2^2}$ as a factor. Continuing in this way, we can successively construct non-nilpotent elements c, c_1, $c_2, \ldots$ in M such that $c_k^2 - c_k$ has the factor $v_1^{2^k}$. Since v_1 is nilpotent we arrive, after a finite number of steps, at a $c_m \in M$ such that $c_m^2 - c_m = 0$. This proves the theorem.

Let N_1 and N_2 be two nilpotent right ideals of **A**. It is easy to see that the sum $N_1 + N_2$ is also nilpotent. Indeed, let $N_1^r = (0)$, $N_2^s = (0)$. In the power $(N_1 + N_2)^{r+s-1}$ each summand contains at least r factors N_1 or s factors N_2. In the first case such a summand has the form

$$\ldots N_1 \ldots N_1 \ldots N_1 \ldots$$

where the factors N_1 are adjacent or the dots stand for factors N_2. Since N_1 is a right ideal we have $N_1N_2 \ldots N_2 \subseteq N_1$. This gives

$$\ldots N_1 \ldots N_1 \ldots N_1 \ldots \subseteq N_1^r \ldots = (0).$$

Similarly, we find that each summand that contains at least s factors N_2 is equal to (0). Thus we obtain $(N_1 + N_2)^{r+s-1} = (0)$. By induction it follows that any sum of finitely many nilpotent right ideals is a nilpotent right ideal.

The sum of all nilpotent right ideals of a ring $\mathbf{A}$ with minimum condition is called the *radical* of $\mathbf{A}$ and is denoted by rad $\mathbf{A}$.

10.2.2. Theorem. *Let* $\mathbf{A}$ *be a ring with minimum condition. Then* rad $\mathbf{A}$ *is a nilpotent two-sided ideal. Moreover,* rad $\mathbf{A}$ *contains every nilpotent left ideal.*

Proof. Clearly, $N = \text{rad}\,\mathbf{A}$ is a right ideal. Suppose that N is not nilpotent. By Theorem 10.2.1, N then contains an idempotent c. Now c is contained in a finite sum of nilpotent right ideals, i.e. in a nilpotent right ideal. So we arrive at the contradiction that c is nilpotent. Therefore N is nilpotent, say $N^m = (0)$.

The product $\mathbf{A}N$ is nilpotent because

$$(\mathbf{A}N)^m = \mathbf{A}(N\mathbf{A})\,(N\mathbf{A})\ldots(N\mathbf{A})\,N \subseteq \mathbf{A}N^m = (0).$$

Since $\mathbf{A}N$ is a right ideal (it is actually a two-sided ideal) we obtain $\mathbf{A}N \subseteq N$ which shows that N is a two-sided ideal.

Finally, let L be any nilpotent left ideal. In the same way as above we conclude that the two-sided ideal $L\mathbf{A}$ is nilpotent. It is easily verified that $L + L\mathbf{A}$ is a right ideal. Since both summands are nilpotent it follows that $L + L\mathbf{A}$ is a nilpotent right ideal. Therefore $L + L\mathbf{A}$ is contained in rad $\mathbf{A}$ and hence $L \subseteq \text{rad}\,\mathbf{A}$.

This completes the proof.

A ring $\mathbf{A}$ with minimum condition is called *semi-simple* if rad $\mathbf{A} = (0)$. Note that semi-simplicity always implies that the minimum condition is satisfied.

If $\mathbf{A}$ is a ring with minimum condition, then so is the residue class ring $\mathbf{A}/\text{rad}\,\mathbf{A}$. This follows from the fact that, due to Theorem 3.7.5 for groups with operators, any right ideal of $\mathbf{A}/\text{rad}\,\mathbf{A}$ is of the form $R/\text{rad}\,\mathbf{A}$ where R is a right ideal of $\mathbf{A}$.

10.2.3. Theorem. *For any ring* $\mathbf{A}$ *with minimum condition the residue class ring* $\mathbf{A}/\text{rad}\,\mathbf{A}$ *is semi-simple.*

Proof. Let $N = \text{rad}\,\mathbf{A}$. As we just observed, every right ideal of $\mathbf{A}/N$ is of the form R/N where R is a right ideal of $\mathbf{A}$. Suppose that R/N is nilpotent, say $(R/N)^k = (0)$. This gives $R^k \subseteq N$. Since N itself is nilpotent, say $N^l = (0)$, we obtain $R^{kl} = (0)$. By the definition of N this implies that $R = N$. Therfore R/N is the zero ideal of $\mathbf{A}/N$ which proves our theorem.

10.3. Semi-simple rings

10.3.1. *Any right ideal* $R \neq (0)$ *of a semi-simple ring* $\mathbf{A}$ *is generated by an idempotent, i.e.* $R = c\mathbf{A}$ *where* $c^2 = c$.

Proof. Since $\mathbf{A}$ is semi-simple, R is not nilpotent. By Theorem 10.2.1, R contains an idempotent. For any idempotent c' of R let $N(c')$ denote the set of all elements x of R such that $c'x = 0$. It is obvious that $N(c')$ is a right ideal in $\mathbf{A}$. Let c be an idempotent in R such that $N(c)$ is minimal. Suppose that $N(c) \neq (0)$, so that, by Theorem 10.2.1, $N(c)$ contains an idempotent c_1. We have $cc_1 = 0$. Putting

$$c_2 = c - c_1c + c_1$$

we find $cc_2 = c = c_2c$, $c_2c_1 = c_1 = c_1c_2$ and hence

$$c_2^2 = c_2(c - c_1c + c_1) = c_2.$$

From $cc_2 = c$ we conclude that $N(c_2) \subseteq N(c)$; moreover, $c_2c_1 = c_1 \neq 0$ but $cc_1 = 0$ shows that $N(c_2)$ is properly contained in $N(c)$. Since this contradicts the choice of c it follows that $N(c) = (0)$.

We show that $R = c\mathbf{A}$. Let r be any element of R. Then $c(r - cr) = 0$ which gives $r - cr = 0$ since $N(c) = (0)$. This shows that c is a left unit element for R and, in particular, $cR = R$. From $c \in R$ it follows that $c\mathbf{A} \subseteq R$. On the other hand, $R \subseteq \mathbf{A}$ implies that $R = cR \subseteq c\mathbf{A}$. This gives $R = c\mathbf{A}$ as was to be proved.

By a *minimal right ideal* of a ring $\mathbf{A}$ we understand a non-zero right ideal R such that R contains no right ideal of $\mathbf{A}$ other than (0) and R. Similarly, a *minimal two-sided ideal* J is a non-zero two-sided ideal that contains no two-sided ideal other than (0) and J.

Our aim is to prove that every semi-simple ring can be represented as a direct sum of finitely many minimal right ideals. The proof is based on the following lemma about direct decompositions of right ideals.

10.3.2. Lemma. *Let* R *be a right ideal of a semi-simple ring* $\mathbf{A}$. *If the right ideal* $R_1 \neq (0)$ *is properly contained in* R, *then there exists a direct decomposition* $R = R_1 \oplus R_2$ *where* R_2 *is a right ideal. Moreover, if* R *is generated by the idempotent* c, *then one can choose* *idempotent generators c_1' and c_2' of R_1 and R_2, respectively, such that*

$$c = c_1' + c_2', \quad c_1'c_2' = c_2'c_1' = 0.$$

Proof. Let $R = c\mathbf{A}$ and $R_1 = c_1\mathbf{A}$ where c and c_1 are idempotents. For any element r of R we set $r_1 = c_1 r$ and $r_2 = r - c_1 r$ so that $r_1 \in R_1$ and $r = r_1 + r_2$. It is obvious that the set of all elements of the form $r - c_1 r$ with $r \in R$ is a right-ideal R_2 in $\mathbf{A}$. We have $R = R_1 + R_2$. To show that the sum is direct we observe that c_1 is a left unit element for R_1 whereas $c_1 r_2 = 0$ for any $r_2 \in R_2$.

Corresponding to $R = R_1 \oplus R_2$ we have $c = c_1' + c_2'$ where $c_1' \in R_1$, $c_2' \in R_2$. From the construction of R_2 it follows that $c_1' = c_1 c$ and $c_2' = c - c_1 c$. Multiplying the last equation by c_1' we obtain

$$c_2'c_1' = cc_1' - c_1cc_1' = c_1' - c_1'c_1' = 0,$$

$$c_1'c_2' = c_1'c - c_1'c_1c = c_1cc - c_1'c_1' = c_1' - c_1'c_1' = 0.$$

It remains to show that c_1' and c_2' generate R_1 and R_2, respectively. For any $r_1 \in R_1$ we have

$$c_1'r_1 = c_1cr_1 = c_1r_1 = r_1,$$

and for any $r_2 \in R_2$ we obtain

$$c_2'r_2 = cr_2 - c_1cr_2 = r_2$$

because $c_1cr_2 = c_1r_2 = 0$. This proves the lemma.

Two idempotents c_1' and c_2' such that

$$c_1'c_2' = c_2'c_1' = 0$$

are said to be *orthogonal*. An idempotent is called *primitive* if it cannot be expressed as the sum of two orthogonal idempotents. It follows from the last lemma that a right ideal of a semi-simple ring is minimal if and only if it is generated by a primitive idempotent.

10.3.3. Theorem. *Every semi-simple ring can be decomposed into a direct sum of finitely many minimal right ideals.*

Proof. Let $\mathbf{A}$ be a semi-simple ring. Since semi-simplicity implies the minimum condition there exists a minimal right ideal R_1 in $\mathbf{A}$.

By Lemma 10.3.2 we have

$$\mathbf{A} = R_1 \oplus R_1'$$

where R_1' is a right ideal. If R_1' is not minimal we can choose a minimal right ideal R_2 contained in R_1'. Again by the last lemma we obtain $R_1' = R_2 \oplus R_2'$ where R_2' is a right ideal. This gives

$$\mathbf{A} = R_1 \oplus R_2 \oplus R_2'.$$

In case R_2' is not minimal we proceed in the same way. After a finite number of steps we arrive at a direct decomposition

$$\mathbf{A} = R_1 \oplus R_2 \oplus \ldots \oplus R_m$$

into minimal right ideals R_i since, otherwise, the right ideals $R_1', R_2', \ldots$ would form an infinite strictly descending chain of right ideals contradicting the semi-simplicity. This proves our theorem.

The next proposition is a converse of the previous theorem.

10.3.4. *If the ring* $\mathbf{A}$ *contains a unit element and is the direct sum of finitely many minimal right ideals, then* $\mathbf{A}$ *is semi-simple.*

Proof. That $\mathbf{A}$ is the direct sum of finitely many, say m, minimal right ideals means that the additive group $\mathbf{A}^+$ with the right operator domain $\mathbf{A}$ is the direct sum of m simple admissible subgroups, i.e. it is completely reducible in the sense of § 3.9. It follows easily that $\mathbf{A}$ satisfies the minimum condition. Indeed, $\mathbf{A}^+$ with the operator domain $\mathbf{A}$ has a composition series of finite length m. Therefore any admissible subgroup R, i.e. every right ideal of $\mathbf{A}$, also has a composition series whose length does not exceed m; this follows by taking the intersections of R with the terms of a composition series of $\mathbf{A}^+$. Hence it follows from Jordan-Hölder's Theorem 3.8.2 that $\mathbf{A}$ satisfies the descending chain condition for right ideals.

It is clear that Theorem 3.9.3 also holds for groups with operators. Let N be the radical of $\mathbf{A}$ and suppose that $N \neq (0)$. By Theorem 3.9.3, $\mathbf{A}$ admits a direct decomposition $\mathbf{A} = N \oplus R$ where R is a right ideal. It follows from Lemma 10.3.2 that the unit element of $\mathbf{A}$ is the sum of two idempotents contained in N and R, respectively. But this contradicts the fact that the nilpotent ideal

N cannot contain an idempotent. So we have $N = (0)$, and our proposition is proved.

Next we deal with two-sided ideals of semi-simple rings.

10.3.5. *If J is a non-zero two-sided ideal in a semi-simple ring* $\mathbf{A}$ *then there exists a unique idempotent c in J such that $J = c\mathbf{A} = \mathbf{A}c$. Moreover, c belongs to the centre of* $\mathbf{A}$.

Proof. Since J is a right ideal it follows from 10.3.1 that there is an idempotent c such that $J = c\mathbf{A}$. Since J is also a left ideal, the set of all $x \in J$ such that $xc = 0$ is a left ideal L of $\mathbf{A}$. From $Lc = 0$ and $cL = L$ we obtain $L^2 = L(cL) = (Lc)\,L = (0)$ so that L is a nilpotent left ideal. Since $\mathbf{A}$ is semi-simple it follows that $L = (0)$. For any element y of J we have $(y - yc)\,c = 0$ so that $y - yc \in L$ and hence $y - yc = 0$. Therefore c is not only a left unit element but also a right unit element for J. Therefore c is the unit element of J and is therefore uniquely determined. From $J \subseteq \mathbf{A}$ it follows that $J = Jc \subseteq \mathbf{A}c$. On the other hand, $c \in J$ implies that $\mathbf{A}c \subseteq J$. This gives $J = \mathbf{A}c$.

Finally, for any $a \in \mathbf{A}$, both ca and ac are contained in J. So we obtain

$$ca = (ca)\,c = c(ac) = ac$$

which shows that c belongs to the centre of $\mathbf{A}$.

This completes the proof. Taking $J = \mathbf{A}$ we obtain the following

10.3.6. Corollary. *Every semi-simple ring contains a unit element.*

We observe that, for any element z of the centre of $\mathbf{A}$, the ideal $z\mathbf{A}$ is evidently two-sided.

Let $\mathbf{A}$ be a ring with minimum condition that contains no two-sided ideals other than (0) and $\mathbf{A}$. Since the radical N of $\mathbf{A}$ is a two-sided ideal we have $N = (0)$ or $N = \mathbf{A}$. Suppose that $N = \mathbf{A}$. Then N^2 is a two-sided ideal which cannot coincide with $\mathbf{A}$ since in our case $\mathbf{A}$ is nilpotent, and the ring that consists of the zero element only has been excluded once for all. This gives $N^2 = \mathbf{A}^2 = 0$ i.e. $\mathbf{A}$ is a zero ring which means, that the product of any two elements of $\mathbf{A}$ is equal to zero. Therefore any subgroup of the additive group $\mathbf{A}^+$ is a two-sided ideal. From the assumption that $\mathbf{A}$ contains no two-sided ideals other than (0) and $\mathbf{A}$ it follows that $\mathbf{A}^+$ is a cyclic group of prime order.

A ring $\mathbf{A}$ that is not a zero ring is called *simple* if it satisfies the minimum condition for right ideals and contains no two-sided ideals other than (0) and $\mathbf{A}$.

As we just saw, every simple ring is semi-simple.

10.3.7. THEOREM. *Any semi-simple ring* $\mathbf{A}$ *is the direct sum*

$$\mathbf{A} = A_1 \oplus \ldots \oplus A_r$$

of unique minimal two-sided ideals A_i. *The* A_i *are simple rings and* $A_iA_j = 0$ *for* $i \neq j$. *Every right ideal of any* A_i *is a right ideal of* $\mathbf{A}$. *Every minimal right ideal of* $\mathbf{A}$ *is contained in some* A_i.

Proof. Let A_1 be a minimal two-sided ideal of $\mathbf{A}$. By Theorem 10.3.5, we have $A_1 = e_1\mathbf{A}$ where e_1 is an idempotent belonging to the centre. Moreover,

$$\mathbf{A} = e_1 \mathbf{A} + (e - e_1) \mathbf{A} \tag{1}$$

where e denotes the unit element of $\mathbf{A}$. Since $e - e_1$ also belongs to the centre, the second summand on the right-hand side is a two-sided ideal. Moreover, the decomposition (1) is direct because e_1 is the unit element for the first summand whereas $e_1(e - e_1) \mathbf{A} = 0$. For any two elements a, b of $\mathbf{A}$ the product $e_1a(e - e_1) b$ is equal to zero since it belongs to the intersection $e_1\mathbf{A} \cap (e - e_1) \mathbf{A} = (0)$. This shows that not only addition but also multiplication is carried out componentwise with respect to the direct decomposition (1). Therefore any one- or two-sided ideal of one of the two summands is also a similar ideal in $\mathbf{A}$. Consequently, $e_1\mathbf{A} = A_1$ is a simple ring and $(e - e_1) \mathbf{A}$ is semi-simple. If $(e - e_1) \mathbf{A}$ is not simple we take a minimal two-sided ideal of $(e - e_1) \mathbf{A}$ and proceed as above. After a finite number of steps we arrive at a direct decomposition

$$\mathbf{A} = A_1 \oplus \ldots \oplus A_r \tag{2}$$

into minimal two-sided ideals A_i. For any two elements $a_i \in A_i$, $a_j \in A_j$, and $i \neq j$ we have $a_ia_j = 0$ since this product belongs to the intersection $A_i \cap A_j = (0)$. This shows that any ideal of an A_i is also an ideal of $\mathbf{A}$. Consequently, the A_i are simple rings.

For any minimal two-sided ideal J of $\mathbf{A}$, not all the intersections $J \cap A_i$ can be equal to (0). Suppose that $J \cap A_j \neq (0)$. Since both J and A_j are minimal, this gives $J \cap A_j = J$. It follows that the minimal two-sided ideals A_i are uniquely determined. For a right ideal R of $\mathbf{A}$, the intersection $R \cap A_j$ is a right ideal of $\mathbf{A}$. If R is minimal, there is a unique index i such that $R \cap A_i = R$ and hence $R \subseteq A_i$. This completes the proof.

Two right ideals of $\mathbf{A}$ are said to be operator isomorphic if they are isomorphic as additive groups with the operator domain $\mathbf{A}$ where the action of $a \in \mathbf{A}$ as an operator is defined as multiplication by a on the right.

10.3.8. *Let*

$$\mathbf{A} = e_1 \mathbf{A} \oplus \ldots \oplus e_r\mathbf{A}$$

be a direct decomposition of the semi-simple ring $\mathbf{A}$ *into minimal two-sided ideals* $e_i\mathbf{A} = A_i$. *Two minimal right ideals* R' *and* R'' *are operator isomorphic if and only if they are contained in one and the same minimal two-sided ideal.*

Proof. Suppose that both R' and R'' are contained in A_i. By Lemma 10.3.2, we have $R' = e'\mathbf{A}$, $R'' = e''\mathbf{A}$ with orthogonal idempotents e', e''. The set of all finite sums of the form

$$a_1e'b_1 + a_2e'b_2 + \ldots + a_me'b_m \quad (a_j, b_j \in \mathbf{A})$$

obviously forms a two-sided ideal of $\mathbf{A}$ contained in A_i. This ideal is distinct from (0), because it contains $e_ie'e_i = e'$. Since A_i is minimal it follows that the ideal in question coincides with A_i. Since $e'' \in A_i$ we have

$$e'' = \bar{a}_1e'\bar{b}_1 + \ldots + \bar{a}_pe'\bar{b}_p$$

with suitable elements $\bar{a}_j$, $\bar{b}_j$ of $\mathbf{A}$. Since e'' is an idempotent, we conclude that there is at least one index k such that

$$e''\bar{a}_ke'\bar{b}_k \neq 0.$$

Therefore, $e''\bar{a}_ke'\bar{b}_k\mathbf{A}$ is a right ideal, which is distinct from (0) and belongs to $e''\mathbf{A} = R''$. As R'' is minimal we have

$$e''\bar{a}_k e'\bar{b}_k \mathbf{A} = R''.$$

Moreover, the right ideal $e'\bar{b}_k\mathbf{A}$ is distinct from (0) and is contained in the minimal right ideal $e'\mathbf{A} = R'$, so that

$$e'\bar{b}_k\mathbf{A} = R'.$$

Hence, we have

$$cR' = R''$$

where $c = e''\bar{a}_k$. The mapping

$$x' \to cx' \quad (x' \in R') \tag{3}$$

is therefore a homomorphism of R' onto R'' as additive groups. The kernel of this homomorphism consists of all $y' \in R'$ for which $cy' = 0$, so that the kernel is a right ideal of $\mathbf{A}$; moreover, the kernel is distinct from R' because $R'' \neq (0)$. Thus, the kernel is (0), hence (3) is an isomorphism. Finally, $c(x'a) = (cx')\,a$ shows that (3) is an operator isomorphism of R' onto R''.

Now, suppose that R' and R'' are contained in distinct two-sided ideals, $R' \subseteq e_i\mathbf{A}$, $R'' \subseteq e_j\mathbf{A}$, $i \neq j$, say. Then, R' and R'' cannot be operator isomorphic. For, since A_i and A_j annihilate each other, we have

$$r'e_i = r', \quad r'e_j = 0$$

for every $r' \in R'$, and

$$r''e_i = 0, \quad r''e_j = r''$$

for every $r'' \in R''$. Thus, both parts of the theorem are proved.

10.4. Simple rings

By a complete matrix algebra of degree n over a skew field $\mathbf{S}$ we understand the ring of all $n \times n$ matrices with elements in $\mathbf{S}$.

10.4.1. WEDDERBURN'S THEOREM. *Every simple ring is isomorphic to a complete matrix algebra over a skew field.*

Proof. A simple ring $\mathbf{A}$ is semi-simple so that $\mathbf{A}$ has a direct decomposition

$$\mathbf{A} = e_{11}\mathbf{A} \oplus e_{22}\mathbf{A} \oplus \dots \oplus e_{nn}\mathbf{A}$$

into minimal right ideals. The $e_{11}, e_{22}, \dots, e_{nn}$ are orthogonal idempotents and

$$e = e_{11} + e_{22} + \dots + e_{nn}.$$

The set of all finite sums

$$a_1e_{11}b_1 + a_2e_{11}b_2 + \dots + a_me_{11}b_m \quad (a_k, b_k \in \mathbf{A}) \tag{1}$$

is a two-sided ideal of $\mathbf{A}$ and distinct from (0), because $e_{11} = e_{11}e_{11}e_{11}$ is such a sum. Thus this set coincides with $\mathbf{A}$. It follows that each e_{ii} has a representation of the form (1), say

$$e_{ii} = \bar{a}_1e_{11}\bar{b}_1 + \dots + \bar{a}_pe_{11}\bar{b}_p.$$

This gives

$$e_{ii} = e_{ii}e_{ii} = \bar{a}_1e_{11}\bar{b}_1e_{ii} + \dots + \bar{a}_pe_{11}\bar{b}_pe_{ii}$$

which shows that for every index i there is an element c_i of $\mathbf{A}$ such that $e_{11}c_ie_{ii} \neq 0$.

After these preliminaries, we show that $S_i = e_{ii}\mathbf{A}e_{ii}$, i.e. the set of all products $e_{ii}ae_{ii}$, $a \in \mathbf{A}$, is a skew field. Clearly, $e_{ii}\mathbf{A}e_{ii}$ is a subalgebra of $\mathbf{A}$ with the unit element e_{ii}. We have to show that each element $e_{ii}ae_{ii} \neq 0$ has an inverse. Now, $e_{ii}ae_{ii}\mathbf{A}$ is a right ideal of $\mathbf{A}$, distinct from (0) and contained in $e_{ii}\mathbf{A}$. Since $e_{ii}\mathbf{A}$ is minimal, we have $e_{ii}ae_{ii}\mathbf{A} = e_{ii}\mathbf{A}$. Thus, there is an element x of $\mathbf{A}$ such that $e_{ii}ae_{ii}x = e_{ii}$. This shows that

$$e_{ii}ae_{ii}e_{ii}xe_{ii} = e_{ii},$$

so that $e_{ii}xe_{ii}$ is the inverse of $e_{ii}ae_{ii}$.

Next, we construct a system of matrix units in $\mathbf{A}$. For each index $i = 2, \dots, n$, we choose an element c_i such that

$$e_{11}c_ie_{ii} = e_{1i} \neq 0.$$

Then, we have

$$e_{11}e_{1i} = e_{1i}e_{ii} = e_{1i} \quad (i = 1, \dots, n).$$

Since $e_{11}\mathbf{A}$ is minimal and $e_{1i}\mathbf{A} \subseteq e_{11}\mathbf{A}$, we conclude that $e_{1i}\mathbf{A} = e_{11}\mathbf{A}$. In particular, we have $e_{1i}d_i = e_{11}$ for suitable $d_i \in \mathbf{A}$ and

$$e_{1i}e_{ii}\,d_ie_{11} = e_{1i}\,d_ie_{11} = e_{11}.$$

For $i = 2, \ldots, n$, we put

$$e_{i1} = e_{ii}\, d_i e_{11},$$

so that

$$e_{ii}e_{i1} = e_{i1}e_{11} = e_{i1} \quad (i = 1, \ldots, n)$$

and

$$e_{1i}e_{i1} = e_{11}.$$

Moreover, we obtain

$$e_{1i}e_{i1}e_{1i}e_{i1} = e_{11}^2 = e_{11}$$

so that $e_{i1}e_{1i} \neq 0$. Since

$$(e_{i1}e_{1i})^2 = e_{i1}e_{1i},$$

$e_{i1}e_{1i}$ is an idempotent of the skew field $S_i = e_{ii}\mathbf{A}e_{ii}$, and since the unit element is the only idempotent of a skew field, we have

$$e_{i1}e_{1i} = e_{ii} \quad (i = 1, \ldots, n).$$

Finally, we put

$$e_{i1}e_{1k} = e_{ik}$$

for $i \neq k$. Then, we obtain

$$e_{ik}e_{km} = e_{i1}e_{1k}e_{k1}e_{1m} = e_{i1}e_{11}e_{1m} = e_{im}$$

and for $l \neq k$

$$e_{ik}e_{lm} = e_{ik}e_{kk}e_{ll}e_{lm} = 0.$$

Thus, the n^2 elements e_{ik} $(i, k = 1, \ldots, n)$ form a system of matrix units in $\mathbf{A}$.

For any element λ_1 of the skew field $S_1 = e_{11}\mathbf{A}e_{11}$, we put

$$\lambda = \lambda_1 + e_{21}\lambda_1 e_{12} + \ldots + e_{n1}\lambda_1 e_{1n}.$$

These elements λ form a skew field $\mathbf{S}$, for one readily verifies that the mapping $\lambda_1 \to \lambda$ is an isomorphism of S_1 onto $\mathbf{S}$. In particular, this isomorphism carries the unit element e_{11} of S_1 into

$$e_{11} + e_{21}e_{11}e_{12} + \ldots + e_{n1}e_{11}e_{1n} = e_{11} + e_{22} + \ldots + e_{nn} = e$$

so that the unit element of $\mathbf{S}$ coincides with the unit element of $\mathbf{A}$.

Every element λ of $\mathbf{S}$ is permutable with every matrix unit e_{ik}; for, we have

$$\lambda e_{ik} = e_{i1}\lambda_1 e_{1k} = e_{ik}\lambda.$$

For an arbitrary element a of $\mathbf{A}$, we can put

$$a = eae = \sum_{i,k=1}^{n} e_{ii}ae_{kk}.$$

The summands on the right-hand side can be written as follows:

$$e_{ii}ae_{kk} = e_{i1}e_{1i}ae_{k1}e_{1k}.$$

Here, $e_{1i}ae_{k1} = e_{11}e_{1i}ae_{k1}e_{11}$ is an element of S_1. The corresponding element of the skew field $\mathbf{S}$ is

$$\alpha_{ik} = e_{1i}ae_{k1} + e_{21}e_{1i}ae_{k1}e_{12} + \dots + e_{n1}e_{1i}ae_{k1}e_{1n}.$$

This gives

$$\alpha_{ik}e_{ik} = e_{ii}ea_{kk}$$

and hence

$$a = \sum_{i,k=1}^{n} \alpha_{ik}e_{ik}.$$

This representation of the elements of $\mathbf{A}$ is unique, for

$$\sum_{i,k=1}^{n} \beta_{ik}e_{ik} = 0$$

implies that

$$\beta_{pq}e_{pq} = e_{pp}\left(\sum_{i,k=1}^{n} \beta_{ik}e_{ik}\right)e_{qq} = 0$$

for $p, q = 1, \dots, n$. This completes the proof.

If $\mathbf{A}$ is a complete matrix algebra of degree n over a skew field $\mathbf{S}$, then the free $\mathbf{S}$-module R_i with the basis

$$e_{i1}, e_{i2}, \dots, e_{in}$$

is a minimal right ideal of $\mathbf{A}$. This follows from the equations

$$e_{ik}e_{pq} = \begin{cases} e_{iq} & \text{if } k = p, \\ 0 & \text{if } k \neq p. \end{cases}$$

(R_i consists of all matrices in which all the entries outside the i-th row are equal to zero.) It is evident that

$$\mathbf{A} = R_1 \oplus \dots \oplus R_n.$$

10.4.2. *If* e_{pq} *and* e'_{pq} $(p, q = 1, \dots, n)$ *are two systems of matrix units in a simple ring* **A**, *then there exists an invertible element* c *of* **A** *such that*

$$ce'_{pq}c^{-1} = e_{pq} \quad (p, q = 1, \dots, n).$$

Proof. As we just observed, every system of matrix units gives rise to a decomposition into a direct sum of simple right ideals. Thus we have

$$\mathbf{A} = R_1 \oplus \dots \oplus R_n = R'_1 \oplus \dots \oplus R'_n$$

corresponding to the e_{ik} and e'_{ik}, respectively. By 10.3.8, R_1 and R'_1 are operator isomorphic, and from the proof of this theorem it follows that there exist elements a, b of **A** such that

$$R_1 = aR'_1, \quad R'_1 = bR_1.$$

Moreover, we can choose a and b in such a way that they induce inverse mappings. Then ab induces the identity automorphism of R_1. This gives $ab = e_{11}$. We set

$$c = \sum_{i=1}^{n} e_{i1}ae'_{1i}, \qquad d = \sum_{i=1}^{n} e'_{i1}be_{1i}$$

and obtain

$$cd = \sum_{i,k=1}^{n} e_{i1}ae'_{1i}e'_{k1}be_{1k} = \sum_{i=1}^{n} e_{i1}ae'_{11}be_{1i}$$

$$= \sum_{i=1}^{n} e_{i1}abe_{1i} = \sum_{i=1}^{n} e_{i1}e_{11}e_{1i} = \sum_{i=1}^{n} e_{ii} = e$$

so that $d = c^{-1}$. We find

$$ce'_{pq}c^{-1} = \sum_{i,k=1}^{n} e_{i1}ae'_{1i}e'_{pq}e'_{k1}be_{1k}$$

$$= e_{p1}ae'_{11}be_{1q} = e_{p1}abe_{1q} = e_{p1}e_{11}e_{1q} = e_{pq}$$

as was to be proved.

We now consider algebras of finite dimension over an algebraically closed field **K**. By Theorem 10.4.1, a simple algebra **A** over **K** is isomorphic to a complete matrix algebra over a skew field **S**. From the proof of this theorem it follows that **A** contains a subfield isomorphic to **S**. Therefore **S** is also a finite dimensional algebra over

K, and **K** is contained in the centre of **S**. Hence, for any element α of **S**, the extension field **K**(α) is algebraic over **K**. Since **K** is algebraically closed this implies that **K**(α) = **K**, i.e. $\alpha \in$ **K**. Thus it follows that **S** = **K**.

By the last result and Theorems 10.3.7 and 10.4.1, any finite-dimensional semi-simple algebra **A** over an algebraically closed field **K** is isomorphic to a direct sum

$$A_1 \oplus \ldots \oplus A_r \tag{2}$$

of complete matrix algebras A_i over **K**. If n is the dimension of **A** and if f_i denotes the degree of the complete matrix algebra A_i, then we obviously have

$$n = f_1^2 + \ldots + f_r^2 .$$

It is easy to see that the centre of a complete matrix algebra of degree f over a field **K** consists of all matrices of the form γI_f, $\gamma \in$ **K**. Hence the centre is isomorphic to **K**. By Theorem 10.3.7, we have $A_i A_j = 0$ for $i \neq j$ so that not only addition but also multiplication is carried out componentwise with respect to the direct decomposition (2). This implies that the centre of **A** is isomorphic to the direct sum of the centres of the A_i. Therefore the centre of **A** is an algebra of dimension r over **A**; it is isomorphic to the direct sum of r copies of **K**.

Summarizing our results we can state the following

10.4.3. Theorem. *Any finite-dimensional semi-simple algebra* **A** *over an algebraically closed field* **K** *is isomorphic to a direct sum*

$$A_1 \oplus \ldots \oplus A_r$$

of complete matrix algebras A_i *over* **K** *where* $A_i A_j = 0$ *for* $i \neq j$. *The number* r *of the direct summands is equal to the dimension of the centre of* **A** *over* **K**.

10.5. Division algebras over the field of the real numbers

By a division algebra we understand an algebra with unit element in which every non-zero element has an inverse.

Our aim is to determine all division algebras of finite dimension.

over the field $\boldsymbol{R}$ of the real numbers. We shall use the fact that every irreducible polynomial in $\boldsymbol{R}[x]$ is of degree one or two. This follows from the Fundamental Theorem of Algebra (cf. § 8.2). Let $f(x)$ be a polynomial of positive degree in $\boldsymbol{R}[x]$. If $f(x)$ has a complex root α, then the conjugate complex number $\bar{\alpha}$ is also a root of $f(x)$. Indeed

$$f(\alpha) = a_n\alpha^n + a_{n-1}\alpha^{n-1} + \ldots + a_0 = 0$$

implies that $\overline{f(\alpha)} = 0$, i.e.

$$\overline{f(\alpha)} = a_n\bar{\alpha}^n + a_{n-1}\bar{\alpha}^{n-1} + \ldots + a_0 = f(\bar{\alpha}) = 0$$

since the coefficients a_i are real numbers. Therefore the linear factors corresponding to complex roots occur in pairs $(x - \alpha)(x - \bar{\alpha})$. Since the product

$$(x - \alpha)(x - \bar{\alpha}) = x^2 - (\alpha + \bar{\alpha})x + \alpha\bar{\alpha}$$

has real coefficients it follows that every polynomial of positive degree in $\boldsymbol{R}[x]$ can be factorized into polynomials of degree one and two with real coefficients.

Since any division algebra $\mathbf{D}$ over $\boldsymbol{R}$ has a unit element e, it contains a subfield isomorphic to $\boldsymbol{R}$, namely the elements γe, $\gamma \in \boldsymbol{R}$. In what follows we shall identify this subfield with $\boldsymbol{R}$ and consequently denote the unit element of $\mathbf{D}$ by 1.

10.5.1. FROBENIUS' THEOREM. *Any division algebra of finite dimension over the field $\boldsymbol{R}$ of the real numbers is isomorphic to one of the following three algebras:*

(i) *the field $\boldsymbol{R}$ itself,*

(ii) *the field of the complex numbers,*

(iii) *the quaternion algebra over $\boldsymbol{R}$* (cf. § 4.1, Example 12).

Proof. Let $\mathbf{D}$ be a division algebra of dimension n over $\boldsymbol{R}$. In case $n = 1$, $\mathbf{D}$ is isomorphic to $\boldsymbol{R}$. Further, assume that $n > 1$.

Let α be any element of $\mathbf{D}$ that is not contained in $\boldsymbol{R}$. Then α is a root of an irreducible polynomial in $\boldsymbol{R}[x]$ of degree greater than 1. As we observed above such a polynomial is of degree 2 and has two conjugate complex roots. This gives

$$\alpha^2 + p\alpha + q = 0$$

where $p, q \in \boldsymbol{R}$ and $q - p^2/4 > 0$. Due to the last inequality there is a number $r \in \boldsymbol{R}$ such that

$$r^2 = q - p^2/4.$$

The element

$$i = \frac{1}{r}(\alpha + p/2)$$

of $\mathbf{D}$ satisfies the equation $i^2 = -1$. If $n = 2$, $\mathbf{D}$ is therefore isomorphic to the field $\boldsymbol{R}(i)$ of the complex numbers.

Now let $n > 2$. As we just saw, $\mathbf{D}$ contains an element i such that $i^2 = -1$. Moreover, $\mathbf{D}$ must contain an element β such that $1, i, \beta$ are linearly independent over $\boldsymbol{R}$. We also know that β is a root of an irreducible quadratic polynomial with coefficients in $\boldsymbol{R}$. In the same way as above we can construct an element j_0 of $\mathbf{D}$ such that $1, i, j_0$ are linearly independent over $\boldsymbol{R}$ and $j_0^2 = -1$.

All elements of $\mathbf{D}$, that are not contained in $\boldsymbol{R}$ are roots of irreducible quadratic polynomials in $\boldsymbol{R}[x]$. In particular, this applies to $i + j_0$ and $i - j_0$. Thus we obtain

$$\begin{aligned} (i + j_0)^2 &= -2 + ij_0 + j_0 i = a(i + j_0) + b, \\ (i - j_0)^2 &= -2 - ij_0 - j_0 i = c(i - j_0) + d \end{aligned} \tag{1}$$

with $a, b, c, d \in \boldsymbol{R}$. Addition yields

$$-4 = (a + c)\, i + (a - c)\, j_0 + (b + d).$$

Since $1, i, j_0$ are linearly independent over $\boldsymbol{R}$ it follows that $a + c = a - c = 0$, i.e. $a = c = 0$. Hence, the first equation (1) gives

$$ij_0 + j_0 i = 2t \tag{2}$$

where $t = \frac{1}{2}(b + 2)$. We set

$$j_1 = j_0 + ti.$$

Clearly, $1, i, j_1$ are linearly independent over $\boldsymbol{R}$. Moreover, $\boldsymbol{R}(j_1)$ is of dimension 2 over $\boldsymbol{R}$ and hence isomorphic to $\boldsymbol{R}(i)$. It follows that $j_1^2 < 0$, say $j_1^2 = -s^2$ where $s \in \boldsymbol{R}$. Then

$$j = \frac{1}{s} j_1$$

satisfies the equation $j^2 = -1$. From (2) we obtain

$$ij + ji = \frac{1}{s}\left(i(j_0 + ti) + (j_0 + ti)\, i\right) = \frac{1}{s}(2t - t - t) = 0,$$

i.e.

$$ij = -ji. \tag{3}$$

We set $k = ij$. Then k cannot be expressed as a linear combination of 1, i, j. For, suppose that, on the contrary,

$$k = a_1 + b_1 i + c_1 j \quad (a_1, b_1, c_1 \in \boldsymbol{R});$$

multiplication by i on the left gives

$$-j = a_1 i - b_1 + c_1 k = a_1 i - b_1 + c_1(a_1 + b_1 i + c_1 j)$$

and hence, due to the linear independence of 1, i, j, $c^2 = -1$ which is impossible. Consequently,

$$1, i, j, k$$

are linearly independent. In case $n = 4$, $\mathbf{D}$ is therefore isomorphic to the quaternion algebra.

Finally, we have to show that n cannot exceed 4. Let us assume that, on the contrary, $n > 4$. Then there exists an element l in $\mathbf{D}$ such that 1, i, j, k, l are linearly independent and $l^2 = -1$. As above we conclude that the expressions

$$il + li = a_2, \quad jl + lj = b_2, \quad kl + lk = c_2$$

are real numbers. We obtain

$$\begin{aligned} lk = l(ij) &= a_2 j - (il)\, j = a_2 j - i(b_2 - jl) \\ &= a_2 j - b_2 i + ijl = a_2 j - b_2 i + kl \\ &= a_2 j - b_2 i + c_2 - lk \end{aligned}$$

so that

$$2lk = a_2 j - b_2 i + c_2.$$

Multiplication by k on the right yields

$$-2l = a_2 i + b_2 j + c_2 k,$$

a contradiction to the linear independence of i, j, k, l.

This completes the proof.

10.6. Representation modules

We recall the definition of a representation module in § 9.3 in a slightly generalized form.

Let V be a vector space of finite dimension n over a field $\mathbf{K}$, and let $\mathbf{A}$ be an arbitrary ring. We assume that V is a right $\mathbf{A}$-module, i.e.

$$\begin{gathered} v(a+b) = va + vb, \quad v(ab) = (va)\,b, \\ (v+w)\,a = va + wa \end{gathered} \tag{1}$$

for arbitrary elements v, w of V and a, b of $\mathbf{A}$. If, moreover,

$$\gamma(va) = (\gamma v)\,a \tag{2}$$

for arbitrary elements $\gamma \in \mathbf{K}$, $v \in V$, $a \in \mathbf{A}$, then V is called a *representation module* for $\mathbf{A}$.

Let $v_1, \ldots, v_n$ be a basis of V. For any $a \in \mathbf{A}$ let $\boldsymbol{v}$ and $\boldsymbol{v}a$ denote the columns with the elements $v_1, \ldots, v_n$ and $v_1 a, \ldots, v_n a$, respectively. We have

$$\boldsymbol{v}a = \Delta(a)\,\boldsymbol{v} \tag{3}$$

where $\Delta(a)$ is an $n \times n$ matrix with elements in $\mathbf{K}$. From (1) and (2) it follows that for arbitrary elements a, b of $\mathbf{A}$

$$\begin{gathered} \boldsymbol{v}(a+b) = \boldsymbol{v}a + \boldsymbol{v}b = \Delta(a)\,\boldsymbol{v} + \Delta(b)\,\boldsymbol{v} = \big(\Delta(a) + \Delta(b)\big)\,\boldsymbol{v}, \\ \boldsymbol{v}(ab) = (\boldsymbol{v}a)\,b = \big(\Delta(a)\,\boldsymbol{v}\big)\,b = \Delta(a)\,(\boldsymbol{v}b) = \Delta(a)\,\Delta(b)\,\boldsymbol{v}. \end{gathered}$$

This shows that the mapping

$$a \to \Delta(a) \quad (a \in \mathbf{A})$$

is a homomorphism of $\mathbf{A}$ into the ring of all $n \times n$ matrices with elements in $\mathbf{A}$. This homomorphism and also the image of $\mathbf{A}$ under this homomorphism are called a *representation* of $\mathbf{A}$ of degree n.

Conversely, if a homomorphism of a ring **A** into the complete matrix algebra of degree n over **A** is given and if to any element a of **A** there corresponds the $n \times n$ matrix $\Delta(a)$, then any vector space of dimension n over **K** becomes a representation module if we define the action of $a \in \mathbf{A}$ on the elements of a basis by (3) and require

$$(\alpha_1 v_1 + \dots + \alpha_n v_n)\, a = \alpha_1(v_1 a) + \dots + \alpha_n(v_n a)$$

to be satisfied. It is then easy to verify that V has all the properties of a representation module.

From the basis $v_1, \dots, v_n$ of V we obtain another basis $w_1, \dots, w_n$ in the form

$$w_i = \sum_{k=1}^{n} \sigma_{ik} v_k \quad (i = 1, \dots, n)$$

where $S = [\sigma_{ik}]$ is an invertible $n \times n$ matrix with elements in **K**. If $\boldsymbol{w}$ denotes the column of $w_1, \dots, w_n$, then the last equations can be written as follows:

$$\boldsymbol{w} = S\boldsymbol{v}.$$

In the new basis we obtain

$$\boldsymbol{w}a = S\boldsymbol{v}a = S\,\Delta(a)\,\boldsymbol{v} = S\,\Delta(a)\,S^{-1}\,\boldsymbol{w} = \Phi(a)\,\boldsymbol{w}.$$

The matrices $\Phi(a) = S\,\Delta(a)\,S^{-1}$ form a representation $\Phi = S\,\Delta S^{-1}$ of **A**. The representations Δ and Φ are said to be *equivalent* and will not be regarded as essentially distinct, since they correspond to the same homomorphism of **A** into the ring of endomorphisms of V and describe the endomorphisms of V only in terms of different bases.

Clearly, two representation modules that are operator isomorphic with respect to **A** as right operator domain yield equivalent representations, and vice versa.

It may happen that a representation module V of **A** contains a submodule T that is admissible with respect to **A**. Let $t_1, \dots, t_m$ with $m < n$ be a basis of such a submodule. By Theorem 3.13.10, we can find suitable $l = n - m$ elements $u_1, \dots, u_l$ of V such that $t_1, \dots, t_m, u_1, \dots, u_l$ is a basis of V. Using this basis, we obtain a representation $S\,\Delta S^{-1}$ whose matrices have the following form:

$$S\,\Delta(a)\,S^{-1} = \begin{bmatrix} \Delta_1(a) & 0 \\ \Delta_3(a) & \Delta_2(a) \end{bmatrix}. \tag{4}$$

0 always denotes a zero matrix of suitable size. In this case 0 has m rows and l columns.

Here, the $m \times m$ matrices $\Delta_1(a)$ form a representation of **A** with T as a representation module. For, if $\boldsymbol{t}$ denotes the column of the elements $t_1, \ldots, t_m$, we have

$$\boldsymbol{t}a = \Delta_1(a)\,\boldsymbol{t}$$

since T is admissible. The matrices $\Delta_2(a)$ also form a representation Δ_2 of **A**. This is easily verified by computing $S\Delta(a)\,S^{-1}S\,\Delta(b)\,S^{-1} = S\,\Delta(ab)\,S^{-1}$, or it can be seen as follows: For the column $\boldsymbol{u}$ of the elements $u_1, \ldots, u_l$ we have

$$\boldsymbol{u}a = \Delta_3(a)\,\boldsymbol{t} + \Delta_2(a)\,\boldsymbol{u}; \tag{5}$$

hence,

$$\boldsymbol{u}a \equiv \Delta_2(a)\,\boldsymbol{u} \pmod{T}.$$

This shows that the factor module V/T is a representation module of Δ_2.

A representation Δ is called reducible if its representation module V has a representation submodule T other than 0 and V. Thus, when Δ is reducible, there exists an equivalent representation $S\,\Delta S^{-1}$ whose matrices are of the form (4). Conversely, if there is an equivalent representation $S\,\Delta S^{-1}$ of the form (4), then every representation module of Δ has a non-trivial representation submodule. For, suppose that the matrices $\Delta_1(a)$ form a representation of degree $m < n$. Then, the first m elements of the basis of V belonging to $S\,\Delta S^{-1}$ are the basis of a non-trivial submodule T of dimension m that is admissible with respect to **A**. The representation Δ is said to split into the constituents Δ_1 and Δ_2.

By contrast, a representation is irreducible if and only if its representation module contains no non-trivial representation submodule. This means that the representation modules regarded as abelian groups with the operator domains **K** and **A** are simple.

We now regard V as an additive abelian group with the operator domains **K** and **A**. As the rank of V is finite, there exists a composition series

$$V = V_k \supset V_{k-1} \cdots \supset V_1 \supset V_0 = 0$$

where each V_{i-1} is a maximal representation submodule of V_i ($i = 1, \ldots, k$) or, in other words, the factor modules V_i/V_{i-1} are simple as additive groups with the operator domains **K** and **A**. We choose a basis of V_1, then we extend it to a basis of V_2; after that, we extend this basis of V_2 to a basis of V_3, etc. until we finally arrive at a basis of V. To this particular basis, there belongs a representation of **A** whose matrices are of the form

$$\begin{bmatrix} \Delta_{11}(a) & 0 & \ldots 0 \\ \Delta_{21}(a) & \Delta_{22}(a) & \ldots 0 \\ \cdots & \cdots & \cdots \\ \Delta_{k1}(a) & \Delta_{k2}(a) & \ldots \Delta_{kk}(a) \end{bmatrix}$$

for every $a \in \mathbf{A}$. For $i = 1, \ldots, k$, the representation Δ_{ii} arrises from the representation module V_i/V_{i-1}, and as these factor modules are simple, it follows that the representations Δ_{ii} are irreducible.

Conversely, each splitting of a representation into irreducible constituents leads to a composition series of the representation module.

By Theorem 3.8.2 for groups with operators, the factor modules of any two composition series of the representation module V are operator isomorphic to within order. Since representations are equivalent if and only if their representation modules are operator isomorphic, our previous arguments yield the following theorem:

10.6.1. THEOREM. *The irreducible constituents of any representation are unique up to equivalence and order.*

Let us assume that V is a representation module of a reducible representation of **A**. Then, there exists a non-trivial representation submodule T. As above, we extend a basis $t_1, \ldots, t_m$ of T by suitable elements $u_1, \ldots, u_l$ to a basis of V. So we obtain a direct decomposition $V = T \oplus U$ where U is the vector space spanned by $u_1, \ldots, u_l$. In general, U is not admissible with respect to the operator domain **A**. Our notation is the same as above, so that (5) shows that U is admissible if and only if $\Delta_3(a) = 0$ for all $a \in \mathbf{A}$. If this condition is satisfied, then U is also a representation module and belongs to the representation Δ_2. We say that Δ *splits completely* into the constituents Δ_1 and Δ_2.

A representation Δ is said to be completely reducible if its representation module V can be decomposed into the direct sum of simple representation submodules:

$$V = V_1 \oplus \ldots \oplus V_r.$$

In this case, there is a representation equivalent to Δ all of whose matrices are of the form

$$\begin{bmatrix} \Delta_1(a) & 0 & \ldots 0 \\ 0 & \Delta_2(a) & \ldots 0 \\ \ldots & \ldots & \ldots \\ 0 & 0 & \ldots \Delta_r(a) \end{bmatrix}$$

where the Δ_i are irreducible representations belonging to the representation modules V_i.

Let $\mathbf{A}$ be a ring with the unit element e. We shall show that it is no loss of generality to assume that e is the identity operator on any representation module for $\mathbf{A}$. Let V be an arbitrary representation module for $\mathbf{A}$, and let Ve denote the image of V under e as a right operator. It is evident that Ve is a subspace of V which is admissible with respect to $\mathbf{A}$ and for which e is the identity operator. Any element v of V can be written as follows:

$$v = ve + (v - ve).$$

So we obtain a direct decomposition

$$V = Ve \oplus (V - Ve). \tag{6}$$

That this decomposition is direct follows from the fact that e as a right operator carries every element of Ve into itself and annihilates every element of $V - Ve$. In particular, $V - Ve$ is admissible with respect to $\mathbf{A}$. By virtue of (6), the representation Δ for which V is a representation module splits completely into the constituents Δ_1 and Δ_0 belonging to Ve and $V - Ve$, respectively. As we observed above, $\Delta_1(e)$ is the identity matrix whereas $\Delta_0(a) = 0$ for each $a \in \mathbf{A}$.

For any subset W of V and any subset B of $\mathbf{A}$ let WB denote the subspace of V generated by all elements wb where $w \in W$, $b \in B$.

Let us suppose that $\mathbf{A}$ is a ring with minimum condition and let N denote the radical of $\mathbf{A}$. If V is a simple representation module for $\mathbf{A}$, then VN is obviously a representation submodule of V so that $VN = V$ or $VN = 0$. The first possibility can readily be excluded; indeed, $VN = V$ implies that $VN^k = V$ for any natural number k which contradicts the fact that $N^m = (0)$ for some natural number m. Hence we have $VN = 0$. This result can be stated as follows.

10.6.2. THEOREM. *In every irreducible representation of a ring* $\mathbf{A}$ *with minimum condition the radical of* $\mathbf{A}$ *is represented by zero matrices.*

Therefore every irreducible representation of $\mathbf{A}$ is actually a representation of $\mathbf{A}/N$.

10.7. Representations of semi-simple algebras

Let $\mathbf{A}$ be an algebra with unit element of finite dimension over a field $\mathbf{K}$. By a representation module of $\mathbf{A}$ we always mean a representation module that is a vector space over $\mathbf{K}$.

Then every right ideal of $\mathbf{A}$ is a representation module for $\mathbf{A}$ because the conditions (1) and (2) of § 10.7 are satisfied by virtue of the distributive and associative laws in $\mathbf{A}$ and the property (ii) of $\mathbf{A}$ in § 10.1. The representation induced by a right ideal is irreducible if and only if the right ideal is minimal.

The representation, or more precisely, the class of equivalent representations, for which $\mathbf{A}$ itself is the representation module is called the *regular representation* of $\mathbf{A}$.

The next theorem is basic for the representations of semi-simple algebras.

10.7.1. THEOREM. *An algebra* $\mathbf{A}$ *with unit element is semi-simple if and only if all representations of* $\mathbf{A}$ *are completely reducible. The representation modules of the irreducible representations of a semi-simple algebra* $\mathbf{A}$ *are operator isomorphic to the minimal right ideals of* $\mathbf{A}$.

Proof. First, suppose that $\mathbf{A}$ is semi-simple. Let

$$\mathbf{A} = e_1\mathbf{A} \oplus \ldots \oplus e_m\mathbf{A}$$

be a direct decomposition into minimal right ideals and let $v_1, \ldots, v_n$ be a basis of a representation module V of $\mathbf{A}$. As we observed in § 10.6 we may assume that the unit element of $\mathbf{A}$ acts as the identity operator on V. So, we have $V = V\mathbf{A}$. Therefore, V is the sum, possibly not direct, of the representation submodules

$$v_i e_k \mathbf{A} \quad (i = 1, \ldots, n;\ k = 1, \ldots, m).$$

If $v_i e_k \mathbf{A} \neq 0$, the mapping

$$e_k a \to v_i e_k a \quad (a \in \mathbf{A})$$

is an operator isomorphism of $e_k\mathbf{A}$ onto $v_i e_k \mathbf{A}$, for, it is clearly an operator homomorphism. The elements $e_k b$ of $e_k\mathbf{A}$ for which $v_i e_k b = 0$ form a right ideal B of $\mathbf{A}$, properly contained in $e_k\mathbf{A}$; since $e_k\mathbf{A}$ is minimal, we have $B = (0)$. Thus, the mapping in question is an operator isomorphism. The representation modules $v_i e_k \mathbf{A}$ that are distinct from 0 are therefore operator isomorphic to minimal right ideals, hence are simple representation modules. Among the modules $v_i e_k \mathbf{A}$, we omit those that are 0. After having arranged the remaining modules in an arbitrary order, we omit those that are contained in the sum of the preceding ones and retain those that are not. Since the modules are simple, we obtain in this way a decomposition of V into a direct sum of suitable modules $v_i e_k \mathbf{A}$. Hence, V, and therefore the corresponding representation of $\mathbf{A}$, are completely reducible, and the simple representation submodules of V are operator isomorphic to minimal right ideals of $\mathbf{A}$.

Conversely, suppose that every representation of $\mathbf{A}$ is completely reducible. Taking the regular representation, we conclude that $\mathbf{A}$ is the direct sum of simple vector spaces over $\mathbf{K}$ that are admissible with respect to multiplication by elements of $\mathbf{A}$ as right factors; in other words, $\mathbf{A}$ is the direct sum of minimal right ideals. Thus, by Theorem 10.3.4, $\mathbf{A}$ is semi-simple. This completes the proof.

10.7.2. THEOREM. *All irreducible representations of a simple algebra are equivalent to each other. The number of inequivalent irreducible representations of a semi-simple algebra* **A** *is equal to the number of minimal two-sided ideals in* **A**.

Proof. Our theorem is an immediate consequence of the Theorems 10.3.7, 10.3.8, and 10.7.1.

Under the additional assumption that the field **K** is algebraically closed we can apply Theorem 10.4.3 to obtain the following corollary of the previous theorem.

10.7.3. COROLLARY. *The number of inequivalent irreducible representations of a finite-dimensional semi-simple algebra* **A** *over an algebraically closed field* **K** *is equal to the dimension of the centre of* **A** *over* **K**.

We now consider representations of finite groups. It is evident that every representation of a group G in a field **K** gives rise to a representation of the group algebra of G over **K** and vice versa; clearly, the representation of G is irreducible if and only if the corresponding representation of the group algebra is irreducible.

First, we prove an important theorem about the complete reducibility of representations of finite groups.

10.7.4. MASCHKE'S THEOREM. *If the characteristic of the field* **K** *is* 0 *or relatively prime to the order of the finite group G, then every representation of G by matrices with elements in* **K** *is completely reducible.*

Proof. It is obviously sufficient to prove the following assertion: If a representation module V for G contains a non-trivial representation submodule T, then V has a direct decomposition

$$V = T \oplus X, \tag{1}$$

where X is a representation module for G.

In any case, we have a direct decomposition

$$V = T \oplus U,$$

where U is not necessarily admissible with respect to G. We shall drove that U can be replaced by another direct summand X that is admissible with respect to G. We use the same notation as in

§ 10.6. Let $u_1, \ldots, u_l$ denote a basis of U; we write $\boldsymbol{u}$ for the column with the elements $u_1, \ldots, u_l$. We have to show that there are elements $x_1, \ldots, x_l$ in V such that the subspace X spanned by $x_1, \ldots, x_l$ is admissible with respect to G and satisfies (1). Writing $\boldsymbol{x}$ for the column of the elements $x_1, \ldots, x_l$ we put

$$\boldsymbol{x} = F\boldsymbol{t} + \boldsymbol{u} \tag{2}$$

and try to determine the $l \times m$ matrix F with elements K such that X is admissible. Note that, for every choice of the matrix F, the elements $x_1, \ldots, x_l$ are linearly independent because $u_1, \ldots, u_l$ are. Moreover, $T \cap X = 0$ for every choice of F. For, from

$$x = \beta_1 x_1 + \ldots + \beta_l x_l$$

it follows by (2) that

$$x \equiv \beta_1 u_1 + \ldots + \beta_l u_l \pmod{T}.$$

Thus, $x \in T \cap X$ implies that

$$\beta_1 u_1 + \ldots + \beta_l u_l \equiv 0 \pmod{T};$$

hence, $\beta_1 = \ldots = \beta_l = 0$, since $T \cap U = 0$. From $T \cap X = 0$ and the fact that X is a subspace of dimension l, it follows that for every matrix F in (2) we have a direct decomposition

$$V = T \oplus X.$$

From

$$\boldsymbol{t}a = \Delta_1(a)\, \boldsymbol{t} \tag{3}$$

and (5) in § 10.6 we obtain

$$\begin{aligned} \boldsymbol{x}a &= (F\boldsymbol{t} + \boldsymbol{u})\, a = F\boldsymbol{t}a + \boldsymbol{u}a = F\,\Delta_1(a)\, \boldsymbol{t} + \Delta_3(a)\, \boldsymbol{t} + \Delta_2(a)\, \boldsymbol{u} \\ &= \big(F\,\Delta_1(a) + \Delta_3(a) - \Delta_2(a)\, F\big)\, \boldsymbol{t} + \Delta_2(a)\,(F\boldsymbol{t} + \boldsymbol{u}) \\ &= \big(F\,\Delta_1(a) + \Delta_3(a) - \Delta_2(a)\, F\big)\, \boldsymbol{t} + \Delta_2(a)\, \boldsymbol{x}. \end{aligned}$$

For X to be admissible, the matrix F must satisfy the condition

$$F\,\Delta_1(a) + \Delta_3(a) - \Delta_2(a)\, F = 0 \text{ for every } a \in G. \tag{4}$$

As Δ is a representation, we have

$$\begin{bmatrix} \Delta_1(a) & 0 \\ \Delta_3(a) & \Delta_2(a) \end{bmatrix} \begin{bmatrix} \Delta_1(b) & 0 \\ \Delta_3(b) & \Delta_2(b) \end{bmatrix} = \begin{bmatrix} \Delta_1(ab) & 0 \\ \Delta_3(ab) & \Delta_2(ab) \end{bmatrix}$$

for any two elements a and b of G. This gives

$$\Delta_1(ab) = \Delta_1(a)\,\Delta_1(b), \quad \Delta_2(ab) = \Delta_2(a)\,\Delta_2(b), \qquad (5)$$
$$\Delta_3(ab) = \Delta_3(a)\,\Delta_1(b) + \Delta_2(a)\,\Delta_3(b).$$

We put $g = |G|$ and

$$F = \frac{1}{g} \sum_{c \in G} \Delta_2(c)^{-1}\,\Delta_3(c) \qquad (6)$$

where c ranges over all elements of G. Then

$$\Delta_2(a)\,F = \frac{1}{g} \sum_{c \in G} \Delta_2(a)\,\Delta_2(c)^{-1}\,\Delta_3(c).$$

If c ranges over all elements of G, then so does $d = ac^{-1}$ for any fixed $a \in G$. By (5), we now obtain

$$\begin{aligned} \Delta_2(a)\,F &= \frac{1}{g} \sum_{d \in G} \Delta_2(d)\,\Delta_3(d^{-1}a) \\ &= \frac{1}{g} \sum_{d \in G} \Delta_2(d)\,\big(\Delta_3(d^{-1})\,\Delta_1(a) + \Delta_2(d^{-1})\,\Delta_3(a)\big) \\ &= \Big(\frac{1}{g} \sum_{d \in G} \Delta_2(d)\,\Delta_3(d^{-1})\Big)\,\Delta_1(a) + \Delta_3(a) \\ &= F\,\Delta_1(a) + \Delta_3(a). \end{aligned}$$

Consequently, the matrix F as defined by (6) satisfies the condition (4), and this proves the theorem.

In view of the previous theorems about representations of semi-simple algebras it is important to know whether a group algebra is semi-simple. This question is answered by the following theorem.

10.7.5. THEOREM. *The group algebra of a finite group G over a field **K** is semi-simple if and only if the characteristic of **K** does not divide $|G|$.*

Proof. If the characteristic of **K** is not a divisor of $|G|$, then it follows from Theorem 10.7.4 that every representation of the group algebra of G over $\mathbf{K}$ is completely reducible. Hence, by Theorem 10.7.1, the group algebra is semi-simple.

Conversely, suppose that the characteristic of $\mathbf{K}$ divides $|G|$. We consider the element

$$s = \sum_{x \in G} x$$

of the group algebra. For any element y of G, we have $sy = s$. For an arbitrary element

$$a = \sum_{y \in G} \alpha_y y \quad (\alpha_y \in \mathbf{K})$$

of the group algebra, we obtain therefore

$$sa = \Big(\sum_{y \in G} \alpha_y\Big) s$$

and hence

$$(sa)^2 = \Big(\sum_{y \in G} \alpha_y\Big)^2 s^2 = \Big(\sum_{y \in G} \alpha_y\Big)^2 |G|\, s = 0.$$

This shows that the set of all elements of the form γs with $\gamma \in \mathbf{K}$ is a nilpotent ideal in the group algebra. Hence, the group algebra is not semi-simple. This proves our theorem.

From now on we shall assume that the field $\mathbf{K}$ is algebraically closed and that its characteristic does not divide $|G|$. Irreducible representations in an algebraically closed field are said to be *absolutely irreducible*.

To find the number of inequivalent absolutely irreducible representations of a finite group G by Corollary 10.7.3 we have to determine the centre of the group algebra.

Let $K_1, \ldots, K_t$ denote the conjugacy classes of G (cf. § 3.11). We form the so-called *class sums*

$$k_i = \sum_{x \in K_i} x \quad (i = 1, \ldots, t)$$

in the group algebra.

10.7.6. *The class sums of G are a basis of the centre of the group algebra of G over any field* $\mathbf{F}$.

Proof. An element

$$z = \sum_{x \in G} \zeta_x x \quad (\zeta_x \in \mathbf{F})$$

belongs to the centre of the group algebra if and only if $y^{-1}zy = z$ for every $y \in G$, i.e.

$$z = \sum_{x \in G} \zeta_x x = y^{-1}zy = \sum_{x \in G} \zeta_x y^{-1}xy = \sum_{x \in G} \zeta_{yxy^{-1}} x .$$

Thus, z belongs to the centre if and only if $\zeta_x = \zeta_{yxy^{-1}}$ for any two elements x, y of G, in other words, if conjugate elements of G have equal coefficients in z. This means that z can be written in the form

$$z = \sum_{i=1}^{t} \zeta_i k_i$$

which proves our proposition.

We now apply the theorems about the structure of semi-simple algebras and their representations to obtain the following results about representations of finite groups.

10.7.7. THEOREM. *Let G be a finite group and $\mathbf{K}$ an algebraically closed field whose characteristic does not divide $|G|$. Representation stands for a representation of G by matrices with elements in $\mathbf{K}$.*

(i) *Every representation of G is completely reducible.*

(ii) *The number of classes of inequivalent absolutely irreducible representations of G is equal to the number of conjugacy classes in G.*

(iii) *Every absolutely irreducible representation class of G occurs among the constituents of the regular representation with a multiplicity equal to its degree.*

(iv) *Among the $|G|$ matrices*

$$\Delta(a) \quad (a \in G)$$

of an absolutely irreducible representation Δ of degree f, there are f^2 linearly independent ones. Thus, only the matrices of the form γI_f, $\gamma \in \mathbf{K}$, are permutable with all the matrices $\Delta(a)$, $a \in G$.

(v) *The degrees $f_1, \ldots, f_t$ of the absolutely irreducible represen-*

tation classes satisfy the equation

$$f_1^2 + \dots + f_t^2 = |G|.$$

Proof. (i) is Theorem 10.7.4; (ii) follows from 10.7.6 and Corollary 10.7.3. The first part of (iii) is a consequence of Theorem 10.7.1. The structure of the group algebra $\mathbf{K}G$ of G over $\mathbf{K}$ is described by Theorem 10.4.3. Let $\mathbf{K}_f$ denote one of the complete matrix algebras that occur as summands in the direct decomposition of $\mathbf{K}G$ according to Theorem 10.4.3 and let f be its degree. As we observed in § 10.4, $\mathbf{K}_f$ is the direct sum of f minimal right ideals each of which is a vector space of dimension f over $\mathbf{K}$. By the second part of 10.3.8, these f minimal right ideals are operator isomorphic so that they yield equivalent absolutely irreducible representations of degree f. Distinct complete matrix algebras that occur as direct summands of $\mathbf{K}G$ yield inequivalent representations. This proves the assertion on the multiplicity in (iii). It follows from Theorem 10.7.2 or can easily be verified by a straightforward calculation that the minimal right ideals of $\mathbf{K}_f$ yield $\mathbf{K}_f$ as its own representation. Thus, the f^2 matrix units of $\mathbf{K}_f$ are linear combinations of the matrices $\Delta(a)$. This implies (iv). By Theorem 10.4.3, $\mathbf{K}G$ is the direct sum of t complete matrix algebras of the degrees $f_1, \dots, f_t$. Thus, the dimension $|G|$ of $\mathbf{K}G$ over $\mathbf{K}$ is also equal to $f_1^2 + \dots + f_t^2$ and this gives (v). This completes the proof.

Let $K_1, \dots, K_t$ be the conjugacy classes of the finite group G. As we observed above, the class sums

$$k_i = \sum_{x \in K_i} x \quad (i = 1, \dots, t)$$

form a basis of the centre of the group algebra $\mathbf{K}G$. If the elements x and y are conjugate in G, then so are x^{-1} and y^{-1}. Therefore the inverses of all elements in some conjugacy class K also form a conjugacy class which will be denoted by K^{-1}. We also write

$$k_i^{(-1)} = \sum_{x \in K_i^{-1}} x = \sum_{y \in K_i} y^{-1}.$$

The product of any two class sums is an element of the centre of $\mathbf{K}G$ and can therefore be represented as a linear combination of the class sums with coefficients that are evidently non-negative integers. It is expedient to write the formulae for the product of class sums in the following form:

$$k_i k_j^{(-1)} = \sum_{l=1}^{t} \gamma_{ijl} k_l .$$

One of the conjugacy classes, say K_1, consists only of the unit element e. We need the coefficient γ_{ij1} which is equal to the multiplicity with which the unit element occurs in the product $k_i k_j^{(-1)}$. It is evident that e does not occur at all unless $i = j$; in case $i = j$, e occurs with the multiplicity g_i where g_i denotes the number of elements in K_i and hence also the number of elements in K_i^{-1}. So we obtain

$$\gamma_{ij1} = g_i \delta_{ij} \quad (i, j = 1, \dots, t) \tag{7}$$

where

$$\delta_{ij} = \begin{cases} 1 & \text{if } i = j, \\ 0 & \text{if } i \neq j. \end{cases}$$

By the Theorems 10.3.5 and 10.3.7 we have the direct decomposition

$$\mathbf{K}G = e_1 \mathbf{K}G \oplus \dots \oplus e_t \mathbf{K}G \tag{8}$$

where the $e_i \mathbf{K}G$ are minimal two-sided ideals. The orthogonal idempotents

$$e^1, \dots, e_t$$

belong to the centre of $\mathbf{K}G$. Therefore the e_i form another basis of the centre of $\mathbf{K}G$. Consequently the class sums can be expressed as linear combinations of the e_i:

$$k_i = \sum_{l=1}^{t} \alpha_l(k_i)\, e_l \tag{9}$$

with coefficients $\alpha_l(k_i)$ in $\mathbf{K}$.

Due to the orthogonality of the e_i we obtain

$$k_i k_j^{(-1)} = \sum_{l=1}^{t} \alpha_l(k_i)\, \alpha_l(k_j^{(-1)})\, e_l = \sum_{m=1}^{t} \gamma_{ijm} \sum_{l=1}^{t} \alpha_l(k_m)\, e_l$$

and hence

$$\alpha_l(k_i)\, \alpha_l(k_j^{(-1)}) = \sum_{m=1}^{t} \gamma_{ijm} \alpha_l(k_m) . \tag{10}$$

By the *character* χ of a representation Δ of degree f, we mean the traces $\chi(a)$ of the matrices $\Delta(a)$. For the unit element e of G we have $\Delta(e) = I_f$ so that $\chi(e) = f$. Equivalent representations Δ and $S\,\Delta S^{-1}$ have the same character since similar matrices $\Delta(a)$ and $S\,\Delta(a)\,S^{-1}$ have the same trace. Later we shall see that the converse is also true, i.e. two representations are equivalent if they have the same character.

For two conjugate elements a and b of G we have $b = c^{-1}ac$ for some $c \in G$ and hence

$$\Delta(b) = \Delta(c)^{-1}\,\Delta(a)\,\Delta(c)$$

so that the two similar matrices $\Delta(a)$ and $\Delta(b)$ have the same trace, i.e. $\chi(a) = \chi(b)$. Thus the character of every representation is a function of the conjugacy classes.

From now on we assume that the field **K** is the field $\boldsymbol{C}$ of the complex numbers.

If the element a of G is of order n we have

$$\Delta(a^n) = \Delta(a)^n = I_f.$$

By a well-known theorem about matrices this implies that the characteristic roots of $\Delta(a)$ are n-th roots of unity. Since the trace of a matrix is equal to the sum of its characteristic roots we obtain

$$\chi(a) = \varepsilon_1 + \ldots + \varepsilon_f$$

where the ε_i are n-th roots of unity. This gives $\varepsilon_i\bar{\varepsilon}_i = 1$ or $\varepsilon_i^{-1} = \bar{\varepsilon}_i$. Now the matrix $\Delta(a^{-1})$ is the inverse of $\Delta(a)$ and hence the characteristic roots of $\Delta(a^{-1})$ are the numbers $\varepsilon_i^{-1} = \bar{\varepsilon}_i$, $i = 1, \ldots, n$. It follows that

$$\chi(a^{-1}) = \varepsilon_1^{-1} + \ldots + \varepsilon_f^{-1} = \bar{\varepsilon}_1 + \ldots + \bar{\varepsilon}_f = \overline{\chi(a)}. \tag{11}$$

So far the character χ has been defined for the elements of G. In a natural way this definition can be extended to arbitrary elements of the group algebra of G over $\boldsymbol{C}$: For an arbitrary element

$$g = \sum_{x \in G} \gamma_x x$$

of the group algebra we set

$$\chi(g) = \sum_{x \in G} \gamma_x \chi(x).$$

By Theorem 10.7.7, there are t absolutely irreducible representations

$$\Delta_1, \ldots, \Delta_t$$

of G whose degrees and characters we denote by $f_1, \ldots, f_t$ and $\chi_1, \ldots, \chi_t$, respectively.

We observe that by the Theorems 10.3.7 and 10.3.8 every absolutely irreducible representation of the group algebra is a representation of a single direct summand in the direct decomposition (8) whereas all other direct summands are represented by zero matrices. We may choose the notation such that Δ_l is a representation of $e_l\boldsymbol{C}G$.

Since the class sums belong to the centre of the group algebra it follows from (iv) of Theorem 10.7.7 that

$$\Delta_l(k_i) = \alpha_l(k_i)\, I \tag{12}$$

where I stands for identity matrix of degree f_l. This gives

$$\chi_l(k_i) = f_l \alpha_l(k_i). \tag{13}$$

The conjugacy class K_i contains g_i elements. For any $a_i \in K_i$ we therefore obtain

$$\chi_l(a_i) = \frac{1}{g_i}\chi_l(k_i) = \frac{f_l}{g_i}\alpha_l(k_i). \tag{14}$$

If any representation Δ contains the absolutely irreducible constituents $\Delta_1, \ldots, \Delta_t$ with the multiplicities $m_1, \ldots, m_t$, respectively, then we write

$$\Delta \sim m_1 \Delta_1 \dotplus \ldots \dotplus m_t \Delta_t.$$

For the character χ of Δ we thus obtain

$$\chi = m_1\chi_1 + \ldots + m_t\chi_t.$$

We now consider the regular representation P of G and denote its character by ϱ. Denoting the column of the $|G| = g$ elements of G by $\boldsymbol{g}$ we obtain for any $a \in G$

$$\boldsymbol{g}a = P(a)\,\boldsymbol{g}$$

where $P(a)$ is a $g \times g$ permutation matrix, i.e. a matrix which has in each row and in each column precisely one entry 1 and zeros elsewhere. To find the trace of $P(a)$ we have to count how often the equation $xa = x$ is satisfied for an element x of G. For $a \neq e$, this equation is not satisfied for any $x \in G$, whereas $xe = x$ for every $x \in G$. So we find the following values of the character ϱ of P:

$$\varrho(a) = \begin{cases} g & \text{if } a = e, \\ 0 & \text{if } a \neq e. \end{cases}$$

By Theorem 10.7.7, we have

$$P \sim f_1 \Delta_1 \dotplus \dots \dotplus f_t \Delta_t$$

which gives

$$\varrho(a) = \sum_{l=1}^{t} f_l \chi_l(a) = \begin{cases} g & \text{if } a = e, \\ 0 & \text{if } a \neq e. \end{cases}$$

In view of (7), (10), and (13) it follows from the last equations that

$$\sum_{l=1}^{t} \chi_l(k_i)\, \chi_l(k_j^{(-1)}) = \sum_{l=1}^{t} f_l^2 \alpha_l(k_i)\, \alpha_l(k_j^{(-1)})$$

$$= \sum_{l=1}^{t} f_l^2 \sum_{m=1}^{t} \gamma_{ijm} \alpha_l(k_m)$$

$$= \sum_{m=1}^{t} \gamma_{ijm} \sum_{l=1}^{t} f_l \chi_l(k_m)$$

$$= \gamma_{ij1} g = g_i g\, \delta_{ij}\,.$$

If a_i is any element of the conjugacy class K_i, then we can use (14) to obtain

$$\sum_{l=1}^{t} \chi_l(a_i)\, \chi_l(a_j^{-1}) = \frac{g}{g_i}\, \delta_{ij}$$

or, by (11),

$$\sum_{l=1}^{t} \chi_l(a_i)\, \overline{\chi_l(a_j)} = \frac{g}{g_i}\, \delta_{ij}.$$

Multiplication of the last equation by $\sqrt{g_i/g}\,\sqrt{g_j/g}$ gives

$$\sum_{l=1}^{t} \sqrt{g_i/g}\; \chi_l(a_i)\, \sqrt{g_j/g}\; \overline{\chi_l(a_j)} = \delta_{ij}\,. \tag{15}$$

The *character matrix* is defined as the $t \times t$ matrix

$$X = \begin{bmatrix} \sqrt{g_1/g}\,\chi_1(a_1) \ldots \sqrt{g_1/g}\,\chi_t(a_1) \\ \cdots\cdots\cdots\cdots\cdots\cdots\cdots\cdots \\ \sqrt{g_t/g}\,\chi_1(a_t) \ldots \sqrt{g_t/g}\,\chi_t(a_t) \end{bmatrix}.$$

If X^* stands for the complex conjugate transpose of X the equation

$$XX^* = I_t \tag{16}$$

is only another form of writing the t^2 equations (15).

Since every matrix commutes with its inverse, (16) implies that

$$X^*X = I_t$$

or, in a more explicit notation

$$\sum_{j=1}^{t} \overline{\sqrt{g_j/g}\,\chi_i(a_j)}\,\sqrt{g_j/g}\,\chi_m(a_j) = \delta_{im},$$

i.e.

$$\frac{1}{g}\sum_{j=1}^{t} g_j\,\overline{\chi_i(a_j)}\,\chi_m(a_j) = \delta_{im}. \tag{17}$$

The equations (15) and (17) are called the *orthogonality relations* of the group characters.

From (16) we conclude that the determinant of the matrix X is distinct from zero. Therefore, the same is true for the matrix

$$\begin{bmatrix} \chi_1(a_1) \;\ldots\; \chi_t(a_1) \\ \cdots\cdots\cdots\cdots\cdots \\ \chi_1(a_t) \;\ldots\; \chi_t(a_t) \end{bmatrix} \tag{18}$$

so that the columns of this matrix are linearly independent.

10.7.8. THEOREM. *Two representations of a finite group G over the field of the complex numbers are equivalent if and only if their characters coincide.*

Proof. We observed above that equivalent representations have the same character.

Conversely let Δ and Δ' be two representations of G whose characters χ and χ' coincide. If

$$\Delta \sim m_1 \Delta_1 \dotplus \ldots \dotplus m_t \Delta_t, \quad \Delta' \sim m_1' \Delta_1 \dotplus \ldots \dotplus m_t' \Delta_t$$

indicate the multiplicities of the absolutely irreducible constituents of Δ and Δ', respectively, then we have

$$\chi = m_1\chi_1 + \ldots + m_t\chi_t, \quad \chi' = m_1'\chi_1 + \ldots + m_t'\chi_t.$$

From $\chi = \chi'$ we conclude that

$$0 = (m_1 - m_1')\,\chi_1(a) + \ldots + (m_t - m_t')\,\chi_t(a)$$

for every $a \in G$. This contradicts the linear independence of the columns of the matrix (18) unless $m_i = m_i'$ for $i = 1, \ldots, t$. Therefore Δ and Δ' are equivalent as was to be proved.

There is even an explicit formula for computing the multiplicities of the absolutely irreducible constituents of a representation in terms of its character.

Since the value of a character is the same for all elements in one and the same conjugacy class it follows from (17) that

$$\sum_{x \in G} \overline{\chi_i(x)}\,\chi_m(x) = g\,\delta_{im}. \tag{19}$$

Now let

$$\Delta \sim m_1 \Delta_1 \dotplus \ldots \dotplus m_t \Delta_t$$

be any representation of G and let χ denote its character so that

$$\chi = m_1\chi_1 + \ldots + m_t\chi_t.$$

From (19) we obtain

$$\sum_{x \in G} \chi(x)\,\overline{\chi_i(x)} = \sum_{j=1}^{t} m_j \sum_{x \in G} \chi_j(x)\,\overline{\chi_i(x)} = g m_i$$

or

$$m_i = \frac{1}{g} \sum_{x \in G} \chi(x)\,\overline{\chi_i(x)}$$

which is the desired formula.

The equations (10) can be written as follows

$$\sum_{m=1}^{t} \left(\gamma_{ijm} - \delta_{im}\alpha_l(k_j^{(-1)})\right) \alpha_l(k_m) = 0.$$

Here we keep the subscript j fixed and consider these equations for $i = 1, \ldots, t$. This gives t linear homogeneous equations for the numbers $\alpha_l(k_1), \ldots, \alpha_l(k_t)$. By (9), not all of these numbers are equal to zero. Therefore the determinant of the coefficients of the linear equations must be equal to zero, i.e.

$$\det\left(\gamma_{ijm} - \delta_{im}\alpha_l(k_j^{(-1)})\right) = 0 \quad (j \text{ fixed}).$$

This means that $\alpha_l(k_j^{(-1)})$ is a characteristic root of the matrix

$$[\gamma_{ijm}] \quad (j \text{ fixed}).$$

Since the γ_{ijm} are integers it follows that the numbers $\alpha_l(k_j^{(-1)})$ are algebraic integers.

For $i = m$, (17) gives

$$\sum_{j=1}^{t} \overline{\chi_i(a_j)}\, g_j \chi_i(a_j) = g$$

and hence, by (13),

$$\sum_{j=1}^{t} \overline{\chi_i(a_j)}\, f_i \alpha_i(k_j) = g$$

or

$$\sum_{j=1}^{t} \overline{\chi_i(a_j)}\, \alpha_i(k_j) = \frac{g}{f_i}\,. \tag{20}$$

The $\chi_i(a_j)$ are sums of roots of unity and hence algebraic integers. As we observed above, the $\alpha_i(k_j)$ are also algebraic integers. Therefore the left-hand side of (20) is an algebraic integer. Hence, g/f_i is a rational number which is an algebraic integer which means that g/f_i is a natural number. Thus we obtain the following

10.7.9. Theorem. *The degree of any absolutely irreducible representation of a finite group G in the field of the complex numbers divides the order of G.*

Exercises

1. Let $\mathbf{T}$ be the algebra of all triangular $n \times n$ matrices with elements in a field. Find the radical of $\mathbf{T}$.

2. Let m be a natural number.
   (i) Find a necessary and sufficient condition on m for $\mathbf{Z}/(m)$ to be semi-simple.
   (ii) In case the condition in (i) is satisfied, represent $\mathbf{Z}/(m)$ as a direct sum of fields. (Cf. § 2.7.)
   (iii) In case the condition in (i) is not satisfied, find the radical of $\mathbf{Z}/(m)$.
3. Let $\mathbf{F}$ be any field and $f(x)$ a polynomial of positive degree with coefficients in $\mathbf{F}$. Answer the questions of Exercise 2 for $\mathbf{F}[x]/(f(x))$ instead of $\mathbf{Z}/(m)$.
4. Let $\mathbf{F}$ be any field and let $f(x)$ be a polynomial with coefficients in $\mathbf{F}$ that is irreducible in $\mathbf{F}[x]$. If $\mathbf{K}$ denotes the algebraic closure of $\mathbf{F}$, find a necessary and sufficient condition for $\mathbf{K}[x]/(f(x))$ to be semi-simple.
5. Let $\mathbf{A}$ be the complete matrix algebra of degree n over any field $\mathbf{F}$. As we observed in § 10.4, the free $\mathbf{F}$-module R_i with the basis $e_{i1}, \ldots, e_{in}$ is a minimal right ideal in $\mathbf{A}$. Verify that R_i yields $\mathbf{A}$ as its own representation.
6. Find the radical of the group algebra of a cyclic group of prime order p over GF(p).
7. Let $\mathbf{A}$ be an algebra over GF(2) with the basis $1, i, j, k$. Suppose that the rule for multiplication of the basis elements is the same as in the quaternion algebra. Find the radical R of $\mathbf{A}$ and determine $\mathbf{A}/R$.
8. Show that the radical of a commutative ring with minimum condition consists precisely of all nilpotent elements.
9. An element r of a ring $\mathbf{A}$ with minimum condition is said to be properly nilpotent if ra is nilpotent for any $a \in \mathbf{A}$.
   (i) Let r be properly nilpotent. Show that ar is nilpotent for any $a \in \mathbf{A}$.
   (ii) Prove that the radical consists precisely of all properly nilpotent elements.
10. Find a representation of degree 2 of a cyclic group of prime order p that is reducible but not completely reducible.

In the following exercises the term representation always means a representation by matrices whose elements belong to the field of the complex numbers.

11. Let G be a finite group and let G' denote the commutator subgroup of G. Prove that the number of representations of degree 1 of G is equal to the index $|G: G'|$.
12. Find the character matrix of the symmetric group S_3 and its absolutely irreducible representations.
13. Let G_m denote the group algebra of a cyclic group of finite order m over $\boldsymbol{C}$. Find a basis of G_m such that the regular representation splits into m constituents of degree 1. (Hint: Consider expressions in G_m similar to Lagrange resolvents.)
14. (i) Show that the characters of the absolutely irreducible representations of a finite abelian group G form a group C with respect to multiplication (i.e. the product of two characters χ_1 and χ_2 is defined by $\chi_1\chi_2(a) = \chi_1(a)\,\chi_2(a)$ for any $a \in G$).
    (ii) Show that C is isomorphic to G. (Hint: First consider the case that G is cyclic, then use the decomposition of a finite abelian group into a direct product of cyclic groups.)
15. Let G be a finite group with a faithful (i.e. isomorphic) absolutely irreducible representation. Prove that the centre of G is cyclic (or trivial).
16. Any symmetric group S_n, $n \geqq 2$, has two representations of degree 1, namely the identity representation and the so-called alternating representation that assigns 1 or -1 to a permutation according as it is even or odd. Find the generating idempotents of the corresponding ideals in the group algebra.
17. Let χ be the character of a representation of a finite group G and let e denote the identity of G. Prove that all elements a of G such that $\chi(a) = \chi(e)$ form a normal subgroup of G.
18. (i) Let

$$\pi = \begin{pmatrix} i \\ i\pi \end{pmatrix} \quad (i = 1, 2, \ldots, n)$$

be a permutation of the natural numbers $1, 2, \ldots, n$. To π we assign the $n \times n$ matrix

$$M_\pi = [\delta_{i\pi,k}]$$

whose element in the i-th row and k-th column is $\delta_{i\pi,k}$

where $\delta_{j,k} = 0$ for $j \neq k$ and $\delta_{k,k} = 1$. Show that these matrices form a representation of S_n.

(ii) Decompose this representation of S_3 into absolutely irreducible constituents.

19. (i) In the notation of § 10.7 let

$$\Delta \sim m_1 \Delta_1 \dotplus \dots \dotplus m_t \Delta_t, \qquad \Delta' \sim m_1' \Delta_1 \dotplus \dots \dotplus m_t' \Delta_t$$

and let χ and χ' denote the characters of Δ and Δ', respectively. Prove that

$$\sum_{a \in G} \chi'(a) \overline{\chi(a)} = |G| \sum_{r=1}^{t} m_r m_r'.$$

(ii) Show that a character χ of a finite group G belongs to an absolutely irreducible representation if and only if

$$\sum_{a \in G} |\chi(a)|^2 = |G|.$$

20. Prove that the values of all characters of the symmetric group are rational integers. (Hint: If a is an element of order m in some symmetric group S_n, then use Theorem 3.10.4 to show that a and a^k are conjugate in S_n if $(k, m) = 1$. On the other hand, show that, for any character χ of S_n, the algebraic integers $\chi(a)$ and $\chi(a^k)$ are conjugate over $\boldsymbol{Q}$.)

11

Lattices

If A and B are subsets of a given set M then so are $A \cap B$ and $A \cup B$. So the set of all subsets of M can be regarded as an algebraic system with the operations $\cap$ and $\cup$. If A and B are subgroups of a given group G, then $A \cap B$ is again a subgroup of G. Apart from trivial cases, however, the set theoretical union of A and B is not a subgroup of G. But if $A \cup B$ is defined as the subgroup generated by A and B, then the set of all subgroups of G can again be considered as an algebraic system with the operations $\cap$ and $\cup$. The same applies to the set of all normal subgroups of G. Similarly, the subfields of a given field form an algebraic system with the operations $\cap$ and $\cup$ where $A \cup B$ means the subfield generated by the subfields A and B. Other examples of such algebraic systems arise when we consider the set of all subspaces of a given vector space or the set of all ideals of a given ring. This suggests the study of algebraic systems with two operations $\cap$ and $\cup$ which have some but in general not all the properties of the set theoretical intersection and union.

11.1. Lattices and partially ordered sets

Let L be a non-empty set in which there are defined two binary operations $\cap$ and $\cup$.

For $a, b \in L$ we call $a \cap b$ the *intersection* or the *meet* of a and b and we call $a \cup b$ the *union* or the *join* of the two elements.[1])

---

[1]) Read a cap b for $a \cap b$ and a cup b for $a \cup b$.

We emphasize that this notation does not refer to any set theoretical meaning of $\cap$ and $\cup$.

L is called a *lattice* if the following laws are satisfied for arbitrary elements a, b, c of L:

$(L_{1\cap})$ $a \cap b = b \cap a$, $(L_{1\cup})$ $a \cup b = b \cup a$,

$(L_{2\cap})$ $(a \cap b) \cap c = a \cap (b \cap c)$, $(L_{2\cup})$ $(a \cup b) \cup c = a \cup (b \cup c)$,

$(L_{3\cap})$ $a \cap (a \cup b) = a$, $(L_{3\cup})$ $a \cup (a \cap b) = a$.

Example 1. For subsets a, b of a given set let $a \cap b$ and $a \cup b$ denote the set theoretical intersection and union, respectively.

Example 2. For $a, b \in \boldsymbol{N}$ define

$$a \cap b = \text{greatest common divisor of } a \text{ and } b,$$
$$a \cup b = \text{least common multiple of } a \text{ and } b.$$

It is easy to verify that the above six postulates are satisfied.

Example 3. Let a and b be subgroups of a given group. Let $a \cap b$ denote the set theoretical intersection and define $a \cup b$ to mean the subgroup generated by a and b, i.e. the set theoretical intersection of all subgroups containing a and b.

Example 4. If a and b are real numbers define

$$a \cap b = \min(a, b), \quad a \cup b = \max(a, b).$$

Due to the commutative laws (L_1) and the associative laws (L_2) the intersection and the union of finitely many elements $a_1, \ldots, a_n$ of L are unambiguously defined and are denoted by

$$\bigcap_{i=1}^{n} a_i, \quad \bigcup_{i=1}^{n} a_i.$$

The postulates for a lattice do not imply the existence of the intersection or the union of infinitely many elements. A lattice is said to be *complete* if the intersection and the union of arbitrary sets of elements exist.

The above postulates consist of pairs $(L_{i\cap})$ and $(L_{i\cup})$ which are interchanged if we interchange the operations $\cap$ and $\cup$. This yields the following

Principle of duality: If any statement can be deduced from the postulates, then the *dual* statement obtained by interchanging $\cap$ and $\cup$ can also be deduced.

A subset S of a lattice L is called a *sublattice* if $a \cap b \in S$ and $a \cup b \in S$ whenever $a \in S$ and $b \in S$.

A mapping

$$\sigma: a \to a\sigma$$

of a lattice L onto a lattice L^* is called a *homomorphism* if

$$(a \cap b)\, \sigma = a\sigma \cap b\sigma, \qquad (a \cup b)\, \sigma = a\sigma \cup b\sigma$$

for any two elements a, b of L. A bijective homomorphism is called an *isomorphism*.

We shall now derive some simple consequences of the axioms.

$$a \cap a = a, \qquad a \cup a = a, \tag{1}$$

$$a \cap b = a \text{ if and only if } b \cup a = b, \tag{2}$$

$$a \cap b = a \cup b \text{ if and only if } a = b. \tag{3}$$

Since the propositions (1) are dual to each other it is sufficient to prove one of them, e.g. the first one. By ($L_{3\cup}$) we have

$$a \cap a = a \cap [a \cup (a \cap b)].$$

Then ($L_{3\cap}$) with $a \cap b$ instead of b gives

$$a \cap [a \cup (a \cap b)] = a$$

which proves the first equation (1).

Suppose that $b \cup a = b$. By ($L_{1\cup}$) this implies that $a \cup b = b$ and hence, by ($L_{3\cap}$),

$$a \cap b = a \cap (a \cup b) = a.$$

Conversely, if $a \cap b = a$, then

$$a \cup b = b \cup a = b \cup (a \cap b) = b.$$

Finally, if $a = b$, then $a \cup b = a \cup a = a$ and $a \cap b = a \cap a = a$ so that $a \cap b = a \cup b$. Conversely, assume that $a \cap b = a \cup b$. Then we obtain

$$a = a \cup (a \cap b) = a \cup (a \cup b) = (a \cup a) \cup b = a \cup b,$$

$$b = b \cup (a \cap b) = b \cup (a \cup b) = a \cup (b \cup b) = a \cup b$$

and hence $a = b$.

Thus we have proved (1), (2), (3).

In § 1.2 we defined a partially ordered set to be one in which an order relation $\leqq$ is defined.

Let P be a partially ordered set and let S be a non-empty subset of P. An element u of P is called an *upper bound* of S if $s \leqq u$ for every $s \in S$. An upper bound u^* of S is said to be the *least upper bound* (l. u. b.) of S if $u^* \leqq u$ for every upper bound u of S. The l. u. b. need not exist but if it exists then it is obviously uniquely determined. In an analogous way we define *lower bounds* and the *greatest lower bound* (g. l. b.).

In an arbitrary lattice L we define a relation $\leqq$ as follows:

$$a \leqq b \text{ if and only if } a \cap b = a.$$

As usual we also write $b \geqq a$ in case $a \leqq b$. By (2) we also have

$$a \leqq b \text{ if and only if } a \cup b = b.$$

We shall now show that $\leqq$ is an order relation on L, i.e. that this relation is reflexive, transitive, and antisymmetric.

By (1) we have $a \leqq a$ for every $a \in L$. Let $a \leqq b$ and $b \leqq c$ so that $a \cap b = a$ and $b \cap c = b$. Then we obtain

$$a \cap c = (a \cap b) \cap c = a \cap (b \cap c) = a \cap b = a$$

and hence $a \leqq c$. Thus the relation $\leqq$ is transitive. If $a \leqq b$ and $b \leqq a$, then $a \cap b = a$ and $b \cap a = b$ and hence $a = b$ so that our relation is antisymmetric.

The dual statement to $a \cap b = a$ is $a \cup b = a$. Therefore we can extend the principle of duality to the order relation by letting $a \leqq b$ and $a \geqq b$ dually correspond to each other.

Next we prove that $a \cap b$ and $a \cup b$ are the g. l. b. and l. u. b., respectively, of a and b. We have

$$(a \cap b) \cap a = (a \cap a) \cap b = a \cap b$$

which means that $a \cap b \leqq a$. By symmetry it follows that $a \cap b \leqq b$. Therefore $a \cap b$ is a lower bound of a and b. Now let l be any lower bound of a and b. From $l \leqq a$, $l \leqq b$, i.e. $l \cap a = l$, $l \cap b = l$ we then obtain

$$l \cap (a \cap b) = (l \cap a) \cap b = l \cap b = l$$

which means $l \leqq a \cap b$. This shows that $a \cap b$ is the g. l. b. of a and b.

Dually one can prove that $a \cup b$ is the l. u. b. of a and b.

We shall summarize the last statements and their converse as follows:

11.1.1. THEOREM. *In an arbitrary lattice the relation $\leqq$ defined by*

$$a \leqq b \text{ if and only if } a \cap b = a$$

is an order relation. Relative to this order

$$a \cap b = \text{g. l. b. of } a \text{ and } b, \tag{4}$$

$$a \cup b = \text{l. u. b. of } a \text{ and } b.$$

Conversely, let P be a partially ordered set in which the g. l. b. and the l. u. b. of any two elements exist. Then P is a lattice with respect to the operations $\cap$ and $\cup$ defined by (4).

Proof. The first part of the theorem has been established above. It remains to prove the converse part. This requires to show that the operations $\cap$ and $\cup$ satisfy the six postulates ($L_{1\cap}$) through ($L_{3\cup}$).

Since the g. l. b. and the l. u. b. of a and b remain unaltered if we interchange a and b we have

$$a \cap b = b \cap a, \quad a \cup b = b \cup a.$$

For any three elements a, b, c of P let $s = b \cap c$ and $t = a \cap s = a \cap (b \cap c)$. Then we have $s \leqq b$, $s \leqq c$ and $t \leqq a$, $t \leqq s$. This gives $t \leqq a$, $t \leqq b$, $t \leqq c$ so that t is a lower bound of a, b, c. If t_0 is any lower bound of a, b, c, then $t_0 \leqq s$ since t_0 is a lower bound of b and c and s is the g. l. b. of b and c. Further t_0 is a lower bound of a and s and hence $t_0 \leqq t$ because t is the g. l. b. of a and s. This shows that $t = a \cap (b \cap c)$ is the g. l. b. of a, b, and c. In a similar way we can prove that $(a \cap b) \cap c$ is the g. l. b. of a, b, and c. This gives

$$a \cap (b \cap c) = (a \cap b) \cap c.$$

Analogously we can prove that ($L_{2\cup}$) holds.

For $a, b \in P$ let $u = a \cup b$ and $v = a \cap u = a \cap (a \cup b)$. Then $a \leqq u$, $b \leqq u$ and $v \leqq a$, $v \leqq u$. Now a is a lower bound of a and u and since v is the g. l. b. of a and u it follows that $a \leqq v$. Due to

$v \leqq a$ we obtain

$$a = v = a \cap (a \cup b)$$

which is ($L_{3\cap}$). In a similar way ($L_{3\cup}$) can be established. This completes the proof.

As a corollary we obtain by induction

$$\bigcap_{i=1}^{n} a_i = \text{g. l. b. of } a_1, a_2, \ldots, a_n;$$

$$\bigcup_{i=1}^{n} a_i = \text{l. u. b. of } a_1, a_2, \ldots, a_n.$$

The following rules are another immediate consequence of the last theorem:

If $a \geqq b$, then

$$a \cap c \geqq b \cap c \quad \text{and} \quad a \cup c \geqq b \cup c.$$

We now derive two inequalities which hold in every lattice. From

$$b \cup c \geqq b$$

it follows that

$$a \cap (b \cup c) \geqq a \cap b$$

and by symmetry we obtain

$$a \cap (b \cup c) \geqq a \cap c.$$

This gives

$$a \cap (b \cup c) \geqq (a \cap b) \cup (a \cap c). \tag{5}$$

The principle of duality yields

$$a \cup (b \cap c) \leqq (a \cup b) \cap (a \cup c). \tag{6}$$

In case $a \geqq c$ we have $a \cap c = c$ so that we obtain the following consequence of (5):

$$\text{If } a \geqq c, \text{ then } a \cap (b \cup c) \geqq (a \cap b) \cup c. \tag{7}$$

If a lattice L contains an element 0 such that $0 \leqq a$ for every $a \in L$, then 0 is called the *null element* of L. It is evident that the null element if it exists is unique. Dually, an element 1 of L such that $1 \geqq a$ for every $a \in L$ is called the *all element*. Again, a lattice need not contain an all element but if an all element exists, then it is unique. If 0 and 1 exist, then we obviously have

$$0 \wedge a = 0, \quad 0 \vee a = a,$$

$$1 \wedge a = a, \quad 1 \vee a = 1$$

for every element a of the lattice.

A finite lattice is one that contains finitely many elements. It is evident that a finite lattice L contains 0 and 1, namely

$$0 = \bigwedge_{a \in L} a, \qquad 1 = \bigvee_{a \in L} a.$$

We introduce the notation $a < b$ to mean that $a \leqq b$ and $a \neq b$. Instead of $a < b$ we also write $b > a$. We say that b *covers* a if there is no element x such that $a < x < b$.

Finite lattices can expediently be described by diagrams. For a finite lattice L such a diagram is constructed as follows. To every element of L we assign a small circle or a dot in a plane. If $a, b \in L$ and $a < b$, then the circle assigned to b is placed above the circle corresponding to a. Further, the circles representing a and b, respectively, are connected by a line in case b covers a. Thus we have $a < b$ if and only if there is an ascending (possibly broken) line from a to b. We give some examples of such diagrams

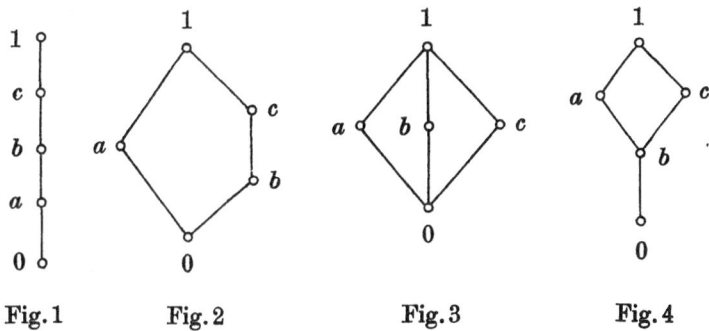

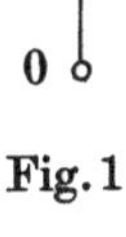

Fig. 1

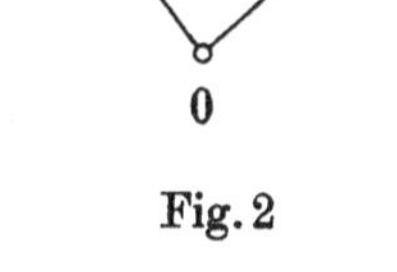

Fig. 2

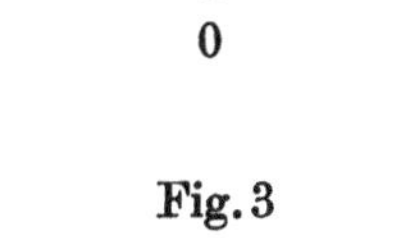

Fig. 3

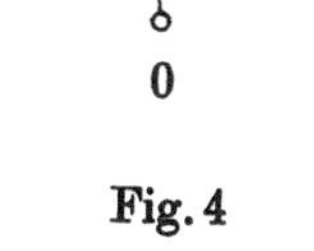

Fig. 4

11.2. Modular lattices

A lattice L is called *modular* if

$$a \geqq c \text{ implies that } a \wedge (b \vee c) = (a \wedge b) \vee c \tag{1}$$

for every $b \in L$.

Not every lattice is modular. For example, the lattice of fig. 2 is not modular, because $c > b$ but

$$c \cap (a \cup b) = c \neq b = (c \cap a) \cup b .$$

This lattice is even characteristic for non-modular lattices.

11.2.1. THEOREM. *A lattice is modular if and only if it contains no sublattice isomorphic to that of fig. 2.*

Proof. If a lattice contains a sublattice isomorphic to that of fig. 2 then it is obviously not modular.

Conversely, if a lattice is not modular, then there exist elements a, b, c such that $a > c$ and $a \cap (b \cup c) \neq (a \cap b) \cup c$. From (7) of § 11.1 it follows that

$$a \cap (b \cup c) > (a \cap b) \cup c .$$

Therefore the lattice contains the sublattice

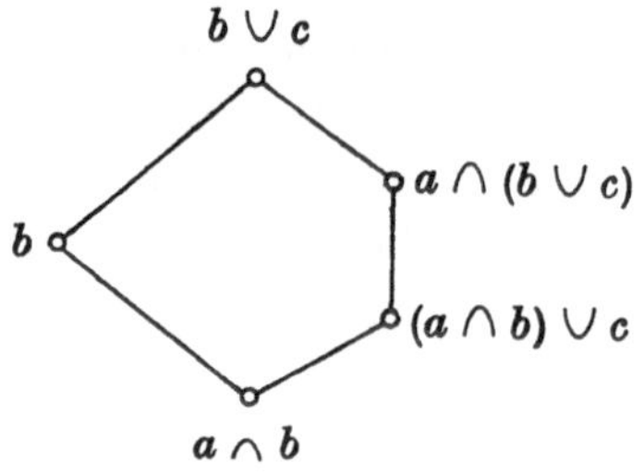

This completes the proof.

As a corollary we obtain a sufficient condition for a lattice to be modular.

11.2.2. COROLLARY. *If for any elements a, b, c of a lattice*

$$a \cap b = a \cap c \quad \text{and} \quad a \cup b = a \cup c$$

imply that $b = c$, then the lattice is modular.

Proof. If the lattice were not modular, then it would contain a sublattice isomorphic to that of fig. 2 in which

$$a \cap b = a \cap c \quad \text{and} \quad a \cup b = a \cup c$$

but $b \neq c$.

The modular lattice of fig. 3 shows that the converse of the corollary does not hold. However, we can prove the following theorem.

11.2.3. *Let a, b, c be elements of a modular lattice such that $b \geqq c$. Then*

$$a \cap b = a \cap c \quad and \quad a \cup b = a \cup c$$

imply that $b = c$.

Proof.

$$b = b \cap (a \cup b) = b \cap (a \cup c) = (b \cap a) \cup c = (c \cap a) \cup c = c.$$

For subgroups U, V of an arbitrary group G we define $U \cap V$ to mean the set theoretical intersection and $U \cup V = \langle U, V \rangle$, the subgroup generated by U and V. As we observed above we thus obtain a lattice, the so-called subgroup lattice of G. If U and V are normal subgroups of G, then so are $U \cap V$ and $U \cup V$. Therefore the subgroup lattice of G contains a sublattice consisting of all normal subgroups of G.

11.2.4. *Let U, V, W be subgroups of an arbitrary group such that $U \subseteq W$ and $UV = VU$. Then*

$$U \cup (V \cap W) = (U \cup V) \cap W.$$

Proof. By (7) of § 11.1 we have

$$U \cup (V \cap W) \subseteq (U \cup V) \cap W.$$

On the other hand, let w be an arbitrary element of $(U \cup V) \cap W$. Since $w \in U \cup V$ and $UV = VU$ the element w can be expressed in the form $w = uv$, $u \in U$, $v \in V$. From $U \subseteq W$ we see that $u^{-1}w = v$ belongs to W so that $v \in V \cap W$. Therefore $w = uv$ is an element of $U \cup (V \cap W)$. This proves our proposition.

Since $UV = VU$ holds for any two normal subgroups U, V we obtain the following result.

11.2.5. Theorem. *The lattice of the normal subgroups of any group is modular.*

We now introduce a notion which plays an important role in the study of modulare lattices.

If a and b are elements of a lattice L such that $a \geqq b$, then the set

$$[a, b] = \{x : x \in L \text{ and } a \geqq x \geqq b\}$$

is called an *interval*. It is easily verified that $[a, b]$ is a sublattice of L.

11.2.6. THEOREM. *For any two elements a, b of a modular lattice the intervals $[a \cup b, a]$ and $[b, a \cap b]$ are isomorphic.*

Proof. We consider the mapping

$$\sigma : x \to b \cap x .$$

If $a \cup b \geqq x \geqq a$, then

$$b = b \cap (a \cup b) \geqq b \cap x \geqq b \cap a$$

so that σ is a mapping of $[a \cup b, a]$ into $[b, a \cap b]$. Let x, $x' \in [a \cup b, a]$ and $b \cap x = b \cap x'$. Then we obtain

$$a \cup (b \cap x) = a \cup (b \cap x') .$$

Since the lattice is modular and $x \geqq a$ as well as $x' \geqq a$ this gives

$$(a \cup b) \cap x = (a \cup b) \cap x' .$$

In view of $x \leqq a \cup b$ and $x' \leqq a \cup b$ the last equation yields $x = x'$. Hence, σ is injective. Now let $y \in [b, a \cap b]$, i.e.

$$b \geqq y \geqq a \cap b .$$

Forming the union with a we obtain

$$a \cup b \geqq a \cup y \geqq a \cup (a \cap b) = a$$

which means that $a \cup y \in [a \cup b, a]$. Since the lattice is modular and $b \geqq y$ we find

$$b \cap (a \cup y) = (b \cap a) \cup y .$$

The element on the right-hand side is equal to y because $y \geqq a \cap b$. This shows that every element of $[b, a \cap b]$ is the image of a suitable element of $[a \cup b, a]$ under σ, i.e. σ is a bijective mapping of $[a \cup b, a]$ onto $[b, a \cap b]$. Finally, σ is order preserving and, hence, is a lattice isomorphism of $[a \cup b, a]$ onto $[b, a \cap b]$, as required.

The last theorem can be regarded as the lattice theoretical analogue of Theorem 3.7.4 (I. Isomorphism Theorem). There are further analogues of theorems on groups, e.g. of the theorems of Schreier and Jordan-Hölder (Theorems 3.8.1 and 3.8.2). To formulate these theorems we introduce a suitable notion of equivalence for intervals.

Two intervals $[p, q]$ and $[r, s]$ of a lattice L are said to be *similar* if there exist elements a and b of L such that one of the intervals is $[a \cup b, a]$ and the other is $[b, a \cap b]$. The intervals $[p, q]$ and $[r, s]$ are called *projective* if there exists a finite sequence of intervals

$$[p, q] = [p_1, q_1], [p_2, q_2], \ldots, [p_n, q_n] = [r, s]$$

such that each pair $[p_i, q_i]$, $[p_{i+1}, q_{i+1}]$ is similar. It follows from Theorem 11.2.6 that projective intervals of a modular lattice are isomorphic.

First we prove an analogue of Zassenhaus' Lemma.

11.2.7. *If* x_1, x_2, y_1, y_2 *are elements of a modular lattice such that* $x_1 \geqq x_2$, $y_1 \geqq y_2$, *then the intervals*

$$[(x_1 \cap y_1) \cup x_2, (x_1 \cap y_2) \cup x_2] \tag{2}$$

and

$$[(x_1 \cap y_1) \cup y_2, (x_2 \cap y_1) \cup y_2] \tag{3}$$

are projective.

Proof. Let

$$a = (x_1 \cap y_2) \cup x_2, \quad b = x_1 \cap y_1.$$

Then

$$a \cup b = (x_1 \cap y_2) \cup x_2 \cup (x_1 \cap y_1) = x_2 \cup (x_1 \cap y_1)$$

because $x_1 \cap y_2 \leqq x_1 \cap y_1$. Further, by (1),

$$\begin{aligned} a \cap b &= [(x_1 \cap y_2) \cup x_2] \cap (x_1 \cap y_1) = (x_1 \cap y_2) \cup [x_2 \cap (x_1 \cap y_1)] \\ &= (x_1 \cap y_2) \cup (x_2 \cap y_1). \end{aligned}$$

Therefore the interval (2) is equal to $[a \cup b, a]$ and, hence, is similar to

$$[b, a \cap b] = [x_1 \cap y_1, (x_1 \cap y_2) \cup \; x_2 \cap y_1)]. \tag{4}$$

The interval (3) is obtained from (2) by interchanging x_i and y_i ($i = 1, 2$). The right-hand side of (4) is invariant under this substitution which shows that the intervals (3) and (4) are similar. Therefore (2) and (3) are projective as was to be proved.

Let a and b be elements of a modular lattice such that $a \geqq b$. We consider finite descending chains

$$a = a_0 \geqq a_1 \geqq \ldots \geqq a_m = b \tag{5}$$

connecting a and b. If any two elements a_i, a_{i+1} are distinct, then we speak of a chain without repetitions. A chain

$$a = a_0' \geqq a_1' \geqq \ldots \geqq a_n' = b$$

is said to be *equivalent* to (5) if $m = n$ and if it is possible to establish a one-to-one correspondence between the intervals $[a_i, a_{i+1}]$ and $[a_j', a_{j+1}']$ such that corresponding intervals are projective. Two equivalent chains connecting a and b with repetitions obviously yield two equivalent chains without repetitions if we remove the repetitions by cancelling suitable elements in both chains.

By a *refinement* of (5) we understand (5) itself and every finite descending chain connecting a and b that is obtained from (5) by inserting additional terms. A proper refinement of (5) is one that contains at least one new element, distinct from all the a_i.

We now use 11.2.7 to prove an analogue of Schreier's Refinement Theorem (Theorem 3.8.1).

11.2.8. Theorem. *If a and b are elements of a modular lattice such that $a \geqq b$, then any two finite descending chains connecting a and b have equivalent refinements.*

Proof. Let

$$a = a_0 \geqq a_1 \geqq \ldots \geqq a_r = b,$$

$$a = b_0 \geqq b_1 \geqq \ldots \geqq b_s = b$$

be any two finite descending chains connecting a and b. We refine these chains as follows. Between a_{i-1} and a_i we insert the $s - 1$ elements

$$a_{i,k} = (a_{i-1} \cap b_k) \cup a_i \qquad (k = 1, \ldots, s - 1)$$

and between b_{k-1} and b_k we insert the $r - 1$ elements

$$b_{i,k} = (b_{k-1} \cap a_i) \cup b_k \qquad (i = 1, \ldots, r - 1).$$

It is easily verified that the chains thus obtained are descending. By 11.2.7, the intervals

$$[a_{i,k-1}, a_{i,k}] = [(a_{i-1} \cap b_{k-1}) \cup a_i, (a_{i-1} \cap b_k) \cup a_i]$$

and

$$[b_{i-1,k}, b_{i,k}] = [(b_{k-1} \cap a_{i-1}) \cup b_k, (b_{k-1} \cap a_i) \cup b_k]$$

are projective as was to be proved.

A finite descending chain

$$a = a_0 > a_1 > \ldots > a_t = b \tag{6}$$

is called a *composition chain* between a and b if each a_i covers a_{i+1}.

If a and b are arbitrary elements of a modular lattice such that $a > b$, then there need not exist composition chains between a and b. But if such chains exist, then there holds a uniqueness theorem analogous to Jordan-Hölder's Theorem in the theory of groups (Theorem 3.8.2).

11.2.9. THEOREM. *Any two composition chains between two given elements of a modular lattice are equivalent.*

Proof. By Theorem 11.2.8, two composition chains between the same elements possess equivalent refinements. On the other hand, however, they do not permit any proper refinement so that they are equivalent without any insertion.

The common number t of the intervals in all composition chains between a and b is called the *length* of the interval $[a, b]$.

Assume that L is a modular lattice with 1 and 0 so that L coincides with the interval $[1, 0]$. In case the length of $[1, 0]$ is finite, L is said to be of finite length and the length of $[1, 0]$ is also called the length or the *dimension* of L.

Let L be a modular lattice of finite length. If a is any element of L, then it is clear that the interval $[a, 0]$ is also of finite length. The length of $[a, 0]$ is called the *length* or the *dimension* of a and is denoted by $l(a)$. If $a, b \in L$ and $a > b$, then it is evident that

$$l(a) = l(b) + \text{length of } [a, b].$$

In particular, for any two elements x, y of L

$$l(x \cup y) - l(x) = \text{length of } [x \cup y, x],$$

$$l(y) - l(x \cap y) = \text{length of } [y, x \cap y].$$

By Theorem 11.2.6, the intervals $[x \cup y, x]$ and $[y, x \cap y]$ are isomorphic so that they have the same length. This gives the following result.

11.2.10. THEOREM. *For any two elements x, y of a modular lattice of finite length*

$$l(x) + l(y) = l(x \cup y) + l(x \cap y).$$

11.3. Distributive lattices

A lattice L is called *distributive* if

$$a \cap (b \cup c) = (a \cap b) \cup (a \cap c) \tag{1}$$

and

$$a \cup (b \cap c) = (a \cup b) \cap (a \cup c) \tag{2}$$

for any three elements a, b, c of L.

It is easy to see that each of the conditions (1) and (2) implies the other. Suppose that (1) is satisfied. Then we obtain

$$\begin{aligned}(a \cup b) \cap (a \cup c) &= [(a \cup b) \cap a] \cup [(a \cup b) \cap c] = a \cap [(a \cup b) \cap c] \\ &= a \cup [(a \cap c) \cup (b \cap c)] \\ &= [a \cup (a \cap c)] \cup (b \cap c) = a \cup (b \cap c)\end{aligned}$$

which shows that (2) holds. Dually it follows that (2) implies (1).

Not every lattice is distributive, e.g. the lattice of fig. 3 in § 11.1 is not.

11.3.1. *Every distributive lattice is modular.*

Proof. If $a \geqq c$, then $a \cap c = c$, hence

$$a \cap (b \cup c) = (a \cap b) \cup (a \cap c) = (a \cap b) \cup c.$$

However, there are lattices that are modular but not distributive, e.g. the lattice of fig. 3. Theorem 11.2.1 gives a characterization of modular lattices by the non-existence of sublattices of a certain type. Our next aim is to derive a similar criterion for a lattice to be distributive.

The inequality

$$(a \cap b) \cup (b \cap c) \cup (c \cap a) \leqq (a \cup b) \cap (b \cup c) \cap (c \cup a) \tag{3}$$

holds in every lattice. Indeed, we have

$$a \cap b \leqq a \cup b, \quad a \cap b \leqq b \cup c, \quad a \cap b \leqq c \cup a.$$

Therefore $a \cap b$ is less than or equal to the right-hand side of (3). By symmetry, the same applies to $b \cap c$ and $c \cap a$ which shows that (3) holds.

11.3.2. *A lattice is distributive if and only if*

$$(a \cap b) \cup (b \cap c) \cup (c \cap a) = (a \cup b) \cap (b \cup c) \cap (c \cup a) \tag{4}$$

for any three elements a, b, c.

Proof. Suppose that (4) holds. If $a \geqq c$, then $a \cap c = c$ and $a \cup c = a$. Therefore (4) gives

$$(a \cap b) \cup (b \cap c) \cup c = (a \cup b) \cap (b \cup c) \cap a$$

or

$$(a \cap b) \cup c = a \cap (b \cup c)$$

which shows that the lattice is modular. If we take the union of a with both sides of (4) we obtain on the left-hand side

$$a \cup (b \cap c).$$

Since the lattice is modular the right-hand side becomes

$$\begin{aligned} & a \cup \{(a \cup b) \cap (b \cup c) \cap (c \cup a)\} \\ &= a \cup \{[(a \cup b) \cap (b \cup c)] \cap (c \cup a)\} \\ &= \{a \cup [(a \cup b) \cap (b \cup c)]\} \cap (c \cup a) \\ &= \{[a \cup (b \cup c)] \cap (a \cup b)\} \cap (c \cup a) \\ &= (a \cup b) \cap (c \cup a). \end{aligned}$$

Therefore the lattice is distributive. Conversely, assume that

$$x \cup (y \cap z) = (x \cup y) \cap (x \cup z). \tag{5}$$

Taking $x = (a \cap b) \cup (c \cap a)$, $y = b$, $z = c$ the left-hand side of (5) becomes

$$(a \cap b) \cup (b \cap c) \cup (c \cap a).$$

Further

$$x \cup y = (a \cap b) \cup (c \cap a) \cup b = (c \cap a) \cup b = (b \cup c) \cap (a \cup b),$$

$$x \cup z = (a \cap b) \cup (c \cap a) \cup c = (a \cap b) \cup c = (b \cup c) \cap (c \cup a)$$

so that the right-hand side of (5) is equal to

$$(a \cup b) \cap (b \cup c) \cap (c \cup a)$$

which shows that (4) is satisfied. This completes the proof.

11.3.3. THEOREM. *A modular lattice is distributive if and only if it contains no sublattice isomorphic to the lattice of fig. 3.*

Proof. If there exists such a sublattice, then it is clear that the lattice is not distributive.

Conversely, assume that the lattice is not distributive. Then it follows from (3) and 11.3.2 that there exist elements a, b, c such that

$$p = (a \cap b) \cup (b \cap c) \cup (c \cap a) < (a \cup b) \cap (b \cup c) \cap (c \cup a) = q.$$

We find that

$$q \cap a = a \cap (b \cup c), \qquad q \cap b = b \cap (c \cup a).$$

Let

$$x = (q \cap a) \cup p, \quad y = (q \cap b) \cup p, \quad z = (q \cap c) \cup p.$$

We obtain

$$\begin{aligned} x \cup y &= [a \cap (b \cup c)] \cup p \cup [b \cap (c \cup a)] \\ &= [a \cap (b \cup c)] \cup [b \cap (c \cup a)] \end{aligned}$$

because

$$\begin{aligned} a \cap b &\leqq a \cap (b \cup c), \\ b \cap c &\leqq b \cap (c \cup a), \\ c \cap a &\leqq a \cap (b \cup c). \end{aligned}$$

Since the lattice is modular it follows that

$$\begin{aligned} x \cup y &= \{[a \cap (b \cup c)] \cup b\} \cap (c \cup a) \\ &= \{(b \cup c) \cap (a \cup b)\} \cap (c \cup a) = q. \end{aligned}$$

Similarly, we find that

$$x \cup z = q, \qquad y \cup z = q.$$

In view of $p < q$ and the modular property of the lattice we can also write

$$x = q \cap (a \cup p), \quad y = q \cap (b \cup p), \quad z = q \cap (c \cup p).$$

Hence the dual argument to the foregoing gives

$$x \cap y = x \cap z = y \cap z = p.$$

This shows that the elements p, q, x, y, and z form a sublattice isomorphic to the lattice of fig. 3 as was to be proved.

As a corollary of the last theorem and of Theorem 11.2.1 we obtain the following characterization of distributive lattices:

11.3.4. THEOREM. *A lattice is distributive if and only if it does not contain a sublattice isomorphic either to the lattice of fig. 2 or to that of fig. 3.*

In case of distributive lattices the Corollary 11.2.2 and 11.2.3 can be sharpened as follows:

11.3.5. THEOREM. *A lattice is distributive if and only if*

$$a \cap b = a \cap c \ \textit{and} \ a \cup b = a \cup c$$

imply that $b = c$.

Proof. If the lattice is not distributive, then it contains a sublattice isomorphic to the lattice of fig. 2 or of fig. 3. In both cases we have $a \cap b = a \cap c$ and $a \cup b = a \cup c$ but $b \neq c$.

Conversely, assume that the lattice is distributive and $a \cap b = a \cap c$, $a \cup b = a \cup c$. Then we have

$$b = b \cup (a \cap b) = b \cup (a \cap c) = (b \cup a) \cap (b \cup c)$$

$$= (a \cup c) \cap (b \cup c) = (a \cap b) \cup c = (a \cap c) \cup c = c$$

as was to be proved.

As is well known the distributive laws (1) and (2) are satisfied in case a, b, c are subsets of a given set and $\cap$, $\cup$ have their set theoretical meaning. It is a remarkable result that conversely every distributive lattice is isomorphic to some lattice of subsets.

11.3.6. Birkhoff's Theorem. *Every distributive lattice is isomorphic to a lattice whose elements are subsets of a certain set and in which $\cap$ and $\cup$ denote the set theoretical intersection and union, respectively.*

To prove this theorem we have to introduce some new notions.

Let L be any lattice. A non-empty subset A of L is called an *ideal* if A satisfies the following conditions:

$$\text{If } a_1, a_2 \in A, \text{ then } a_1 \cap a_2 \in A. \tag{6}$$

$$\text{If } a \in A, \text{ then } a \cup x \in A \text{ for every } x \in L.^{1)} \tag{7}$$

Since $a \cup x = x$ if and only if $a \leqq x$ it is clear that (7) is equivalent to the following condition:

$$\text{If } a \in A \text{ and } x \geqq a, \text{ then } x \in A. \tag{7'}$$

If a is an arbitrary element of L, then the set

$$(a) = \{x : x \in L \text{ and } x \geqq a\}$$

is an ideal. It is called the *principal ideal* generated by a. Indeed, if $x_1 \in (a)$ and $x_2 \in (a)$, that is $x_1 \geqq a$ and $x_2 \geqq a$, then $x_1 \cap x_2 \geqq a$ so that $x_1 \cap x_2 \in (a)$. Further, if $x \in (a)$ and $y \in L$, then $x \cup y \geqq x \geqq a$ which means that $x \cup y \in (a)$. In view of (7'), every ideal that contains a also contains (a).

If A and B are ideals in L, let $\langle A, B\rangle$ denote the ideal generated by A and B, i.e. the set theoretical intersection of all ideals that contain A as well as B.

11.3.7. Lemma. *The following statements are equivalent:*

(i) $x \in \langle A, B\rangle$,

(ii) *There exist $a \in A$ and $b \in B$ such that $x \geqq a \cap b$.*

Proof. Let X be the set of all $x \in L$ with the following property: there exist $a \in A$ and $b \in B$ such that $x \geqq a \cap b$.

If a and b are arbitrary elements of A and B, respectively, then $a \geqq a \cap b$ and $b \geqq a \cap b$, so that $a \in X$ and $b \in X$. Hence $A \subseteq X$, $B \subseteq X$.

---

1) It is evident that the notion of an ideal does not coincide with its dual notion. This suggests to use the term intersection-ideal. But since the dual of an intersection-ideal, which would then be called a union-ideal, does not occur in our arguments we shall simply speak of an ideal. The term ideal refers to a rather formal analogy to ideals in rings.

If $x_1, x_2 \in X$, then $x_1 \geqq a_1 \cap b_1$ and $x_2 \geqq a_2 \cap b_2$ where $a_1, a_2 \in A$ and $b_1, b_2 \in B$. It follows that

$$x_1 \cap x_2 \geqq (a_1 \cap b_1) \cap (a_2 \cap b_2) = (a_1 \cap a_2) \cap (b_1 \cap b_2) = a_3 \cap b_3$$

where $a_3 = a_1 \cap a_2$ belongs to A and $b_3 = b_1 \cap b_2$ belongs to B. This implies that $x_1 \cap x_2 \in X$. Further, if $x \in X$ and $y \in L$, then $x \geqq a \cap b$ for some $a \in A$ and $b \in B$; hence

$$x \cup y \geqq x \geqq a \cap b$$

so that $x \cup y \in X$. This shows that X is an ideal. Due to $A \subseteq X$ and $B \subseteq X$ we have $\langle A, B\rangle \subseteq X$.

Conversely, let $x \in X$. Then $x \geqq a \cap b$ for some $a \in A$, $b \in B$. Since $a \cap b \in \langle A, B\rangle$ it follows that

$$x = (a \cap b) \cup x$$

is contained in $\langle A, B\rangle$. Hence, $X \subseteq \langle A, B\rangle$ so that $X = \langle A, B\rangle$ as was to be proved.

In the case of a distributive lattice the last lemma can be sharpened as follows:

11.3.8. Lemma. *If A and B are ideals of a distributive lattice, then the following statements are equivalent.*

(i) $x \in \langle A, B\rangle$,

(ii) *There exist $a_1 \in A$, $b_1 \in B$ such that $x = a_1 \cap b_1$.*

Proof. In view of Lemma 11.3.7 it remains to prove the following proposition: If A and B are ideals of a distributive lattice and $x \geqq a \cap b$ for some elements $a \in A$, $b \in B$, then there exist elements $a_1 \in A$, $b_1 \in B$ such that $x = a_1 \cap b_1$. Now, if $x \geqq a \cap b$, then

$$x = x \cup (a \cap b) = (x \cup a) \cap (x \cup b) = a_1 \cap b_1$$

where $x \cup a = a_1$ and $x \cup b = b_1$ belong to A and B, respectively. This completes the proof.

An ideal P is called a *prime ideal* if

$$x \cup y \in P \textit{ implies that } x \in P \textit{ or } y \in P \text{ (or both)}.$$

If A is an arbitrary ideal, then $x \in A$ and $y \in A$ imply that $x \cap y \in A$. Conversely, if $x \cap y \in A$, then both

$$x = x \cup (x \cap y) \text{ and } y = y \cup (x \cap y)$$

belong to A. This gives

$$x \in A \textit{ and } y \in A \text{ if and only if } x \cap y \in A.$$

Moreover, by (7),

$$x \in A \textit{ or } y \in A \textit{ implies that } x \cup y \in A.$$

This shows that a prime ideal P can be characterized as follows: P is a non-empty subset of L such that

$$x \in P \textit{ and } y \in P \text{ if and only if } x \cap y \in P, \tag{8}$$

$$x \in P \textit{ or } y \in P \text{ if and only if } x \cup y \in P. \tag{9}$$

Proof of Theorem 11.3.6. Let L be a distributive lattice. In what follows we have to distinguish between the operations $\cap$, $\cup$ defined in L and the set theoretical intersections and unions of certain sets of prime ideals in L. Therefore we shall use $\Cap$ and $\Cup$ to denote the set theoretical intersection and union, respectively.

For $x \in L$ let $\Pi(x)$ denote the set of all prime ideals in L that contain x. Our first aim is to prove that

$$\Pi(x \cap y) = \Pi(x) \Cap \Pi(y), \tag{10}$$

$$\Pi(x \cup y) = \Pi(x) \Cup \Pi(y). \tag{11}$$

The following statements are equivalent:

(i) $P \in \Pi(x \cap y)$,

(ii) $x \cap y \in P$,

(iii) $x \in P$ and $y \in P$,

(iv) $P \in \Pi(x)$ and $P \in \Pi(y)$,

(v) $P \in \Pi(x) \Cap \Pi(y)$.

That (i) and (ii) are equivalent follows from the definition of $\Pi(x \cap y)$. The equivalence of (ii) and (iii) is due to (8). The definition of $\Pi(x)$ and $\Pi(y)$ shows that (iii) and (iv) are equivalent. The equivalence of (iv) and (v) is obvious.

The equivalence of (i) and (v) means that (10) holds.

Dually, (11) can be proved by showing that the following statements are equivalent:

(i) $P \in \Pi(x \cup y)$,

(ii) $x \cup y \in P$,

(iii) $x \in P$ or $y \in P$,

(iv) $P \in \Pi(x)$ or $P \in \Pi(y)$,

(v) $P \in \Pi(x) \cup \Pi(y)$.

To establish the equivalence of (i) through (v) is just as easy as above and can be left to the reader.

(10) and (11) show that the mapping

$$x \rightarrow \Pi(x) \tag{12}$$

for every $x \in L$ is a lattice homomorphism of L into the lattice of the subsets of the prime ideals in L where the operations in the latter lattice have their set theoretical meanings.

It remains to show that this mapping is an isomorphism in case L is distributive.

If x and y are elements of L such that $x \leqq y$, then $x \cap y = x$ so that

$$\Pi(x) \cap \Pi(y) = \Pi(x \cap y) = \Pi(x),$$

i.e. $\Pi(x) \subseteq \Pi(y)$.

To establish that (12) is an isomorphism we shall prove that conversely, $\Pi(x) \subseteq \Pi(y)$ implies $x \leqq y$. To do this we show that in case $x \leqq y$ is false there exists a prime ideal M such that $x \in M$ and $y \notin M$.

Let Λ denote the set of all ideals in L that contain x but do not contain y. Λ is not empty because $(x) \in \Lambda$. With respect to set theoretical inclusion Λ is partially ordered. Let Σ be an arbitrary chain in Λ and let K denote the set theoretical union of all the ideals that belong to Σ. We shall prove that K is an ideal and that K itself belongs to Σ. Suppose that $k_1, k_2 \in K$. Then there exist ideals R_1, R_2 in Σ such that $k_1 \in R_1$, $k_2 \in R_2$. Since Σ is a chain one of the ideals R_1, R_2 contains the other, $R_1 \subseteq R_2$ say. Then both k_1 and k_2 belong to R_2. Since R_2 is an ideal we have $k_1 \cap k_2 \in R_2$ and hence $k_1 \cap k_2 \in K$. Further, if z is any element of L, then $k_1 \cup z \in R_2$ which implies that $k_1 \cup z \in K$. This shows that K is an ideal. It is obvious that K is an upper bound of all the ideals in Σ. Thus it follows from Zorn's Lemma that there exists a maximal ideal M in Λ. We have $x \in M$, $y \notin M$ because all ideals in Λ have this property.

To complete the proof we show that M is a prime ideal. Assume that, on the contrary, M is not a prime ideal. Then there exist elements a, b of L such that $a \notin M, b \notin M$ and $a \cup b \in M$. We have to show that this leads to a contradiction.

Since M is maximal in Λ, the ideals $\langle M, (a)\rangle$ and $\langle M, (b)\rangle$ do not belong to Λ which means that

$$y \in \langle M, (a)\rangle, \quad y \in \langle M, (b)\rangle.$$

Consequently it follows from Lemma 11.3.8 that there are elements $m_1, m_2 \in M$, $u \in (a)$, $v \in (b)$ such that

$$y = m_1 \cap u = m_2 \cap v.$$

This gives

$$\begin{aligned} y = y \cup y &= (m_1 \cap u) \cup (m_2 \cap v) \\ &= [m_1 \cup (m_2 \cap v)] \cap [u \cup (m_2 \cap v)] \qquad (13) \\ &= [m_1 \cup (m_2 \cap v)] \cap (u \cup m_2) \cap (u \cup v). \end{aligned}$$

From $m_1 \in M$ it follows that $m_1 \cup (m_2 \cap v) \in M$, and $m_2 \in M$ implies that $u \cup m_2 \in M$. Finally, since $a \cup b \in M$ and $u \geqq a$, $v \geqq b$ we have $u \cup v \geqq a \cup b$ so that $u \cup v \in M$. This shows that the right-hand side of (13) belongs to M contradicting the hypothesis that y is not contained in M.

This completes the proof.

Let L be any lattice with 1 and 0 and let $a \in L$. An element b of L is said to be a *complement* of a if

$$a \cap b = 0 \quad \text{and} \quad a \cup b = 1.$$

Of course, a is also a complement of b. An element of an arbitrary lattice need not have a complement and, on the other hand, an element may have several complements. For example, the element b in the lattice of fig. 4 has no complement whereas the element a in the lattice of fig. 2 or of fig. 3 has two complements, namely b and c.

The elements 1 and 0 are complements of each other.

11.3.9. Theorem. *Every element of a distributive lattice with* 1 *and* 0 *has at most one complement.*

Proof. Suppose that b_1 and b_2 are complements of a. Then we have

$$a \wedge b_1 = a \wedge b_2 = 0, \quad a \vee b_1 = a \vee b_2 = 1.$$

Since the lattice is distributive Theorem 11.3.5 gives $b_1 = b_2$ as was to be proved.

A distributive lattice with 1 and 0 in which every element has a complement is called a *Boolean lattice.*

Due to the last theorem the complement of any element a of a Boolean lattice is uniquely determined and may therefore be denoted by a'. We have

$$a \wedge a' = 0, \quad a \vee a' = 1, \quad (a')' = a.$$

11.3.10. De Morgan's Laws. *In every Boolean lattice*

$$(a \wedge b)' = a' \vee b', \qquad (a \vee b)' = a' \wedge b'.$$

Proof.

$$\begin{aligned}(a \wedge b) \vee (a' \vee b') &= (a \vee a' \vee b') \wedge (b \vee a' \vee b')\\ &= (1 \vee b') \wedge (1 \vee a') = 1 \wedge 1 = 1,\\ (a \wedge b) \wedge (a' \vee b') &= (a \wedge b \wedge a') \vee (a \wedge b \wedge b')\\ &= (0 \wedge b) \vee (a \wedge 0) = 0 \vee 0 = 0.\end{aligned}$$

Thus $a' \vee b'$ is the complement of $a \wedge b$ which proves the first law. The second law is proved dually.

Let B be a Boolean lattice. In B we define two new operations, addition and multiplication, as follows

$$a + b = (a \wedge b') \vee (a' \wedge b), \tag{14}$$
$$ab = a \wedge b.$$

We shall prove that B is a commutative ring with unit element with respect to these new operations.

By $(L_{1\wedge})$ and $(L_{2\wedge})$ multiplication is commutative and associative. Moreover, for every $a \in B$

$$1a = a, \quad 0a = 0.$$

We also observe that $a^2 = aa = a$ for every $a \in B$. We refer to this property by saying that every element of B is idempotent.

Addition is commutative because it follows from $(L_{1\wedge})$ and $(L_{1\vee})$ that

$$b + a = (b \wedge a') \vee (b' \wedge a) = (a \wedge b') \vee (a' \wedge b) = a + b.$$

To prove that additon is associative we proceed as follows:

$$\begin{aligned}(a+b)+c &= [(a+b)\cap c'] \cup [(a+b)'\cap c] \\ &= \{[(a\cap b')\cup(a'\cap b)]\cap c'\} \cup \{(a\cap b')'\cap(a'\cap b)'\cap c\} \\ &= \{(a\cap b'\cap c')\cup(a'\cap b\cap c')\} \cup \{(a'\cup b)\cap(a\cup b')\cap c\} \\ &= (a\cap b'\cap c')\cup(a'\cap b\cap c')\cup(a'\cap b'\cap c)\cup(a\cap b\cap c).\end{aligned}$$

The right-hand side is symmetric with respect to a, b, c. This gives

$$(a+b)+c = (b+c)+a = a+(b+c)$$

so that addition is associative. We also have

$$a+0 = (a\cap 0')\cup(a'\cap 0) = (a\cap 1)\cup 0 = a$$

and

$$a+a = (a\cap a')\cup(a'\cap a) = 0\cup 0 = 0.$$

The last equation shows that any $a \in B$ has an additive inverse, namely a itself. Thus the elements of B form an abelian group with respect to addition. Moreover, every element of this group is of order $\leqq 2$.

Finally, we establish the distributive law:

$$\begin{aligned}ab+ac &= [(ab)\cap(ac)'] \cup [(ab)'\cap(ac)] \\ &= [(a\cap b)\cap(a'\cup c')] \cup [(a'\cup b')\cap(a\cap c)] \\ &= (a\cap b\cap c')\cup(b'\cap a\cap c) \\ &= a\cap[(b\cap c')\cup(b'\cap c)] = a(b+c).\end{aligned}$$

Hence, B is a ring.

It turns out that the commutative law of multiplication in B and the property $a + a = 0$ are consequences of the fact that every element of B is idempotent. Indeed,

$$a+b = (a+b)^2 = a^2+ab+ba+b^2 = a+ab+ba+b$$

from which it follows that

$$ab+ba = 0. \tag{15}$$

Taking $a = b$ we find that $a^2 + a^2 = 0$, that is, $a + a = 0$ for every $a \in B$. Therefore we have $ab + ab = 0$ so that (15) gives $ab = ba$.

By a *Boolean ring* we understand a ring with unit element in which every element is idempotent. As we just saw every Boolean ring is commutative and every element is of order ≤ 2 in its additive group.

We shall show that every Boolean ring B becomes a Boolean lattice if we define the operations $\cap$, $\cup$ and the complement as follows:

$$a \cap b = ab, \quad a \cup b = a + b + ab, \quad a' = 1 + a. \tag{16}$$

The postulates ($L_{1\cap}$), ($L_{1\cup}$), ($L_{2\cap}$) are satisfied because the ring B is associative and commutative. We find that

$$\begin{aligned}(a \cup b) \cup c &= (a \cup b) + c + (a \cup b)\, c \\ &= a + b + ab + c + ac + bc + abc.\end{aligned}$$

Since the right-hand side is symmetric with respect to a, b, c and since $\cup$ is commutative we obtain

$$(a \cup b) \cup c = (b \cup c) \cup a = a \cup (b \cup c)$$

so that ($L_{2\cup}$) holds. ($L_{3\cap}$) and ($L_{3\cup}$) follow by a simple calculation:

$$\begin{aligned}a \cap (b \cup a) &= a(a + b + ab) \\ &= a^2 + ab + a^2 b = a + ab + ab = a, \\ a \cup (b \cap a) &= a \cup ab = a + ab + a^2 b = a.\end{aligned}$$

Therefore B is a lattice. This lattice is distributive because

$$\begin{aligned}a \cap (b \cup c) &= a(b + c + bc) = ab + ac + abc \\ &= ab + ac + abac = ab \cup ac = (a \cap b) \cup (a \cap c).\end{aligned}$$

Finally, $a' = 1 + a$ is the complement of a since

$$\begin{aligned}a \cup a' &= a + a' + aa' = a + 1 + a + a(1 + a) \\ &= 1 + a + a + a + a = 1, \\ a \cap a' &= aa' = a(1 + a) = a + aa = a + a = 0.\end{aligned}$$

Thus B is a Boolean lattice.

We shall now show that the construction of a Boolean lattice out of a Boolean ring and the construction of a Boolean ring out of a Boolean lattice are inverses of each other. In other words, if we start from a Boolean ring, define a Boolean lattice by (16) and then use (14) to define a ring, then we arrive again at the initial ring.

As to multiplication, this is obvious. Let $a + b$ denote the addition in the Boolean ring from which we start. After having constructed the Boolean lattice and out of it again a Boolean ring we arrive at the addition

$$a \oplus b = (a \cap b') \cup (a' \cap b).$$

By (14) and (16) we obtain

$$\begin{aligned} a \oplus b &= (ab') \cup (a'b) = ab' + a'b + ab'a'b \\ &= a(1 + b) + (1 + a)b + 0 = a + b + ab + ab = a + b \end{aligned}$$

as was to be shown.

11.3.11. **Theorem.** *There is a one-to-one correspondence between Boolean lattices and Boolean rings established by*

$$a \cap b = ab,$$

$$a \cup b = a + b + ab, \quad a + b = (a \cap b') \cup (a' \cap b),$$

$$a' = 1 + a.$$

Exercises

1. Show that the lattice of Example 2 in § 11.1 is distributive.
2. Determine the ideals and the prime ideals in the lattice of Example 2 in § 11.1.
3. Let L be a finite lattice in which $[a, a \cap b]$ and $[a \cup b, b]$ are isomorphic for all $a, b \in L$. Prove that L is modular.
4. Prove that a lattice is distributive if and only if
$$a \cup (b \cap c) \geqq (a \cup b) \cap c$$
for all a, b, c.
5. Let V be a vector space over an arbitrary field. If S and T are subspaces of V define $S \cap T$ to denote the set theoretical intersection and $S \cup T = S + T$. Is the lattice of all subspaces of V modular? Is it distributive?
6. A lattice is said to be complemented if it contains 1 and 0 and if every element has at least one complement; it is called relatively complemented if every interval $[a, b]$ is complement-

ed, that is, if for every $x \in [a, b]$ there exists at least one $y \in [a, b]$ such that $x \cup y = a$, $x \cap y = b$. Prove that every complemented modular lattice is relatively complemented.

7. Draw diagrams of the subgroup lattices of the symmetric group S_3 and of the alternating group A_4. Which of these lattices are modular and which are distributive?
8. Give an example to show that non-isomorphic groups may have isomorphic subgroup lattices.
9. Show that the subgroup lattice of a finite abelian group is distributive if and only if the group is cyclic.
10. Find all finite abelian groups whose subgroup lattice is a chain.
11. Let G and H be finite cyclic groups. Find a necessary and sufficient condition for the subgroup lattices of G and H to be isomorphic.
12. Show that Jordan-Hölder's Theorem on groups with an operator domain Ω follows from its analogue on lattices in case Ω contains all inner automorphisms. Explain the reason for the assumption about Ω.
13. Let $\varphi: x \to \varphi(x)$ be a mapping of a lattice L into a lattice L^*. We consider the following conditions on φ:

(1) $x \leqq y$ implies that $\varphi(x) \leqq \varphi(y)$,

(2) $\varphi(x \cap y) = \varphi(x) \cap \varphi(y)$,

(3) $\varphi(x \cup y) = \varphi(x) \cup \varphi(y)$.

φ is called a lattice homomorphism if it satisfies all the three conditions. If φ satisfies (1), (2), or (3), then we speak of an isotone mapping, an intersection-homomorphism, or a union-homomorphism, respectively.

(i) Show that either of the conditions (2) and (3) implies (1).

(ii) Give an example of a mapping φ that satisfies (1) but neither (2) nor (3).

(iii) Give examples of mappings φ that satisfy precisely one of the conditions (2) or (3).

(Hint: Consider suitable mappings of the lattice of all subsets of a set of two elements onto a chain with two or with four elements.)

14. Let φ be a lattice homomorphism of L onto L^* (cf. the definition in Exercise 13).

    (i) Assume that L^* has an all element 1^*. Show that the elements x of L such that $\varphi(x) = 1^*$ form an intersection ideal.

    (ii) Assume that L^* has a null element and describe its inverse image.

15. Let L and L^* be lattices. A bijective mapping $\varphi\colon x \to \varphi(x)$ of L onto L^* is called a dual isomorphism if

$$\varphi(x \cap y) = \varphi(x) \cup \varphi(y), \quad \varphi(x \cup y) = \varphi(x) \cap \varphi(y).$$

    Use this notion to state the Principal Theorem of the Galois Theory.

16. A lattice L is called a metric lattice if there exists a real valued function ψ, defined for all $x \in L$, such that

$$\psi(x) + \psi(y) = \psi(x \cup y) + \psi(x \cap y) \text{ for any } x, y \in L$$

    and

$$\psi(x) > \psi(y) \text{ whenever } x > y.$$

    (i) Prove that every metric lattice is modular.

    (ii) Prove that a metric lattice L is distributive if and only if, for any $x, y, z \in L$,

$$\psi(x \cup y \cup z) - \psi(x \cap y \cap z)$$
$$= \psi(x) + \psi(y) + \psi(z) - \psi(x \cap y) - \psi(y \cap z) - \psi(z \cap x).$$

    (iii) Prove that the lattice of all the subspaces of a finite-dimensional vector space over any field is a metric lattice.

17. Let a and b be subsets of a given set m and let a' denote the complement of a in m, that is, $a \cap a' = \emptyset$, $a \cup a' = m$. Describe the elements of the so-called symmetric difference

$$(a \cap b') \cup (a' \cap b)$$

    (cf. (14) in § 11.3) in terms of their membership of a and/or b.

18. Show that in every Boolean ring

$$x + y = (x \cap y) + (x \cup y).$$

Bibliography

The following is a list of books which we suggest to the reader for further studies in algebra. In view of the abundance of the pertinent literature this list contains only a small selection of suitable works. We would like to point out that the classification by topics is by no means strict, books of different classes may rather overlap considerably.

General

[1] Bourbaki, N. *Eléments de mathématique.* Livre II, Algèbre, Hermann, Paris.

[2] Jacobson, N. *Lectures in abstract algebra,* I, II, III. Van Nostrand, Princeton (1965, 1953, 1964).

[3] Landau, E. *Foundations of analysis.* Chelsea Publ. Comp., New York (1951).

[4] MacLane, S. and Birkhoff, G. *Algebra.* Mac Millan Comp., New York (1967).

[5] Mitchell, B. *Theory of categories.* Academic Press, New York (1965).

[6] Northcott, D. G. *An introduction to homological algebra.* Univ. Press., Cambridge (1960).

[7] Zariski, O. and Samuel, P. *Commutative algebra,* I, II. Van Nostrand, Princeton (1958, 1960).

Linear Algebra

[8] Kaplansky, I. *Linear algebra and geometry.* A Second Course. Allyn and Bacon, Boston (1969).

[9] Lang, S. *Linear algebra*. Addison-Wesley Publ. Comp., Reading (1968).

[10] Schwerdtfeger, H. *Introduction to linear algebra and the theory of matrices*. P. Noordhoff Ltd., Groningen (1962).

Number Theory and Theory of Algebraic Numbers

[11] Artin, E. *Algebraic numbers and algebraic functions*. Gordon and Breach, New York (1967).

[12] Borevich, Z. I. and Shafarevich, I. R. *Number theory*. Academic Press, New York (1966).

[13] Hardy, G. and Wright, E. *An introduction to the theory of numbers*. Clarendon Press, Oxford (1960).

[14] Weiss, E. *Algebraic number theory*. McGraw-Hill Book Comp., New York (1963).

[15] Weyl, H. *Algebraic theory of numbers*. Princeton Univ. Press, Princeton (1940).

Theory of Groups

[16] Curtis, C. W. and Reiner, I. *Representation theory of finite groups and associative algebras*. Interscience, J. Wiley & Sons, New York (1962).

[17] Fuchs, L. *Abelian groups*. Pergamon Press, New York (1960).

[18] Hall, M. Jr. *The theory of groups*. Mac Millan Comp., New York (1959).

[19] Kaplansky, I. *Infinite abelian groups*. Univ. of Michigan Press, Ann Arbor (1956).

[20] Kurosh, A. G. *The theory of groups*, I, II. Chelsea Publ. Comp., New York (1960).

[21] Scott, W. R. *Group theory*. Prentice Hall, Englewood Cliffs (1964).

[22] Wielandt, H. *Finite permutation groups*. Academic Press, New York (1964).

Theory of Rings

[23] Artin, E. and Nesbitt, C. and Thrall, R. *Rings with minimum condition*. Univ. of Michigan Press, Ann Arbor (1964).

[24] Jacobson, N. Structure of rings. *Amer. Math. Soc. Coll. Publ. Vol. 37*, Providence (1956).
[25] Jacobson, N. *Lie algebras.* Interscience Press, New York (1962).
[26] Jans, J. P. *Rings and homology.* Holt, Rinehart, and Winston, New York (1964).
[27] Northcott, D. G. *Ideal theory.* Cambridge Univ. Press, Cambridge (1963).

Theory of Lattices

[28] Birkhoff, G. Lattice theory. *Amer. Math. Soc. Coll. Publ. Vol. 25*, New York (1948).
[29] Rutherford, D. E. *Introduction to lattice theory.* Oliver & Boyd, Edinburgh (1964).

Index

Adjunction 183
Algebra 116, 336
—, complete matrix 117
—, division 119, 352
Algebraic closure 200
— system 11
Ascending chain condition 315
Associate 129
Associative law 8, 11, 13, 113
Automorphism 52, 121
— group 52
—, inner 52
—, outer 53

Basis of module 108, 293
— of vector space 100
Birkhoff's Theorem 396
Block, elementary 309
Bound, greatest lower 382
—, least upper 382
—, lower 382
—, upper 6

Cardano's formula 259
Cauchy's Theorem 91
Cayley group table 46
Cayley-Hamilton's Theorem 315
Centre of group 57
Centre of ring 120
Chain 5
Character of permutation 86
— of representation 369
Characteristic 145, 146
Chinese Remainder Theorem 35
Combination, linear 99, 292
Commutative law 12, 113
Commutator 74
Commute 41
Companion matrices 308
Compatible 30
Complex in group 56
Congruence 29, 124
—, right 59
Conjugacy class 93
Constituent of representation 358
Constructions by ruler and compass 261
Coset 60
Cycle 83

De Morgan's Laws 401
Degree, inseparable 208
— of extension field 188
— of permutation 83
— of polynomial 152
— of representation 305, 356

Degree, reduced 206
—, separable 206
Derivative of polynomial 162
Descending chain condition 315
Dimension of algebra 117
— of vector space 104
Discriminant 171
Distributive law 11, 113
Division algorithm 22
— ring 115
Divisor 22
—, elementary 290, 311
—, greatest common 24, 131
— of zero 119
—, trivial 23, 130

Eisenstein's Theorem 158
Element, algebraic 185
—, generating 49, 184
—, maximal 5
—, nilpotent 338
—, primitive 184, 203
—, separable 202
—, transcendental 185
Elements, conjugate of field 200
—, — of group 52, 65
Embed 53
Endomorphism 94
— ring 95
Epimorphism 66
Equation, cubic 257
—, pure 251
—, quartic 259
Equivalence class 4
Euclidean algorithm 26
Euler's function 36
Exponent of inseparability 207
Extension field 183
— —, algebraic 185
— —, Galois 216
— —, normal 201
Extension field, purely inseparable 210
— —, separable 203
— —, simple 184
— of isomorphism 193

Factor, direct 77
— group 68
— module 292
Fermat's Theorem 61
Field 12, 115, 183
—, algebraically closed 198
—, cyclotomic 226
—, fixed 213
—, formally real 269
— of quotients 142
—, ordered 267
—, perfect 166
—, prime 144
—, real closed 270
Four-group 48
Frobenius' Theorem 353

Galois field 235
— group 216, 244
— resolvent 244
Gauss' Lemma 155
Gauss' Theorem 158
Gaussian integers 115, 139
Group 12, 42
—, abelian 13, 44
—, additive of ring 114
— algebra 118
—, alternating 86
—, completely reducible 79
—, cyclic 48, 61
—, derived 74
—, finite 45
—, general linear 13
— of type p^∞ 48
—, p- 93

Group, simple 65
—, soluble 74
—, symmetric 48

Hilbert's Basis Theorem 321
Homomorphism 66, 121
—, natural 68, 124
— Theorem for Groups 69
— — for Rings 124

Ideal 122
—, left 337
—, maximal 126
—, minimal 341
—, nilpotent 338
— of lattice 396
—, prime 128
—, principal 127
—, right 337
—, two-sided 337
Idempotent 338
—, primitive 342
Idempotents, orthogonal 342
Identity 13, 42
—, left 42
— permutation 9
—, right 42, 43
Image 6
—, inverse 7
Indeterminate 151
Index 61
Integer 1, 20
—, algebraic 321
Integral domain 128
Intersection 2
Interval 388
Inverse 42
— of mapping 7
—, right 43
Isomorphism 51, 120, 187
— Theorems 69, 70

Jordan-Hölder's Theorem 74
Jordan normal form 309

Kernel of homomorphism 67, 122

Lagrange resolvent 252
Lagrange's Theorem 61
Lattice 379
—, Boolean 401
—, distributive 392
—, modular 385
Length of cycle 83
Linearly dependent 99, 187
— independent 99, 188, 211, 293

Mapping 6
—, bijective 7
—, injective 7
—, linear 106, 303
—, one-to-one 7
— onto 7
—, surjective 7
Maschke's Theorem 363
Matrices, equivalent 285
—, similar 304
Matrix, unimodular 285
Maximum condition 315
Minimum condition 316
Möbius' function 228
Module 108, 291
—, cyclic 292
—, finitely generated 292
—, free 109, 293
—, primary 302
Monomorphism 66
Multiple, least common 29

Newton's relations 175
Noetherian ring 320
Norm 224
Normalizer 93

Nullity 313
Number, algebraic 186
—, complex 1
—, natural 1
—, rational 1
—, real 1
—, trancendental 186

Operation, binary 10
Operator domain 96
— homomorphism 97
Orbit 81
Order, complete 5
— ideal 293
— of element 58
— of group 45
—, partial 5
— relation 5
Orthogonality relations of characters 373

Permutation 8
—, even 87
— group 81
— —, intransitive 82
— —, transitive 82
—, odd 87
Polynomial 151
—, characteristic 310
—, constant 152
—, cyclotomic 227
—, elementary symmetric 171
—, general 247
—, irreducible 154
—, minimal 185
—, monic 152
—, normal 225
—, prime 154
—, primitive 155
—, separable 202
—, symmetric 171
Preimage 6
Prime divisor 23
— element 130
— number 23
—, relatively 26, 132
Principal ideal ring 131
Product, Cartesian 3
—, direct 76
— of mappings 8

Quaternion algebra 119

Radical of algebra 340
Rank of module 109, 294
Refinement 72
Relation, antisymmetric 3
—, binary 3
—, defining 49
—, equivalence 4
—, order 5
—, reflexive 3
—, symmetric 3
—, transitive 3
Remainder 22
Representation 305, 356
—, absolutely irreducible 366
—, completely reducible 360
—, irreducible 358
— module 305, 356
—, permutational 89
—, reducible 358
—, regular 83, 361
Representations, equivalent 305
Representative of equivalence class 5
Residue class 30, 123
— —, prime 34
— — ring 33, 123
Resolvent, cubic 260
Resultant 178
Ring 11

Ring, Boolean 403
—, commutative 114, 127
—, division 115
—, euclidean 137
—, principal ideal 131
—, semi-simple 340
—, simple 126, 345
—, unique factorization 136
Root, characteristic 310
—, multiple 161
— of polynomial 161
— of unity 225
— —, primitive 227
—, primitive mod p 232
—, simple 161

Schreier's Refinement Theorem 73
Semigroup 40
—, abelian 41
— algebra 117
—, cancellative 53
—, commutative 41
—, cyclic 42
Series, composition 73
—, subnormal 72
Set, empty 1
—, well ordered 21
Skew field 115
Solubility by radicals 251
Span 99
Splitting field 195
Stabilizer subgroup 82
Subgroup 56
—, admissible 96
—, commutator 74
—, normal 65
—, Sylow 91
Subgroup, trivial 56
Subring 119
Subset 2
—, proper 2
Subsets, disjoint 2
Subspace 99
Sum, direct 79, 292
Sylow's Theorem 91
Sylvester's determinant 178

Torsion module 301
Trace 224
Transformation in group 52
Transposition 84
Transversal 5, 60
Type of permutation 84

Union 2
Unique factorization theorem 27, 134
Unit 129
— element 12, 13, 115

Valuation 275
—, archimedean 279
—, non-archimedean 279
—, p-adic 277
—, trivial 276
Valuations, equivalent 278
Vector space 14, 98

Wedderburn's Theorem 238, 347
Wilson's Theorem 238

Zassenhaus' Lemma 71
Zero 113
— ring 114
Zorn's Lemma 6